Logic Synthesis

Other Computer Engineering Books of Interest

Logic Synthesis

Srinivas Devadas

Abhijit Ghosh

Kurt Keutzer

McGraw-Hill, Inc.

New York San Francisco Washington, D.C. Auckland Bogotá
Caracas Lisbon London Madrid Mexico City Milan
Montreal New Delhi San Juan Singapore
Sydney Tokyo Toronto

Library of Congress Cataloging-in-Publication Data

Devadas, Srinivas.
Logic synthesis / Srinivas Devadas, Abhijit Ghosh, Kurt Keutzer.
p. cm. — (McGraw-Hill series on computer engineering)
Includes bibliographical references and index.
ISBN 0-07-016500-9
1. Integrated circuits—Very large scale integration—Design—Data processing.
2. Logic design—Data processing. 3. Computer-aided design.
I. Ghosh, Abhijit. II. Keutzer, Kurt William. III. Title.
IV. Series: Series on computer engineering.
TK7874.D478 1994
621.39'5—dc20 93-41524
CIP

1 2 3 4 5 6 7 8 9 0 DOC/DOC 9 0 9 8 7 6 5 4

ISBN 0-07-016500-9

The sponsoring editor for this book was Stephen S. Chapman.
Printed and bound by R. R. Donnelley & Sons Company.

Contents

Preface

The dramatic increase in designer productivity over the past decade in the area of *very large scale integrated* (VLSI) circuit design is the direct result of the development of sophisticated *computer-aided design* (CAD) tools. Today, designers routinely describe the functionality of a circuit in a *hardware description language* (HDL) (which is a high-level description akin to a programming language) and use synthesis tools to produce optimized circuit layouts that can be sent off to a chip manufacturer to be fabricated on a silicon integrated circuit.

The two major areas in VLSI synthesis that have enabled vastly improved design turnaround times are logic synthesis and layout synthesis. The logic synthesis process consists of the translation of the input HDL description into a gate-level circuit and the optimization of the gate-level circuit. An optimized layout is produced for the final gate-level circuit by the layout synthesis process. Both the logic and layout synthesis process require the solution of difficult combinatorial optimization problems. In this book we focus solely on the logic synthesis process.

Switching and automata theory form the cornerstones of logic synthesis. Combinatorial problems associated with optimizing switching circuits abound in logic synthesis. In order to meet the demands of the designers, logic optimization systems have to be versatile and efficient. Versatility implies that the system should be able to target a variety of design parameters such as circuit area, delay, power dissipation, and testability. Efficiency implies that the system should be able to produce near-optimum or at least acceptable results for large VLSI circuits with reasonable CPU time expenditure.

The development of logic optimization systems that are versatile and efficient posed one of the major challenges for CAD in the 1980s. Today, thanks to a large amount of research and developmental effort such systems are in wide use among integrated circuit designers. This book focuses on describing the basic principles of logic design as well as the practical aspects of engineering a logic synthesis system.

Very few individuals will themselves undertake to implement a logic synthesis system, but we feel that understanding the core principles of logic synthesis will be of use to a number of communities.

One community consists of educators in computer science and electrical engineering. In general an educator is looking for material that both disciplines the intellect of the student as well as prepares the student for practical problems the student is likely to encounter. In particular, the modern educator searching to find material that is relevant to the logical design of VLSI circuits has been faced with choosing either the classical works on switching and automata theory or anthologies of collected articles on logic synthesis and optimization methods. A large body of theory as well as many practical algorithms and methodologies to design logic circuits have been developed by researchers in logic synthesis, but so far theoretical and practical results have only been documented in numerous articles. This book attempts to fill the gap between the classical books written in the 1970s and modern logic optimization articles.

Another community consists of integrated circuit designers who are presently using logic synthesis to design integrated circuits. For these designers the former skills of handcrafting transistor-level layout or manually entering the schematics of a carefully designed gate-level implementation of a circuit are being replaced by the skill of writing efficient HDL models of integrated circuits. In order to develop the skill of writing HDL models of circuits that will result in efficient implementations, it is necessary for a designer to understand the basic principles underlying logic synthesis and optimization. One of the recurrent obstacles for a hardware designer using synthesis is the assumption that two functionally equivalent HDL models will produce similar circuits after logic synthesis and optimization. While we do not present a primer on HDL model development, it is our hope that by making the designer understand the capabilities and limitations of logic optimization software the designers will be able to build more efficient circuits within a synthesis framework. It is our hope that the discipline of logic synthesis will play just as central a role to the training and education of circuit designers as the discipline of compilers plays in the education and training of software developers.

A final community we wish to address are those fellow researchers in CAD who are working in logic synthesis or a related area. Research in logic synthesis is a very satisfying enterprise because improvements in algorithms can immediately translate into smaller or faster circuits, and smaller and faster circuits have substantial commercial impact. Researchers wishing to get up to speed in logic synthesis have also been forced to rely on the classical works on switching and automata theory or on the anthologies of collected articles on logic synthesis and optimization. Here we hope to provide these researchers with a self-contained reference book that covers most of the principal synthesis and optimization techniques.

This book is organized into ten chapters. We provide an introduction to synthesis, verification, and testing in Chapter 1. The translation of an HDL model into a netlist of gates is introduced in Chapter 2. While integrated circuits implement sequential circuits, the most successful logic synthesis and optimization techniques have focused on the combinational portions. Therfore, in the remainder of the book we focus on the combinational portions of the circuits. The core algorithms for logic optimization were initially developed on two-level circuits and most easily understood in that context. For these reasons we will introduce these circuits first in Chapter 3. In the following chapters, we deal with the problems of minimizing two-level circuits so as to improve area and speed. We focus on testing a two-level circuit under various models of faulty behavior. We show how logic transformations applied to minimize the two-level circuit's area affect the testability of the circuit. While logic optimization techniques were first developed on two-level circuits, multilevel circuits are of much greater practical importance. In Chapters 6 through 9 we deal with the problems of synthesizing multilevel combinational logic circuits for minimal area, maximal speed, and high testability. Strong relationships between the area, speed, and testability of a circuit are highlighted throughout the book. We summarize the state-of-the-art in logic synthesis in Chapter 10.

Circuit representations and data structures cut across all facets of design such as synthesis, testing, and verification. At the combinational or sequential circuit level, Boolean functions are manipulated in various ways. The search for more efficient representations of Boolean functions is ceaseless, mainly because discovering such representations can have a significant impact on synthesis, testing, and verification problems. In this book we describe commonly used representations for combinational circuits and their advantages and disadvantages when applied to particular problems in synthesis and test.

Since we intend this book to be useful to CAD researchers, educators, and VLSI designers, we have included considerable detail in the description of the various algorithms. Because we cannot comprehensively present the full panorama of logic optimization techniques, we present those techniques that have proven to be most useful in practice.

Acknowledgements

Over the years, several people have helped to deepen our understanding of VLSI synthesis, test generation, and synthesis for testability. We thank Jonathan Allen, Pranav Ashar, Robert Brayton, Michael Bryan, Gaetano Borriello, Raul Camposano, Steve Carlson, Tim Cheng, Aart De Geus, Giovanni De Micheli, Alfred Dunlop, Gary Hachtel, Niraj Jha, Charles Leiserson, Michael Lightner, Bill Lin, Tony Ma, Sharad Malik, Rick McGeer, Richard Newton, Paul Penfield, Sudhakar Reddy, Richard Rudell, Alexander Saldanha, Alberto Sangiovanni-Vincentelli, Fabio Somenzi, Kanwar Jit Singh, Albert Wang, Ruey-sing Wei, Jacob White, Tom Williams, and Wayne Wolf.

Some of the problems in this book have been taken or modified from Robert Brayton's class notes. Material modified from a variety of other sources is acknowledged by citations in the text.

Lastly, we thank Sulochana Devadas, Eliane Setton, Cate Hunter, and the rest of our families for their continual patience and encouragement.

Acknowledgments

Logic Synthesis

Chapter 1

Introduction

Electronic systems are ubiquitous in modern society and *very large scale integrated* (VLSI) circuits are at the heart of most modern electronic systems. These VLSI circuits contain thousands to millions of transistors and interconnections within a very small area. The design of such circuits is a complicated and time-consuming process, and the complexity of the process is constantly increasing. At the core of the VLSI design problem is the fact that semiconductor processing doubles the number of transistors that can be implemented on an integrated circuit every 2 or 3 years. Imagine the challenges to aeronautical engineering if the horsepower of rocket engines doubled every 2 years. The *computer-aided design* (CAD) tool developer must try to find a way to enable the VLSI designer to effectively harness the capabilities offered by similar improvements in semiconductor processing. These improvements in semiconductor processing are not new, and as a result paradigm shifts in VLSI design methodologies are common. To manage increases in complexity human beings use techniques such as abstraction and hierarchy. To manage the complexity of the integrated circuit design process the view of the circuit has moved from the transistor level to the gate level and from the gate level to the register-transfer level. As a result in less than 10 years we have seen the core skill of an integrated circuit designer change from the ability to handcraft a compact and efficient transistor-level layout to manually entering the schematics of a carefully designed gate-level implementation of a circuit. The importance of each of these skills is now becoming secondary to the skill of writing an efficient *hardware description language* (HDL) model of an *integrated circuit* (IC).

1.1 Digital Integrated Circuits

Before discussing design techniques, let us begin by understanding the object to be designed, a *digital circuit.* The circuit is called *digital* because it works on discrete (binary) values in distinction to *analog circuits* which utilize continuous values. The use of the term *circuit* probably has its origins with the term *switching circuit.* Switching circuits were used to control electromechanical relays in telecommunication networks. These circuits were originally strictly *combinational circuits*, meaning that they had no memory elements and they were first formalized in terms of Boolean algebra in a classic paper by Shannon [64]. The study of combinational switching circuits has been an active area of research for over three decades (see e.g., [45, 52, 71]), but combinational circuits alone are not of great utility. To be of significant practical use for computation a combinational circuit needs to be augmented by memory elements that retain the state of a circuit. Such a circuit is called a *sequential circuit.*

The first sequential circuits were built from vacuum tubes and required one vacuum tube for each combinational or sequential circuit element. The ability to build sequential circuits from discrete transistor components resulted in a significant advance in the size and reliability of sequential circuits that could be built. Another tremendous advance came in 1958 when Jack Kilby and Robert Noyce fabricated a number of discrete devices on a single piece of germanium. The result was the first integrated circuit. There have been many significant technological advances since the advent of the integrated circuit, but from a designer's perspective the primary change has been an exponential increase in both the number of transistors available and the speed of the transistors. It is the need to cope with this continual increase in design complexity that is the primary driver for changes in design technology, and logic synthesis is the latest response to managing design complexity.

1.2 IC Design Methodology

Two conflicting forces drive the IC design process: circuit quality and time to market. To get to market early with a very poor quality chip is not a good strategy. Neither is it a good strategy to miss the market entirely while carefully crafting an IC. Typically, an IC designer is trying to reach a market window with an IC design that has a competitive performance within that window. At any point in time IC design systems are trying to provide a designer with design tools that allow for sufficient design productivity to realize time-to-market constraints while producing circuits that are efficient enough to realize performance constraints. If the semiconductor processing of integrated circuits were a slowly improving process, then we would expect to see a relatively stable IC design methodology with slow incremental improvements in individual tools. As we discussed in Section 1.1, semiconductor processing is actually undergoing exponential improvements, and rather than a single stable IC design methodology we see rapidly changing paradigm shifts.

1.3 Transistor-Level Layout

The first codified IC design methods were very focused on transistor-level design closely coupled to layout. The first design approaches emphasized only transistor-level layout entry and transistor-level simulation. These transistor-level methods evolved into the "silicon compiler" approach to IC design that has its roots in the famous Mead-Conway book on VLSI design [53]. It is useful to remember that by the time the Mead-Conway book was published integrated circuits of over 100,000 transistors had already been designed. The silicon compiler based IC design methodology emphasized software generators that created dense transistor layouts. Other key tools in this methodology were: layout editors, for detailed manual design; design rule checkers, to check for legal layout configurations; transistor-level simulators; and layout compactors. Books which reflect this era in design are [69] and [42].

Module generators are still in use today, and they have been found to be most useful for datapaths in medium-to-high performance ICs. Datapath portions in the highest-performance ICs are still typically layouts handcrafted at the transistor level. Handcrafting is justified by the reusability of these layouts, since most circuits have arithmetic units like adders and multipliers in them.

1.4 Gate-Level Entry

Transistor-level design, even when accelerated by module generators, is a time-consuming process. In time, as both the number of transistors and the individual die-size increased designers began to look for ways to accelerate the design process, even if circuit density and speed were sacrificed. Board-level designers using *transistor-transistor logic* (TTL) parts had long benefited from the existence of a library of precharacterized modules. The introduction of *gate arrays* and *standard cells* brought comparable benefits to the IC designers. A gate array is an array of transistors and routing channels which can be configured into an IC through a metalization process during semiconductor fabrication. The metalization phases are used for cell definition, such as defining a NOR cell, and for interconnecting the cells. The electrical characteristics of cells after metalization have been carefully defined and are embodied in a databook. Standard cells are combinational and sequential logic gates whose electrical characteristics have been carefully defined and embodied in a library. Standard cells are similar to gate arrays in that they are precharacterized in a databook, but they offer additional degrees of freedom since they go through all the mask steps of semiconductor processing. This IC technology is supported by automated place-and-route systems. These systems take as input a netlist of cells from the predefined cell library and automatically place and route them in rows and columns.

Design using standard cells and gate arrays raised the level of abstraction from the transistor level to the gate level. The primary design-entry method in the gate-level IC design methodology is gate-level schematic entry by means of a schematic editor. Other key tools in this methodology were: gate-level simulators, for simulating the entered netlist; automatic place-and-route tools, for placing and routing the netlist; and layout editors, for performing final layout edits.

The gate-level design methodology was in use at major corporations since the 1970s and many designers had already migrated to gate-level design by the time transistor-level design was being popularized in academic circles. Corporations such as AT&T, IBM, and NEC used internally developed CAD tools to support their own semiconductor processing.

Three simultaneous developments contributed to a dramatic increase in the popularity of gate-level design. One of these was the evolution of independent CAD vendors offering gate-level schematic entry and simulation tools. The deployment of these tools was significantly affected by the development of engineering workstations. With these two developments an independent engineer could, with a moderate investment, design an IC. Once having entered a netlist of an IC and verified its functionality through simulation the question still remained: Where should the circuit be manufactured? The third enabling development was the emergence of independent semiconductor vendors offering semiconductor processing for gate arrays and standard cells. With these three components in place there was a significant increase in IC design, particularly *application specific integrated circuit* (ASIC) design.

Gate-level schematic entry probably still remains the dominant design methodology in the IC design industry although few would claim that it will remain so.

1.5 Initial Use of Logic Optimization

The first widespread use of logic synthesis technology made rather modest use of logic optimization techniques. Once a designer has created a gate-level netlist targeted for one cell library, the designer might later wish to convert the design into the library of another semiconductor processing vendor. Reentering the design would be too time consuming, cell libraries can be very different, and even logically identical cells can have different speed and area characteristics. For these reasons a manual translation of a netlist from one cell library to another could eliminate the advantages of moving to another library in the first place. One of the first applications of logic synthesis, or more properly *logic optimization*, was to remap a netlist of cells from one semiconductor cell library to another. While this is a very limited use of logic optimization, the need for this capability gave logic synthesis tools their first commercial foothold.

A designer using logic synthesis for the remapping of an IC design also often found a significant improvement in the design due to logic optimization. This demonstrated the potential for giving logic synthesis and optimization a role in the gate-level design methodology. After entering a netlist in a gate-level schematic editor, logic optimization could be used to optimize the resulting netlist and map it into a target library. The value of the tool was measured by the speed or area improvements in resulting circuitry.

1.6 Emergence of Synthesis-Based Design

While logic optimization was finding its first commercial use for remapping, designers at major corporations such as IBM had already been demonstrating the viability of a top-down design methodology based on logic synthesis. At these corporations internal simulation languages, such as BDLCS or ESIM [34], were coupled with synthesis systems, such as LSS [22] or CONES [67], that translated the simulation model into a gate-level netlist. Designers at IBM had demonstrated the utility of this synthesis-based design methodology on thousands of real industrial ICs. Based on this industrial experience a broader, commercial use of a logic synthesis methodology seemed viable, but a key requirement was a standard HDL that was both simulatable and synthesizable. This need was filled by VERILOG [68].

Designers who were faced with designing circuits of over 50,000 gates were already in desperate need of a productivity improvement, and other designers who were already using VERILOG as a netlist language were attracted to synthesis strictly for its logic optimization capabilities. Both these groups were quick to realize the advantages of designing by entering a simulation model in a HDL. In this way logic synthesis and optimization moved from a minor tool in a gate-level schematic based design methodology to the cornerstone of a highly productive IC design methodology.

1.7 A Logic Synthesis Design Methodology

Having presented a historical context for the synthesis design methodology we will now overview the methodology itself. Synthesis of a digital circuit involves an interwoven series of translation, optimization, and mapping steps. These steps in a typical synthesis process are shown in Figure 1.1.

Arriving at an initial product-level specification for an IC depends at least as much on marketing issues as on engineering issues. The engineering-oriented aspect of circuit design begins with a specification of the required circuit functionality and performance. This specification will cover issues such as die size (final IC area), clock rate (speed), required fault coverage (reliability), power dissipation, and pin count. The bulk of the specification will deal with the logical functionality required of the integrated circuit. For an IC implementing an industry standard, such as a protocol, the functional specification may be defined by a standards document. For an IC that is a "knock-off" of a competitor's chip the functional specification may be all the data available on the competitor's chip. Most commonly, an IC is an enhancement of an "in-house" IC, and the functional specification will be to retain backward compatability with the prior generation while enhancing the speed, reducing the area, or perhaps adding a new feature set.

1.7.1 Behavioral Modeling

For complex ICs, such as a high-performance microprocessor, a *behavioral model* of the IC will probably be developed. Modeling a circuit's functionality at the *behavioral level* means modeling the functionality correctly, but without regard to exact clock-cycle by clock-cycle behavior. In signal-processing applications such models are useful for verifying the functionality of algorithms. In microprocessor development such models are useful for gathering information on instruction usage and for supporting the development of assemblers and other tools in the software environment for the microprocessor.

This behavioral model can be expressed in a number of different languages. A number of different languages have been suggested for supporting behavioral simulation models. These languages include ISPS [5], ELLA [54], VHDL [58], and VERILOG [68]. Nevertheless, most commonly used behavioral models are written in the C programming language. The advantages of using the C language are that it has an excellent programming environment and the speed of simulation is very fast. Speed of a behavioral simulator is of preeminent importance if it is to run the millions of vectors required for software development or for instruction-set analysis.

Before the behavioral model can be implemented as an IC, the behavioral model must be evolved into a *register-transfer level* (RTL) model representing the *microarchitecture* of the circuit. Currently the most popular hardware design languages for describing RTL models are VERILOG and VHDL. A register-transfer level model is a model in which the operations of a sequential circuit are described as synchronous transfers between functional units such as arithmetic-logic units and register files. These transfer functions are under the control of an independent controller and are synchronized to a clock. Two significant transformations must take place in turning a behavioral-level model into a register-transfer model. One transformation is that the behavior of the IC must be *scheduled* into synchronous clock-cycle by clock-cycle behavior. Another transformation is that the various operations defined in the behavioral model must be mapped onto resources available in the actual IC. This latter step is termed *resource allocation* and is a step that binds the resources to the available functional units. For example, if an integer multiply occurs in the behavioral model, then the register-transfer level model must describe precisely whether the multiply is pipelined over a number of cycles or accomplished in a single clock cycle. The register-transfer level model must also describe exactly which functional units perform the multiplication.

The translation of a behavioral model into an RTL model is typically performed manually. The development of an RTL model that will result in an efficient IC is among the most important skills of a contemporary IC designer. Behavioral synthesis tools have been developed which automate the translation of a behavioral model to an RTL model [31, 39, 43, 56, 70, 72, 73]. Although these behavioral synthesis tools have been in existence for over 15 years, there are significant barriers to their use. One barrier is that the microarchitecture produced by behavioral synthesis tools is usually one of a relatively small set of datapath oriented microarchitectures. These microarchitectures are most useful for arithmetic intensive applications such as signal processing and are not widely applicable to other control dominated applications such as protocol processsing. Another barrier is that these arithmetic intensive applications often have the greatest performance constraints, and these are hard to meet with automatically generated architectures. Finally, the question arises how much design productivity improvement will be gained by synthesizing from a behavioral model instead of an RTL model. Whatever the potential of behavioral synthesis in the future, almost all RTL models are presently generated manually.

1.7.2 Register-Transfer Level Modeling

Presently the register-transfer level description is most often entered textually in a HDL such as VHDL or VERILOG. This model is then simulated for functional correctness. Once the functional correctness is assured, the model is translated into a mixture of combinational logic gates and sequential memory elements using translation tools. In current practice, in the remainder of the process the synthesized sequential memory elements remain untouched and only the combinational portion is optimized. This translation is the first step in logic synthesis. As we mentioned in the Preface, the term *logic synthesis* is used to denote the entire translation and optimization process. The translation step is relatively straightforward and only a few papers have been published on this aspect [59, 63]. The next step in logic synthesis is logic optimization.

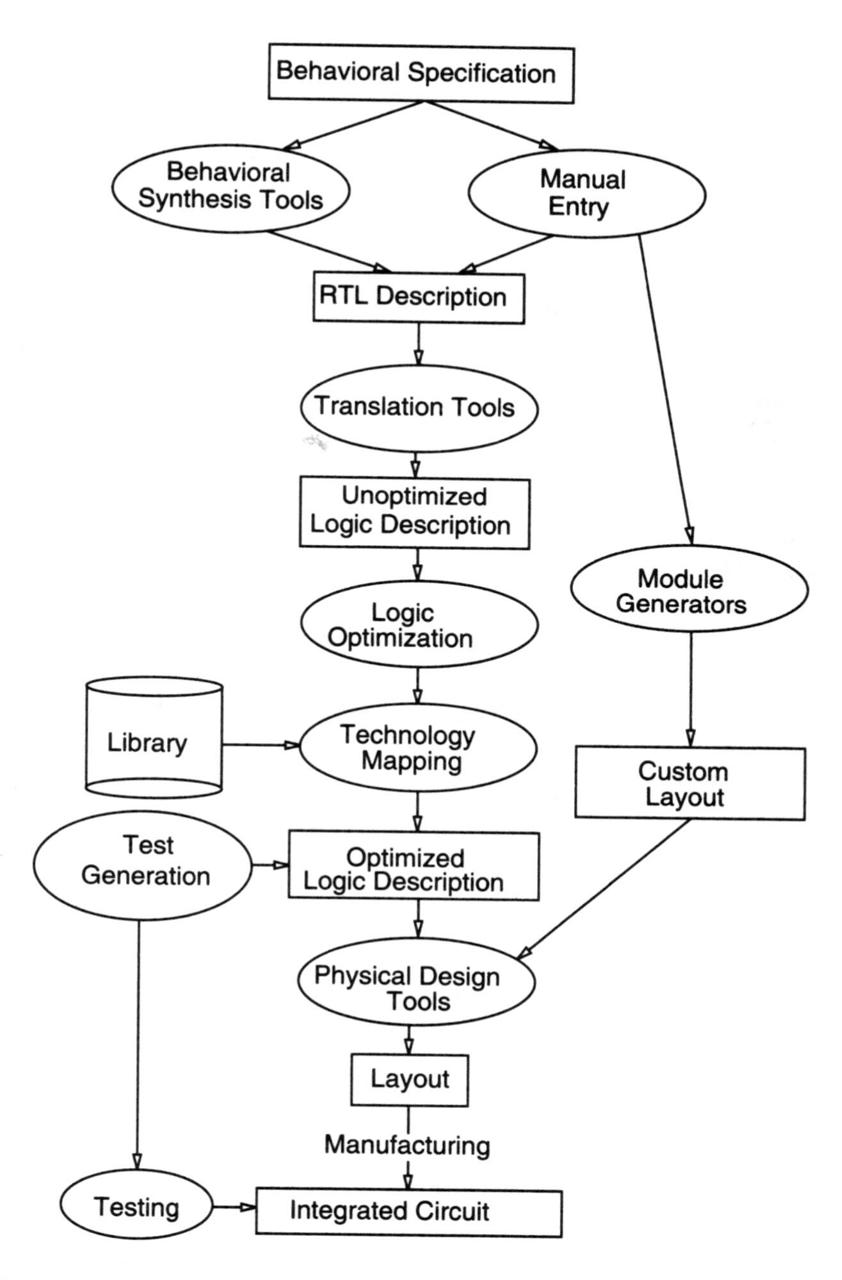

Figure 1.1: A typical synthesis design flow

Because the translation from the HDL is straightforward, the resulting gate level description often is a suboptimal logic implementation. The task of logic optimization tools [4, 8, 9, 10, 24] is to transform this description into a closer-to-optimal implementation of the circuit in terms of area and speed. A goal of logic optimization is to minimize the area while meeting the speed constraints. Another objective is to improve the testability of the circuit [6, 26, 27, 30]. The combinational logic implementation can either be a two-level implementation or a multilevel implementation.

1.7.3 Two-level Logic Optimization

There are a variety of two-level logic implementations. The most common one is the sum-of-products implementation, where the first level of logic corresponds to AND gates and the second level to OR gates. NOR-NOR structures, NAND-NAND structures, AND-XOR structures, and OR-AND structures are also possible. Two-level logic is typically implemented as a *programmable logic array* (PLA) [33] in a NOR-NOR form followed by inverters at the outputs. PLAs have the advantage of being very structured and are therefore amenable to automated logic and layout synthesis. A disadvantage of two-level representations (from which the implementation is derived) is that many regular functions have a minimum two-level representation whose size grows exponentially with the number of inputs to the function (e.g., parity functions and adders). However, PLAs and two-level logic circuits can efficiently implement control logic. Further, optimized two-level sum-of-products representations are often used as a starting point for multilevel synthesis. For these reasons, the problem of two-level logic optimization is of great interest.

1.7.4 Multilevel Logic Optimization

Because multilevel logic can often result in a faster or smaller implementation of a function than two-level logic, synthesis of multilevel logic has received considerable attention over the past decade (e.g., [8, 9]). Efficient algebraic optimization methods were proposed [12] and successfully implemented in the MIS-II program [9]. The program BOLD [8] uses Boolean optimization methods that exploit external and internal "don't-care" conditions (roughly speaking, these are conditions under which the particular value of a signal does not matter). The program SOCRATES [7] uses a rule-based approach combined with an algorithmic approach for area and timing optimization. A comprehensive treatment of state-of-the-art in multilevel logic optimization can be found in [11]. It is of great interest to analyze the algebraic and Boolean transformations, used in the various multi-level logic optimization programs, from the standpoint of testability. Constraining these transformations can result in highly testable circuits. However, constraining logic optimization may adversely affect the area and speed of the resulting design. Thus, there are complex tradeoffs involved in testability-driven logic synthesis.

1.7.5 Technology Mapping

The result of logic optimization is an optimized gate-level (also called logic-level) netlist with combinational subcircuits reintegrated with sequential memory elements. This netlist is composed of generic components such as NORs or NANDs. The next step is to efficiently map this netlist into a library of gates available from a semiconductor vendor [23, 44]. This step is called *technology mapping.* Simply translating a netlist of generic components into a cell library is not a challenging process. The real challenge lies in maximally utilizing the components in the library such that the resulting netlist realizes its area, speed, and testability goals. For these reasons technology mapping recapitulates many of the problems faced in higher level logic optimization except that the optimizations are more locally directed.

1.7.6 Physical Design

So far we have described a design flow strictly within a logic synthesis methodology. Some circuitry, such as ROMs or RAMs, are not at all suited for being generated by logic synthesis. For these portions the most common automated design approach is to use *module generators.* Module generators [42, 69] use a combination of layout language and programming language constructs to automatically generate a layout. Module generators are also useful for generating medium-to-high performance datapaths. Datapath portions in the highest-performance ICs are still typically layouts handcrafted at the transistor level. The result of these programs are fixed portions of custom layout.

To produce the final layout the netlists generated by logic synthesis are integrated with the modules from module generators and the result is placed and routed using placement and routing tools. A good survey of the tools required is given in [57] and a large amount of research literature has grown around the topic as well (e.g., [18, 60, 62, 66]). This phase produces the final layout-level description of the circuit. From this final layout a mask can be generated, and this mask is used to manufacture the final product.

In a typical design process, a designer might have to iterate over all the steps of synthesis before an acceptable final circuit is produced. This is because constraints imposed on the design may not be satisfied in the first pass. Information from any level can be fed back to a higher level to enable the tools working at that level to come to better decisions in order to satisfy the design constraints. For example, a logic optimization tool might not correctly estimate the capacitances of wires in a combinational netlist. After layout the precise values of the capacitances can be calculated, back-annotated on the netlist, and used by logic optimization. The synthesis process can be made fully automatic. However, designer intervention and insight is often necessary to produce high-quality circuits. The synthesis process involves the solution of many optimization problems, most of which are conjectured to be computationally intractable. Therefore, most tools use heuristics to obtain solutions.

1.8 Verification

It is not enough that an integrated circuit be designed and implemented, it must work correctly when deployed in a system. For an integrated circuit to work correctly it has to be designed, implemented, and manufactured correctly. A treatment of the verification issues involved in a logic synthesis design flow is beyond the scope of this book; therefore, in this section we will briefly review the problems and give pointers to relevant work on these problems.

1.8.1 Design Verification

The first verification problem encountered is design verification. Design verification involves the checking of the initially entered design of a circuit for correctness against a specification of desired behavior. In a synthesis methodology the initial design is the HDL model described at the register-transfer level. The correctness of this HDL model is most often checked by simulating the HDL model with meaningful input stimuli and checking the output for the expected behavior. A simulation environment targeted for such use is described in [41].

While simulation is currently the most common technique for design verification, the use of formal methods is increasing. An automata-oriented method for verifying properties is presented in

[40]. The problem of verifying the consistency of a behavioral description and a HDL description at the register-transfer level has been the subject of extensive investigation (e.g., [2, 14, 25]). This is currently one of the most active areas of research in computer-aided verification.

1.8.2 Implementation Verification

In Figure 1.1 there are as many as six consecutive steps in translating a HDL model into an IC. The resulting IC will only work correctly if each of the steps has been performed without error; however, each tool in a step is embodied in hundreds-of-thousands of lines of code and may have bugs. Errors may also be introduced because of the misuse of an automatic tool. For these reasons it is necessary to verify, at each design step, that the original and resulting descriptions of a circuit are consistent. *Implementation verification* ensures that the manual design or automatic synthesis process is correct, i.e., the mask-level description correctly implements the specification.

Algorithms for verifying the equivalence of a HDL description against a logic-level description for combinational and sequential circuits are presented in [28, 49]. Logic verification algorithms used to verify the equivalence of two different logic-level descriptions are presented in [20, 28, 49, 51]. Algorithms for verifying the equivalence of logic-level descriptions and layout (layout verification) can be found in [15, 16, 17].

1.9 Manufacture Testing

Even if a circuit is designed and implemented correctly, the resulting manufactured circuits may not function correctly. This is because the complex manufacturing process is not perfect and some defects are introduced. The types of defects depend on the technology. Across various technologies, the most common types of defect during manufacturing are short-circuits, open-circuits, open bonds, open interconnections, bulk shorts, shorts due to scratches, shorts through dielectric, pin shorts, cracks, and missing transistors [13]. Also, the larger the circuit in terms of area the greater the probability of it having a defect. It is necessary to separate the bad circuits from the good ones after manufacturing. It has been shown that the cost of detecting a faulty component is lowest before the component is packaged and becomes a part of a larger system. Therefore, from the point of view of economics testing is a very important aspect of any VLSI manufacturing system.

1.9.1 Fault Detection

There are two aspects of testing: one is fault detection and the other is fault diagnosis. In fault detection only the presence of a fault is detected, whereas in diagnosis the exact location of the fault has to be identified. The testing process involves the application of test patterns to the circuit and comparing the response of the circuit with a precomputed expected response. Any discrepancy constitutes an error, the cause of which is said to be a *physical fault* [13]. Such faults, for digital circuits, can be classified as *static* or *dynamic*. A static fault is one which causes the logic function of a circuit element (elements) or an input signal to be changed to some other function. These faults are also called *logic* faults. Dynamic faults alter the magnitude of a circuit parameter, causing a change in some factor such as circuit speed, current, or voltage levels. These faults are also called *parametric* faults.

Testing must be performed throughout the life of a circuit, since faults may be introduced in the circuit during assembly, storage, and operation. The most commonly occurring faults during storage and operation are due to temperature, humidity, aging, vibration, radiation, and voltage or current stress [13].

Generation of test patterns is a very important problem and has been under investigation for a long time [1, 21, 35, 38, 46, 48, 55, 61, 65]. Test generation may be performed at various levels during the design process. The average case complexity of test generation (empirical), the fault model, and the fault coverage obtained depend on the representation used.

1.9.2 Fault Models

An important issue in testing is the fault model used in test generation. Physical faults are often modeled as logic faults (i.e., faults that alter the logic function being implemented). By doing so, the problem of fault analysis becomes a technology-independent problem. In addition, tests derived for logic faults may be useful for physical faults whose effect on circuit behavior is not well understood or too complex to be analyzed otherwise. The main requirement for the choice of the fault model is that the model should be able to capture the change in functionality caused by most of the commonly occurring physical defects in the circuit.

There are a number of fault models which are used to describe the most common types of physical faults which might occur in a fabricated circuit. These models can be broadly classified into two categories — static (logic) fault models and dynamic (parametric) fault models.

Static fault models are models for those types of physical faults which affect the logic functionality of the circuit and can be detected independent of delays in the circuit. The static fault models considered in this book are the single stuck-at, multiple stuck-at, and bridging fault models.

Dynamic fault models are models for those types of physical faults which do not necessarily affect the static operation of the circuit and thus can only be detected by tests which either measure delay through a circuit or which are applied at the same speed at which the circuit is specified to operate. The three dynamic fault models considered in this book are the gate delay fault, transistor stuck-open fault, and path delay fault models.

The fault model used most often in practice today is the *single stuck-at* fault model, where a single gate input or output in the circuit gets stuck at a 1 or 0 value. We will describe the fault models used in this book in detail in Section 5.2.

Another important issue in testing is fault diagnosis. It is not only important to identify the presence of a fault, but also to locate the fault and find a reason for the existence of the fault. Fault location, which is one aspect of fault diagnosis, is used to debug circuits and fix manufacturing errors. Tests for one fault can simultaneously detect other faults in the circuit. Two faulty circuits might also have identical response for a particular test pattern. Therefore, a test set has to be derived which not only identifies all the faults, but can also help in locating the fault from the analysis of the response. The more comprehensive fault models, described in detail in Section 5.2, have greater diagnosability. For example, if a path delay fault is detected, it is possible to identify the path in the circuit that has become slower. However, if a gate delay fault is detected, it not always possible to identify the gate in the path that has become slower.

Tests are applied to circuits using *automatic test equipment* (ATE). This equipment is usually very expensive and has a relatively short lifetime before it becomes obsolete. The amount of time that each circuit requires for testing determines the testing cost and is therefore very important. This time is determined by the amount of time required to apply the test vectors and the time required to compare the data with the expected response. Testing time is the main motivation for deriving test sets that are relatively small in size.

1.10 Synthesis For Testability

Synthesis for testability addresses two broad issues to improve and ease the testing of circuits. The first problem attacked by synthesis is the automatic synthesis of design for testability circuitry such as scan latches and scan chains. This may seem like a small problem but it greatly eases the problem of generating this circuitry.

The second problem addressed in synthesis for testability is actually modifying the detailed implementation of the circuitry to improve the testability of the circuitry. This is a much more challenging process for synthesis and has been the focus of a great deal of research.

Synthesis algorithms that produce optimal (to be defined later) two-level or multilevel combinational logic circuits can produce fully testable combinational circuits under the single stuck-at fault model [6]. However, removal of redundant single stuck-at faults (i.e., faults for which there are no tests) based on stuck-at fault test generation algorithms is currently the best way of obtaining fully testable combinational logic designs.

Test generation for multifaults and delay faults is a harder problem than test generation for single stuck-at faults. The large number of multifaults and path delay faults in a circuit make explicit consideration of each fault impossible (the number of faults is an exponential function of the number of gates and wires in the circuit). Synthesis for testability approaches can alleviate the burden on test pattern generators by making circuits more testable. Furthermore, constrained synthesis approaches have been developed which can guarantee full multifault or path delay fault testability. Also, the synthesis process can provide insights that can make the subsequent task of test generation easier. The disadvantage of such approaches is that the constraints imposed during logic optimization may adversely affect circuit area and/or speed. However, the penalties can be kept low in many cases by judiciously applying constraints during the synthesis process.

Generating tests for sequential circuits is more difficult than for combinational circuits. Some current algorithms for test generation under the single stuck-at fault model, comprehensively described elsewhere [37], provide means for handling large circuits. However, some circuits have redundant faults, and a significant fraction of the test generation time could be spent in the identification of redundant faults while generating tests for such circuits. The task of test generation will be greatly simplified if circuits are synthesized to be fully testable.

Optimal sequential logic synthesis strategies that result in fully single stuck-at fault testable interacting sequential circuits are described in [30, 29]. These methods rely on extracting a set of don't-care conditions for the circuit and using the don't-cares during combinational logic optimization to derive a fully testable implementation of the circuit. The primary issue in this approach to synthesis for testability is the efficient derivation of the don't-care conditions. For sequential circuits, these don't-cares have been traditionally obtained from the *state transition graph* (STG) [30]. Recently, STG traversal techniques based on *binary decision diagrams* (BDDs) [20] have been used to derive the set of don't-cares as in [47]. The use of register-transfer-level descriptions to derive the don't-cares is described in [36, 37].

Methods for the synthesis of sequential circuits such that they are testable for multiple stuck-at fault and delay fault models are considerably less developed than the corresponding methods for single stuck-at faults. Some approaches have been presented that have tackled the problem (e.g., [3, 19]). A standard scan-design methodology as in [32] can be used to test sequential circuits under the multiple stuck-at fault model, and an enhanced scan-design methodology as in [50] can be used to test sequential circuits under the delay fault model.

1.11 Outline

We introduce terminology regarding Boolean algebra and two-level combinational logic circuits in Chapter 3. Representations and implementations of two-level circuits are presented. We describe the problem of two-level Boolean optimization in Chapter 4 and give details of the state-of-the-art procedures for exact and heuristic two-level Boolean optimization. In Chapter 5, we focus on the test generation and synthesis for testability problems for two-level circuits. For each of the single stuck-at, multiple stuck-at, hazard-free robust path delay, general robust path delay, gate delay, and stuck-open fault models, we give conditions for faults in a circuit to be testable. We describe necessary modifications to the two-level synthesis procedure presented in Chapter 4 to target improved or full testability under the different fault models. Finally, test generation methods for the various fault models are presented.

Chapters 6, 7, and 9 focus on multilevel combinational logic circuits. Representations and implementations of multilevel combinational logic circuits are presented in Chapter 6. State-of-the-art in multilevel logic synthesis is described in Chapter 7. In Chapter 9, we focus on test generation and synthesis for testability for multilevel circuits. For each chosen fault model, we give conditions for faults in a circuit to be testable. We describe necessary modifications to the multilevel synthesis procedures presented in Chapter 7 to target improved or full testability under the different fault models. Finally, test generation methods for the various fault models are presented.

Conclusions and directions for future work are presented in Chapter 10.

REFERENCES

[1] V. D. Agrawal, K-T. Cheng, and P. Agrawal. CONTEST: A Concurrent Test Generator for Sequential Circuits. In *Proceedings of the* 25th *Design Automation Conference*, pages 84–89, June 1988.

[2] K. Apt and D. Kozen. Limits for Automatic Verification of Finite State Concurrent Systems. In *Information Processing Letters*, pages 307–309, 1986.

[3] P. Ashar, S. Devadas, and A. R. Newton. Multiple Fault Testable Sequential Machines. In *Proceedings of the International Conference on Circuits and Systems*, pages 3118–3121, May 1990.

[4] P. Ashar, S. Devadas, and A. R. Newton. Optimum and Heuristic Algorithms for a Problem of Finite State Machine Decomposition. *IEEE Transactions on Computer-Aided Design of Integrated Circuits*, 10(3):296–310, March 1991.

[5] M. R. Barbacci, G. E. Barnes, R. G. Cattell, and D. P. Siewiorek. The ISPS Computer Description Language. Technical report, Dept. of EECS, CMU, Pittsburgh, PA, August 16, 1979.

[6] K. Bartlett, R. K. Brayton, G. D. Hachtel, R. M. Jacoby, C. R. Morrison, R. L. Rudell, A. Sangiovanni-Vincentelli, and A. R. Wang. Multilevel Logic Minimization Using Implicit Don't-Cares. *IEEE Transactions on Computer-Aided Design of Integrated Circuits*, 7(6):723–740, June 1988.

[7] K. Bartlett, W. Cohen, A. J. De Geus, and G. D. Hachtel. Synthesis of Multilevel Logic under Timing Constraints. *IEEE Transactions on Computer-Aided Design of Integrated Circuits*, CAD-5(4):582–595, October 1986.

[8] D. Bostick, G. D. Hachtel, R. Jacoby, M. R. Lightner, P. Moceyunas, C. R. Morrison, and D. Ravenscroft. The Boulder Optimal Logic Design System. In *Proceedings of the International Conference on Computer-Aided Design*, pages 62–65, November 1987.

[9] R. Brayton, R. Rudell, A. Sangiovanni-Vincentelli, and A. Wang. MIS: A Multiple-Level Logic Optimization System. *IEEE Transactions on Computer-Aided Design of Integrated Circuits*, CAD-6(6):1062–1081, November 1987.

[10] R. K. Brayton, G. D. Hachtel, C. McMullen, and A. Sangiovanni-Vincentelli. *Logic Minimization Algorithms for VLSI Synthesis.* Kluwer Academic Publishers, Norwell, MA, 1984.

[11] R. K. Brayton, G. D. Hachtel, and A. L. Sangiovanni-Vincentelli. Multilevel Logic Synthesis. *Proceedings of the IEEE*, 78(2):264–300, February 1990.

[12] R. K. Brayton and C. McMullen. The Decomposition and Factorization of Boolean Expressions. In *Proceedings of the International Symposium on Circuits and Systems*, pages 49–54, Rome, May 1982.

[13] M. A. Breuer and A. D. Friedman. *Diagnosis and Reliable Design of Digital Systems.* Computer Science Press, Woodland Hills, CA, 1976.

[14] M. C. Browne and E. M. Clarke. A High Level Language for the Design and Verification of Finite State Machines. In *IFIP WG 10.2 International Workshop: From HDL Descriptions to Guaranteed Correct Circuit Designs*, pages 269–292, 1986.

[15] R. E. Bryant. Symbolic Verification of MOS Circuits. In *Proceedings of the 1985 Chapel Hill Conference on VLSI*, pages 419–438, December 1985.

[16] R. E. Bryant. Algorithmic Aspects of Symbolic Switch Network Analysis. *IEEE Transactions on Computer-Aided Design of Integrated Circuits*, CAD-6(4):618–633, July 1987.

[17] R. E. Bryant, D. Beatty, K. Brace, K. Cho, and T. Sheffler. COSMOS : A Compiled Simulator for MOS Circuits. In *Proceedings of the* 24th *Design Automation Conference*, pages 9–16, June 1987.

[18] J. Burns, A. Casotto, M. Igusa, F. Marron, F. Romeo, A. Sangiovanni-Vincentelli, C. Sechen, H. Shin, G. Srinath, and H. Yaghutiel. MOSAICO: An Integrated Macro-Cell Layout System. In *Proceedings of the VLSI-87 Conference*, Vancouver, Canada, August 1987.

[19] K-T. Cheng, S. Devadas, and K. Keutzer. Delay-Fault Test Generation and Synthesis for Testability Under a Standard Scan Design Methodology. *IEEE Transactions on Computer-Aided Design of Integrated Circuits*, 12(8):1217–1231, August 1993.

[20] O. Coudert, C. Berthet, and J. C. Madre. Verification of Sequential Machines Using Boolean Functional Vectors. In *IMEC-IFIP International Workshop on Applied Formal Methods for Correct VLSI Design*, pages 111–128, November 1989.

[21] H. Cox and J. Rajski. A Method of Fault Analysis for Test Generation on Fault Diagnosis. *IEEE Transactions on Computer-Aided Design of Integrated Circuits*, 7(7):813–833, July 1988.

[22] J. Darringer, W. Joyner, L. Berman, and L. Trevillyan. Logic Synthesis through Local Transformations. IBM *Journal of Research and Development*, 25(4):272–280, July 1981.

[23] E. Detjens, G. Gannot, R. Rudell, A. Sangiovanni-Vincentelli, and A. Wang. Technology Mapping in MIS. In *Proceedings of*

the International Conference on Computer-Aided Design, pages 116–119, November 1987.

[24] S. Devadas. Optimizing Interacting Finite State Machines Using Sequential Don't Cares. *IEEE Transactions on Computer-Aided Design of Integrated Circuits*, 10(12):1473–1484, December 1991.

[25] S. Devadas and K. Keutzer. An Automata-Theoretic Approach to Behavioral Equivalence. *INTEGRATION, the VLSI Journal*, 12(2):109–129, December 1991.

[26] S. Devadas and K. Keutzer. Synthesis of Robust Delay-Fault Testable Circuits: Theory. *IEEE Transactions on Computer-Aided Design of Integrated Circuits*, 11(1):87–101, January 1992.

[27] S. Devadas and H-K. T. Ma. Easily Testable PLA-Based Finite State Machines. *IEEE Transactions on Computer-Aided Design of Integrated Circuits*, 9(6):604–611, June 1990.

[28] S. Devadas, H-K. T. Ma, and A. R. Newton. On the Verification of Sequential Machines at Differing Levels of Abstraction. *IEEE Transactions on Computer-Aided Design of Integrated Circuits*, 7(6):713–722, June 1988. Addendum in May 1989 issue.

[29] S. Devadas, H-K. T. Ma, and A. R. Newton. Redundancies and Don't Cares in Sequential Logic Synthesis. *Journal of Electronic Testing: Theory and Applications*, 1(1):15–30, February 1990.

[30] S. Devadas, H-K. T. Ma, A. R. Newton, and A. Sangiovanni-Vincentelli. Irredundant Sequential Machines via Optimal Logic Synthesis. *IEEE Transactions on Computer-Aided Design of Integrated Circuits*, 9(1):8–18, January 1990.

[31] S. Devadas and A. R. Newton. Algorithms for Hardware Allocation in Datapath Synthesis. *IEEE Transactions on Computer-Aided Design of Integrated Circuits*, 8(7):768–781, July 1989.

[32] E. B. Eichelberger and T. W. Williams. A Logic Design Structure for LSI Testability. In *Proceedings of the* 14th *Design Automation Conference*, pages 462–468, June 1977.

[33] H. Fleisher and L. I. Maissel. An Introduction to Array Logic. IBM *Journal of Research and Development*, 19(3):98–109, March 1975.

[34] E. Frey. ESIM: A Functional Level Simulation Tool. In *Proceedings of the International Conference on Computer-Aided Design*, pages 48–53, November 1984.

[35] H. Fujiwara and T. Shimono. On the Acceleration of Test Generation Algorithms. *IEEE Transactions on Computers*, C-32(12):1137–1144, December 1983.

[36] A. Ghosh, S. Devadas, and A. R. Newton. Sequential Logic Synthesis for Testability using Register-Transfer Level Descriptions. In *Proceedings of International Test Conference*, pages 274–283, September 1990.

[37] A. Ghosh, S. Devadas, and A. R. Newton. *Sequential Logic Testing and Verification*. Kluwer Academic Publishers, Norwell, MA, 1991.

[38] P. Goel. An Implicit Enumeration Algorithm to Generate Tests for Combinational Logic Circuits. *IEEE Transactions on Computers*, C-30(3):215–222, March 1981.

[39] L. J. Hafer and A. Parker. Register-Transfer Level Digital Design Automation : The Allocation Process. In *Proceedings of the* 15th *Design Automation Conference*, pages 213–219, June 1978.

[40] Z. Har'El and R.P. Kurshan. Software for Analysis of Coordination. In *Proceedings of the International Conference on System Science and Engineering*, pages 382–385, 1988.

[41] D. Hill and D. Coelho. *Multi-Level Simulation for VLSI Design*. Kluwer Academic Publishers, Norwell, MA, 1989.

[42] D. Hill, D. Shugard, J. Fishburn, and K. Keutzer. *Algorithms for VLSI Layout Synthesis*. Kluwer Academic Publishers, Norwell, MA, 1989.

[43] C. Y. Hitchcock III and D. E. Thomas. A Method of Automatic Data Path Synthesis. In *Proceedings of the* 20th *Design Automation Conference*, pages 484–489, June 1983.

[44] K. Keutzer. DAGON: Technology Mapping and Local Optimization. In *Proceedings of the* 24th *Design Automation Conference*, pages 341–347, June 1987.

[45] Z. Kohavi. *Switching and Finite Automata Theory.* Computer Science Press, New York, NY, 1978.

[46] T. Larrabee. Efficient Generation of Test Patterns Using Boolean Difference. In *Proceedings of the International Test Conference*, pages 795–801, August 1989.

[47] B. Lin, H. Touati, and A. R. Newton. Don't Care Minimization of Multilevel Sequential Logic Networks. In *Proceedings of the International Conference on Computer-Aided Design*, pages 414–417, November 1990.

[48] H-K. T. Ma, S. Devadas, A. R. Newton, and A. Sangiovanni-Vincentelli. Test Generation for Sequential Circuits. *IEEE Transactions on Computer-Aided Design of Integrated Circuits*, 7(10):1081–1093, October 1988.

[49] J-C. Madre and J-P. Billon. Proving Circuit Correctness Using Formal Comparison between Expected and Extracted Behaviour. In *Proceedings of the 25th Design Automation Conference*, pages 205–210, June 1988.

[50] Y. K. Malaiya and R. Narayanswamy. Testing for Timing Failures in Synchronous Sequential Integrated Circuits. In *Proceedings of the International Test Conference*, pages 560–571, October 1983.

[51] S. Malik, A. R. Wang, R. K. Brayton, and A. Sangiovanni-Vincentelli. Logic Verification Using Binary Decision Diagrams in a Logic Synthesis Environment. In *Proceedings of the International Conference on Computer-Aided Design*, pages 6–9, November 1988.

[52] E. J. McCluskey. *Introduction to the Theory of Switching Circuits.* McGraw-Hill, New York, NY, 1965.

[53] C. Mead and L. Conway. *Introduction to VLSI Systems.* Addison-Wesley, Reading, MA, 1980.

[54] J. D. Morison, N. E. Peeling, and T. L. Thorp. ELLA: Hardware Description or Specification? In *Proceedings of the International Conference on Computer-Aided Design*, pages 54–56, November 1984.

[55] S. Nitta, M. Kawamura, and K. Hirabayashi. Test Generation by Activation and Defect-Drive (TEGAD). *INTEGRATION Journal*, 3(1):2–12, March 1985.

[56] A. Parker, D. Thomas, D. Siewiorek, M. Barbacci, L. Hafer, G. Leive, and J. Kim. The CMU Design Automation System. In *Proceedings of the* 16th *Design Automation Conference*, pages 73–79, June 1979.

[57] B. Preas and M. Lorenzetti, editors. *Physical Design Automation of VLSI Systems*. Benjamin-Cummings, Menlo Park, CA, 1982.

[58] IEEE Press. *IEEE Standard VHDL Language Reference Manual*. The IEEE, Inc., New York, NY, 1987.

[59] D. Ravenscroft and M. R. Lightner. Functional Language Extractor and Boolean Cover Generator. In *Proceedings of the International Conference on Computer-Aided Design*, pages 120–123, November 1986.

[60] J. Reed, A. Sangiovanni-Vincentelli, and M. Santamauro. A New Symbolic Channel Router: YACR2. *IEEE Transactions on Computer-Aided Design of Integrated Circuits*, CAD-4(3):208–219, July 1985.

[61] M. Schulz, E. Trischler, and T. Sarfert. SOCRATES : A Highly Efficient Automatic Test Pattern Generation System. *IEEE Transactions on Computer-Aided Design of Integrated Circuits*, 7(1):126–137, January 1988.

[62] C. Sechen and A. Sangiovanni-Vincentelli. The TimberWolf Placement and Routing Package. In *Proceedings of the 1984 Custom Integrated Circuit Conference*, pages 522–527, Rochester, NY, May 1984.

[63] R. Segal. BDSYN : Logic Description Translator; BDSIM : Switch-Level Simulator. Master's thesis, University of California, Berkeley, CA, May, 1987. UCB ERL Memo No. M87/33.

[64] C. E. Shannon. A Symbolic Analysis of Relay and Switching Circuits. *Transactions of the AIEE*, 57:713–723, 1938.

[65] S. Shteingart, A. W. Nagle, and J. Grason. RTG: Automatic Register Level Test Generator. In *Proceedings of the* 22nd *Design Automation Conference*, pages 803–807, June 1985.

[66] C. B. Shung, R. Jain, K. Rimey, R. W. Brodersen, E. Wang, M. B. Srivastava, B. Richards, E. Lettang, L. Thon, S. K. Azim, P. N. Hilfinger, and J. Rabaey. An Integrated CAD System for Algorithmic-Specific IC Design. *IEEE Transactions on Computer-Aided Design of Integrated Circuits*, 10(4):447–463, April 1991.

[67] C. E. Stroud, R. R. Munoz, and D. A. Pierce. CONES: A System for Automated Synthesis of VLSI and Programmable Logic from Behavioral Models. In *Proceedings of the International Conference on Computer-Aided Design*, pages 428–431, November 1986.

[68] D. E. Thomas and P. Moorby. *The Verilog Hardware Description Language.* Kluwer Academic Publishers, Norwell, MA, 1991.

[69] S. Trimberger. *An Introduction to CAD for VLSI.* Kluwer Academic Publishers, Norwell, MA, 1987.

[70] C-J. Tseng and D. P. Siewiorek. Automated Synthesis of Data Paths in Digital Systems. *IEEE Transactions on Computer-Aided Design of Integrated Circuits*, CAD-5(4):379–395, July 1986.

[71] S. H. Unger. *Asynchronous Sequential Switching Circuits.* John Wiley and Sons, New York, NY, 1969.

[72] R. A. Walker and D. E. Thomas. Behavioral Transformation for Algorithmic Level IC Design. *IEEE Transactions on Computer-Aided Design of Integrated Circuits*, 8(10):1115–1128, October 1989.

[73] G. Whitcomb and A. R. Newton. Abstract Data Types and High-Level Synthesis. In *Proceedings of the* 27th *Design Automation Conference*, pages 680–685, June 1990.

Chapter 2

Translation from HDL Descriptions

2.1 Introduction

Hardware description languages (HDLs) provide *integrated circuit* (IC) designers with the capability of describing the functionality of a design at a higher, more abstract level of representation than a gate-level representation. Logic synthesis provides an automated means of generating an optimized gate-level circuit from a HDL description. The input to logic synthesis is a *register-transfer level* (RTL) description, written in some HDL, of the circuit that is to be designed. As mentioned in the Chapter 1, RTL descriptions are distinct from behavioral descriptions in that the RTL description has a definite structure while a behavioral description does not. The output of logic synthesis is an optimized gate-level circuit.

Logic synthesis is composed of two steps:

- **Translation**: Conversion of the given RTL description into an unoptimized gate-level description.

- **Optimization**: Technology-independent as well as technology-dependent transformation of the initial gate-level description into a netlist of gates implementable in a target technology; the resulting implementation must meet area, speed, and testability requirements.

Optimization ensures that design quality is not being compromised for increased designer productivity. The focus of this book is on the optimization step; however, in this chapter we will describe the translation step which acts as the bridge between two different levels of abstraction, namely the RTL and the gate level.

In order to perform synthesis from any HDL description a synthesis policy is needed. The purpose of the policy is to enable the designer to predictably produce high quality silicon. The productivity gains afforded by a HDL synthesis methodology are lost if a chip that does not meet the area or speed constraints is synthesized. We describe a language independent synthesis policy in Section 2.2.

Many HDLs exist today (e.g., [1, 3]) but it appears that VHDL [4] is emerging as the industry standard. For this reason, we will focus on the VHDL hardware description language in this chapter, and describe the translation of an RTL description written in VHDL to a gate-level circuit. In Section 2.3 we give an overview of VHDL and of synthesis from RTL descriptions written in VHDL using several examples taken from [2]. These examples illustrate both the constructs supported in VHDL and the translation process. For the rest of this chapter it is assumed that the reader is familiar with VHDL. To gain further familiarity see [2, 4].

2.2 A Policy for Synthesis from a HDL

To ensure successful synthesis a synthesis policy that ensures high quality results is needed. The major requirement is a powerful optimization step following the translation process. The techniques we will describe in Chapters 3 through 9 can be used to optimize for area, speed, or testability. However, there are two other important components of the design policy, namely the description style and the supported language constructs.

2.2.1 Description Style

The style used for design description is a useful mechanism for controlling the translation step in logic synthesis. Different HDL descriptions of the same functionality can yield dramatically different unoptimized networks. It is possible that a badly written HDL description will result in a large unoptimized network, and the optimization steps will not be able to improve it sufficiently for the circuit to meet its constraints. Thus, the final circuit quality is dependent both on the style of the description used and the quality of the optimization methods used.

In practice, there are no strict style guidelines that are followed; rather, a designer is familiarized with description styles for specific examples that are effective starting points for the synthesis process. There are thus proven styles of descriptions for representative hardware modules.

The two areas in which a designer can exert the most influence using style as a control mechanism are in utilizing design hierarchy and taking advantage of predefined blocks or modules. Hierarchy allows a complex design to be partitioned into smaller, more manageable parts. If the given hierarchical HDL description corresponds to a good partitioning of the design, the task of the optimization step is eased considerably. Most design libraries have large building blocks like adders, multipliers, etc. These highly optimized blocks provide the designer with a resource of high quality parts that can ease the entire design task and the task of optimization.

2.2.2 Supported Language Constructs

The language constructs used to support the level of description required for synthesis from RTL descriptions form another important component of the synthesis policy. These constructs correspond to the designer's vocabulary for design description.

Most HDLs were created for simulation purposes and not for synthesis. Synthesis tools have to be able to work around the simulation specific constructs of the language and focus on the portion of the design description that relates to circuit function. Thus, not all constructs in a given HDL are supported by the synthesis process.

Typically, some constructs in the HDL (e.g., logical operators, if and case statements, and static expressions) are fully supported. Others (e.g., overloaded operators, assignment statements and variable declarations) may be constrained to be used in a certain way in order for the translation step to be successful. Finally, some simulation-specific constructs (e.g., concurrent assertion statement and disconnection specification) in the description may be ignored in the translation step.

2.3 Synthesis Examples from VHDL

VHDL is a strongly typed language with a rich set of constructs that enable the description of networks at many different levels of abstraction. The network described could be a collection of gates, chips, or

```
entity VHDL is
   port(
          A, B, C : in BIT;
          Z : out BIT
   );
end VHDL;

architecture VHDL_1 of VHDL is
begin
  Z <= (A and B) or C;
end VHDL_1;
```

Figure 2.1: Entities and architectures

boards. VHDL has many different language constructs that allow one language to be used for the entire design process. We will give examples of VHDL descriptions that use various constructs and illustrate the translation process of deriving gate-level representations from the given VHDL descriptions.

2.3.1 Entities and Architectures

The example of Figure 2.1 illustrates the translation of basic structural constructs of VHDL. The entity section defines the name of a VHDL design. The port declarations correspond to a signal interface to the design. The architecture section declares the body of the design. In this case, the body is a simple combinational logic expression relating `Z` to `A`, `B`, and `C`.

The translation of the above example begins with the instantiation of the input and output ports. The architecture declaration defines the implementation, which in this case is a Boolean expression over `A`, `B`, and `C`. The resulting gate-level circuit is shown in Figure 2.2.

2.3.2 Bit Vectors and Sequential Statements

Consider the VHDL description of Figure 2.3. The description illustrates the use of VHDL's vectored data types. The ports `A` through `C` are 3 bits wide as is port `Z`.

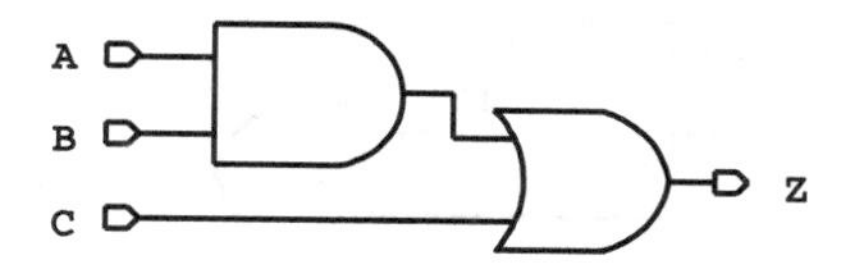

Figure 2.2: Gate-level circuit for an entities and architectures example

```
entity VHDL is
   port(
         A, B, C : in BIT_VECTOR(1 to 3);
         Z : out BIT_VECTOR(1 to 3)
   );
end VHDL;

architecture VHDL_1 of VHDL is
begin
  process (A, B, C)

   variable TEMP : BIT_VECTOR(1 to 3);

  begin
    TEMP := A and B;
    Z <= TEMP or C;
  end process
end VHDL_1;
```

Figure 2.3: Bit vectors and sequential statements

VHDL descriptions may contain processes denoted by the **begin process** and **end process** declarations. Processes can contain sequential statements that cause logic to be cascaded together. The sequential statements correspond to the assignment to the TEMP and Z variables in Figure 2.3.

The gate-level circuit corresponding to the VHDL description is shown in Figure 2.4. During translation, multiple-bit busses are created for signals and variables whose types must be represented by more than one bit. The declared process contains sequential statements that cause a cascaded AND-OR structure in the derived circuit.

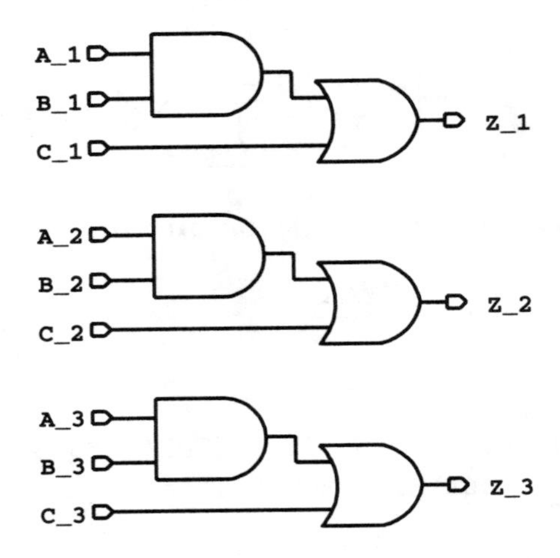

Figure 2.4: Gate-level circuit for a bit vectors and sequential statements example

```
entity VHDL is
   port(
         A, B : in INTEGER range 0 to 7;
         C : out INTEGER range 0 to 7
   );
end VHDL;

architecture VHDL_1 of VHDL is
begin
  C <= A + B;
end VHDL_1;
```

Figure 2.5: Arithmetic operators

2.3.3 Arithmetic Operators

A useful feature enabled by synthesis from HDL descriptions is the support for higher-level arithmetic operators. An example of arithmetic addition over integers is shown in Figure 2.5. Operators other than add such as subtract, compare, increment, and multiply are also supported in most synthesis systems.

The first step in the translation process determines that 3 bits are required for the operands `A`, `B`, and `C`. The `+` operator is

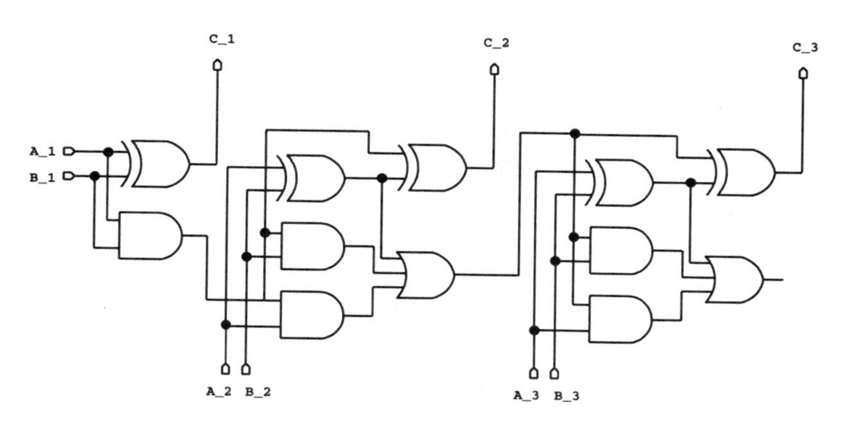

Figure 2.6: Gate-level circuit for an arithmetic operators example

recognized as addition and a 3-bit ripple-carry adder is instantiated as shown in Figure 2.6. The circuit of Figure 2.6 is the starting point for the optimization step which may, for instance, convert the ripple-carry adder into a carry look-ahead adder for greater speed.

2.3.4 If-Then-Else Statements

If-then-else statements are used in HDLs to conditionally execute (sequential) statements. An example of an if-then-else in VHDL is shown in Figure 2.7.

For the purposes of synthesis the if-then-else translates into a multiplexor whose control input is the signal or variable in the `if` clause. Assignments to signals or variables in the `then` or `else` clauses will be multiplexed. The gate-level circuit corresponding to the VHDL description of Figure 2.7 is shown in Figure 2.8.

The case statement in VHDL also implies a multiplexor like the if-then-else statement.

2.3.5 Wait Statements

Wait statements in VHDL are used to signify locations of flip-flops or registers. An example of the use of a wait statement is given in Figure 2.9.

The use of a wait statement signifies to the translation process that the value of the signals must be stored. Registers are inserted wherever they are required to create hardware whose behavior

```
entity VHDL is
   port(
         A, B, USE_B : in BIT;
         Z : out BIT
   );
end VHDL;

architecture VHDL_1 of VHDL is
begin
  process (A, B, USE_B) begin
    if (USE_B = '1') then
      Z <= B;
    else
      Z <= A;
    end if;
  end process;
end VHDL_1;
```

Figure 2.7: If-then-else statement

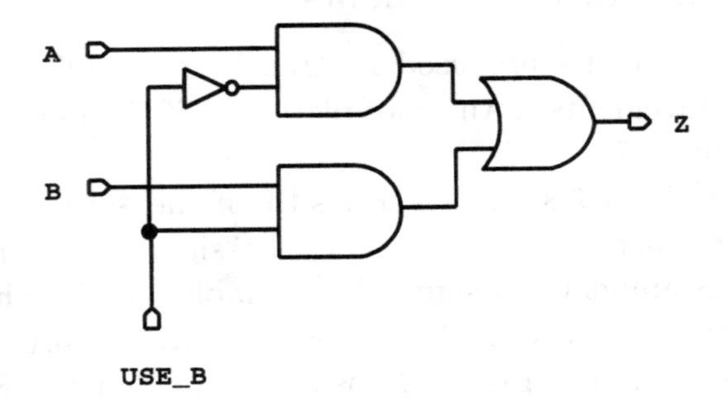

Figure 2.8: Gate-level circuit for an if-then-else statement example

matches that of the VHDL description. Since the signal `TOGGLE` has been assigned after the `wait`, this means that the assignment can only take place at a clock edge, implying that the value of the signal `TOGGLE` is preserved across clock cycles. Therefore, the translation mechanism assigns a flip-flop to the signal `TOGGLE`. This signal is complemented if the signal `ENABLE` is a 1 and remains what it was if `ENABLE` is 0. This means that at the clock edge the new value

```
entity VHDL is
   port(
         ENABLE : in BIT;
         CLOCK : in BIT;
         TOGGLE : buffer BIT
   );
end VHDL;

architecture VHDL_1 of VHDL is
begin
  process begin
    wait until CLOCK'event and CLOCK = '1';
    if (ENABLE = '1') then
      TOGGLE <= not TOGGLE;
    end if;
  end process;
end VHDL_1;
```

Figure 2.9: Wait statement

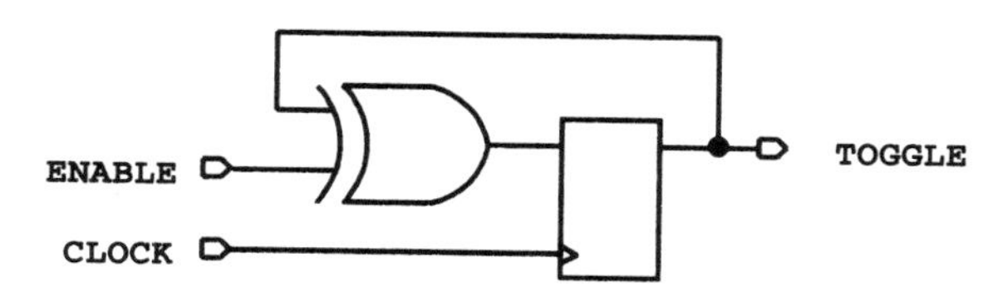

Figure 2.10: Gate-level circuit for a wait statement example

of TOGGLE is the XOR of the old value of TOGGLE and ENABLE. The gate-level circuit corresponding to Figure 2.9 is shown in Figure 2.10.

Most *very large scale integrated* (VLSI) circuits have many registers. Thus, support for wait statements is crucial for a complete synthesis process. After translation the synthesis process focuses on all the register-bounded combinational logic blocks that are generated and optimizes them. In addition, sequential optimization techniques can also be used.

2.3.6 Parameterizable Circuits

We end with a final, relatively large example that illustrates some

```
package MATH is
   function add(L, R : BIT_VECTOR)
            return BIT_VECTOR;
end MATH;

package body MATH is
   function add(L, R : BIT_VECTOR)
            return BIT_VECTOR is
     variable carry : BIT
     variable A:BIT_VECTOR(L'length-1 downto 0);
     variable B, sum:BIT_VECTOR(L'length-1 downto 0);
   begin
     A := L;
     B := R;
     carry := '0';

     for i in 0 to A'left loop
       sum(i) := A(i) xor B(i) xor carry;
       carry := (A(i) and B(i)) or (A(i) and carry)
                 or (carry and B(i));
     end loop
     return sum;
  end;
end MATH;

use work.MATH.all;
entity test is
  port(ARG1, ARG2 : in BIT_VECTOR(1 to 2);
       RESULT : out BIT_VECTOR(1 to 2));
end test

architecture BEHAVIOR of test is
begin
  RESULT <= add(ARG1, ARG2);
end BEHAVIOR
```

Figure 2.11: Parameterizable adder

language constructs and some aspects of translation not seen in the previous examples.

VHDL provides the designer with the capability to create parameterizable functions, i.e., functions which can be used for operands of any bit width. In Figure 2.11 we have an example of a parameterizable adder.

A VHDL package is a library of commonly used arithmetic or control functions and is normally stored in a separate file. The `MATH` package stores the function `add`. In the architecture construct the function `add` is invoked with the appropriate arguments. The translator will create hardware for a function whenever that function is called. Functions can be called several times and can be nested.

The description of the `add` function contains array indices. These are particularly useful in conjunction with `for` loops and can be used to iterate sections of logic.

The `L` and `R` arguments of the `add` function are declared with an unconstrained array type `BIT_VECTOR`. When an unconstrained array type is used for an argument to a function, the actual constraints of the array are taken from the values that are passed to the function. In this example the arguments to the function, `ARG1` and `ARG2`, are declared as `BIT_VECTOR(1 to 2)`. This causes the function `add` to work on 2-bit arrays.

Within the function `add` two temporary variables `A` and `B` are declared. These variables are created to be the same length as `L` and `R` and have uniform constraints from `L'length-1 downto 0`. By assigning arguments `L` and `R` to `A` and `B` the array arguments are normalized to a known index value. Once the arguments are normalized, a ripple carry adder can be easily created by using a `for` loop.

Throughout the function `add` there are no explicit references to a fixed array length. Instead the VHDL constructs `'left` and `'length` are used. This allows the function to work on arrays of any length.

The gate-level circuit obtained after translating the above VHDL description will correspond to a 2-bit ripple carry adder.

2.3.7 Greatest Common Divisor Circuit

We illustrate how large sequential circuits containing both control and arithmetic logic can be compactly specified in VHDL using the example of Figure 2.12.

The circuit computes the greatest common divisor of two given numbers corresponding to the `xi` and `yi` inputs. The inputs

```
entity VHDL is
   port(
         xi, yi : in BIT_VECTOR(1 to SIZE);
         gcd : out BIT_VECTOR(1 to SIZE);
         x, y : buffer BIT_VECTOR(1 to SIZE);
         rst : buffer BIT;
   );
end VHDL;

architecture VHDL_1 of VHDL is
begin
  process begin
    variable gtr, equ : BIT;

    wait until CLOCK'event and CLOCK = '1';
    if (rst = '1') then
      x <= xi;
      y <= yi;
      rst <= '0';
    else
      gtr := x > y;
      if (gtr = '1') then
        x <= x - y;
      else
        equ := x = y;
        if (equ = '1') then
          gcd <= x;
          rst <= '1';
        else
          y <= y - x;
        end if;
      end if;
    end if;
  end process;
end VHDL_1;
```

Figure 2.12: Greatest common divisor computation

have a bit-width SIZE. The greatest common divisor is computed at the output gcd.

There are three sets of registers in the circuit corresponding to `x`, `y`, and `rst`. Registers `x` and `y` store the intermediate values of the computation. Register `rst` is used to signify the end of a computation and the beginning of another computation (on a new pair of numbers).

Initially `rst` stores the value 1. The inputs `xi` and `yi` are sampled and stored in the registers `x` and `y` and `rst` is set to 0. In the next clock cycle `x` and `y` are compared to see if the value stored in `x` is greater than the value stored in `y`. If so, the new value of `x` becomes the old value of `x` minus the old value of `y`. The value of `y` is unchanged. If not, we check if the values of `x` and `y` are equal. If they are equal, the computation is complete, and the value of `x` is the greatest common divisor. (Since the computation is complete, the value of `rst` is set to 1.) If not, the value of `y` is updated to be the old value of `y` minus the value of `x` and the computation will continue in the next clock cycle.

The sequential logic circuit corresponding to the VHDL description of Figure 2.12 is shown in Figure 2.13. There are three sets of registers generated by the **`wait`** statement, several multiplexors generated by the **`if`** statements, and combinational blocks generated by the arithmetic operators.

The sequential circuit corresponding to Figure 2.13 can be optimized for area, speed, and testability. The structure and size of the combinational logic portion of the sequential circuit is the predominant determiner of the various parameters. Therefore, in the next few chapters we will focus on the optimization of combinational logic for area, speed, and testability.

2.4 Dependence on Input Description Style

A limitation of current logic synthesis systems is that a description style has to be enforced on the input RTL description in order to produce quality results. Further, there is no formal definition of such a description style. Examples that are effective starting points constitute the definition of the style. If a description is badly written, the final optimized circuit is unacceptable from an area or speed standpoint in many cases.

For example, a description of an adder-subtracter can be written as shown in Figure 2.14. The translation step will produce a circuit consisting of an adder, a subtracter, and a multiplexor with

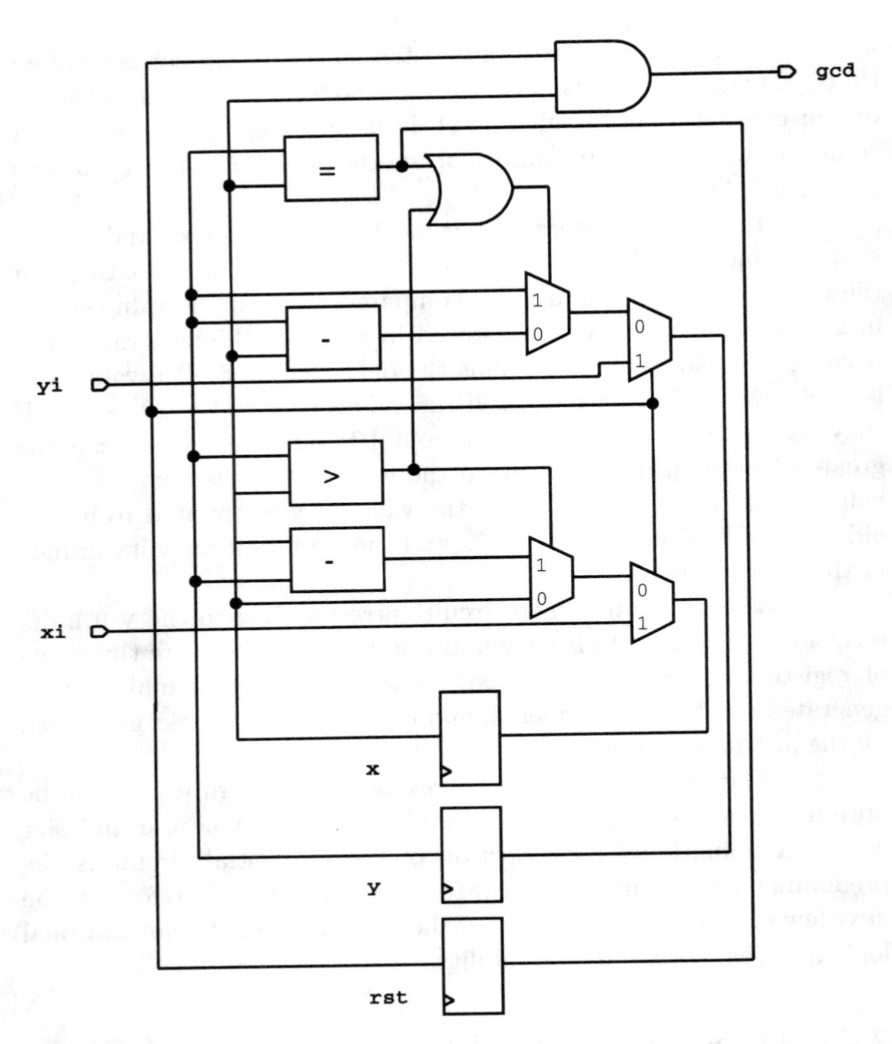

Figure 2.13: Gate-level circuit for a greatest common divisor example

control input `SUB` that chooses between `c` and `d`. The initial circuit will have, approximately, twice the area of an adder. Consider the equivalent description in Figure 2.15.

The translation step will produce an adder with multiplexors at each input to the adder as the initial circuit. This circuit will have an area approximately equal to that of an adder.

Running the optimization targeting minimum area on the circuit obtained from the first description may recover some fraction

```
c := a + b;
d := a - b;
if (SUB = '1') then
  result <= d;
else
  result <= c;
end if;
```

Figure 2.14: Add-subtract computation

```
if (SUB = '1') then
  b := - b;
end if;
result <= a + b;
```

Figure 2.15: Alternate add-subtract computation

of the area, but in general the optimized circuit obtained from the first description will not be as small as even the initial circuit corresponding to the second description. This is because the multiplexors at the outputs of the original circuit corresponding to Figure 2.14 have to be pushed across many levels of logic all the way to the circuit inputs during the optimization.

One approach to solving this problem is to increase the power of the optimization step so it can recover from badly written input. Another solution is to precisely define the characteristics of a "good" RTL description that is suitable as a starting point for logic synthesis and then to manipulate the given RTL description into this form.

Problems

1. Draw a gate-level circuit corresponding to the VHDL description of Figure 2.16. How many flip-flops does the circuit have?

2. Write a VHDL description for a parameterizable adder-subtracter

```
entity VHDL is
   port(
         ENABLE : in BIT;
         CLOCK : in BIT;
         COUNT : in BIT_VECTOR(1 to 2)
   );
end VHDL;

architecture VHDL_1 of VHDL is
begin
  process begin
    wait until CLOCK'event and CLOCK = '1';
    if (ENABLE = '1') then
      COUNT <= COUNT + 1;
    end if;
  end process;
end VHDL_1;
```

Figure 2.16: VHDL description with a wait statement

function `add_sub` using the description of Figure 2.11 as a starting point. The function `add_sub` will have an extra argument `ADD : BOOLEAN`. If `ADD` is 1, the addition operation should be performed on `L` and `R`, else `R` should be subtracted from `L` to produce the result. (Such a description can be found in [2].)

REFERENCES

[1] M. R. Barbacci, G. E. Barnes, R. G. Cattell, and D. P. Siewiorek. The ISPS Computer Description Language. Technical report, Dept. of EECS, CMU, Pittsburgh, PA, August 16, 1979.

[2] S. Carlson. *Introduction to HDL-Based Design Using VHDL*. Synopsys, Inc., Mountain View, CA, 1991.

[3] J. D. Morison, N. E. Peeling, and T. L. Thorp. ELLA: Hardware Description or Specification? In *Proceedings of the International Conference on Computer-Aided Design*, pages 54–56, November 1984.

[4] IEEE Press. *IEEE Standard VHDL Language Reference Manual.* The IEEE, Inc., New York, NY, 1987.

Chapter 3

Two-Level Combinational Circuits

3.1 Introduction

Modern approaches to logic optimization began with the study of two-level combinational circuits, and many algorithms are easier to understand on two-level circuits than on their multilevel counterparts. Furthermore, optimized two-level sum-of-products representations are often used as a starting point for multilevel synthesis. For these reasons, we begin with two-level combinational logic circuits in this chapter. Two-level logic circuits may be represented and implemented in many ways. The most common implementation is obtained directly from the sum-of-products representation, an example of which is shown in Figure 3.1(a). The first level of gates are AND gates and the second level of gates are OR gates. Note that, strictly speaking, the inverters on the primary inputs make this a three-level circuit.

The function of Figure 3.1(a) can be reexpressed in product-of-sums form and implemented as the circuit shown in Figure 3.1(c). A sum-of-products implementation can be directly converted into an equivalent NAND-NAND implementation by replacing all the AND gates and OR gates by NAND gates. A NAND-NAND implementation of the function of Figure 3.1(a) is shown in Figure 3.1(b). Similarly, a product-of-sums implementation can be directly converted into a NOR-NOR implementation as shown in Figure 3.1(d).

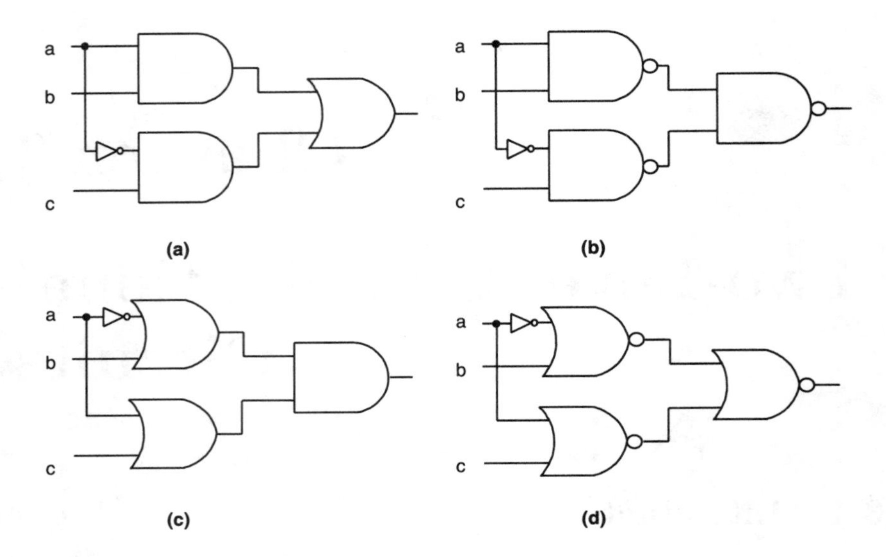

Figure 3.1: Two-level logic implementations

3.2 Terminology

A *completely specified Boolean function* f of N input variables, $x_1, \cdots, x_N$, and of M output variables, $f_1, \cdots, f_M$, is a mapping $f : B^N \rightarrow B^M$ where $B^N = \{0,1\}^N$ and $B^M = \{0,1\}^M$. B^N can be modeled as a binary N-cube. For each output f_i of f, the *ON-set* can be defined to be the set of input values x such that $f_i(x) = 1$. Similarly, the *OFF-set* is the set of input values x such that $f_i(x) = 0$. A function in which $M = 1$ is a *single-output function*, and a function with $M > 1$ is a *multiple-output function*.

The *complement* of a completely specified logic function f, denoted as $\overline{f}$, is a completely specified logic function such that the *ON*-set of $\overline{f}$ is equal to the *OFF*-set of f and the *OFF*-set of $\overline{f}$ is equal to the *ON*-set of f.

An *incompletely specified Boolean function* f of N input variables, $x_1, \cdots, x_N$, and M output variables, $f_1, \cdots, f_M$, is a mapping $f : B^N \rightarrow Y^M$ where $B^N = \{0,1\}^N$ and $Y^M = \{0,1,X\}^M$, such that for some $x \in B^N$, $f_i(x) = X$, where X signifies a don't-care, i.e., the value of the function $f_i(x)$ does not matter. The outputs of an incompletely specified Boolean function f can assume the *don't-care* (DC) value in addition to the logic values 0 and 1. The *DC-set* is the set of input values x such that $f_i(x) = X$, where X or $-$ represents the don't-care value.

A *literal* is a Boolean variable or its complement. A *cube* is a conjunction of literals. A cube has two parts, an *input part* and an *output part*. The input part is defined to be a set of literals and is interpreted as a product of literals. For example, if $\{a, b, c\}$ is the input part of a cube, then it is interpreted as the product term $a \cdot b \cdot c$. The input part of the cube may be abbreviated as abc. The output part of the cube is the set of outputs to whose *ON*-set the cube belongs. In the case of single-output functions, the output is clear, and we will not always specify the output part of the cube. We only specify whether a cube is part of the *ON*-set, *OFF*-set, or *DC*-set.

A *cube* can also be written in bit-vector notation. A cube of a Boolean function f with N inputs and M outputs is written as $c = [c_1, \cdots, c_N, c_{N+1}, \cdots, c_{N+M}]$. For $1 \leq i \leq N$, c_i is 0 if variable x_i appears complemented in c, c_i is 1 if variable x_i appears uncomplemented in c, and c_i is $-$ or X if x_i does not appear in c. For $N + 1 \leq i \leq N + M$, c_i is 0 if c belongs to the *OFF*-set of the output $i - N$ of f, c_i is 1 if c belongs to the *ON*-set of output $i - N$ of f, and c_i is $-$ if c belongs to the *DC*-set of output $i - N$ of f. For single-output functions, we may not write the $N + 1^{th}$ bit of the cube if the function is fully specified.

A *minterm* is a cube in which every variable in the Boolean functions appears. The minterm may be interpreted as a *vertex* in the Boolean N-cube. For example, a single-output function with three inputs a, b, and c can be viewed as a physical cube as shown in Figure 3.2(a), with each of the minterms being represented by the labeled vertices. A function f is shown in Figure 3.2(b), with f(000, 011, 100, 101, 110) = 1, $f(010) = X$, and $f(001, 111) = 0$.

A minterm m_1 is said to *dominate* minterm m_2 (denoted by $m_1 \succ m_2$) if for each position that m_2 has a 1, m_1 also has a 1.

Minterms and cubes may be used to represent the values of a set of input variables, e.g., $x\overline{y}z$ is shorthand for $x = 1,\ y = 0$, and $z = 1$. Therefore, there is a natural correspondence between an input vector or input stimulus, a minterm, and a vertex in the N-cube. This correspondence may be extended to cubes where unspecified values in the function are assumed to be undefined values. Thus, if a circuit C has inputs $v,\ w,\ x,\ y$, and z then applying the cube $x\overline{y}z$ to C is shorthand for applying $v = X,\ w = X,\ x = 1,\ y = 0$, and $z = 1$. (Here X is used to denote an unknown value.)

For single-output functions, the *distance* δ between a cube q and a cube r is defined as the cardinality of the set: $\{l | (l \in q) \wedge (\bar{l} \in$

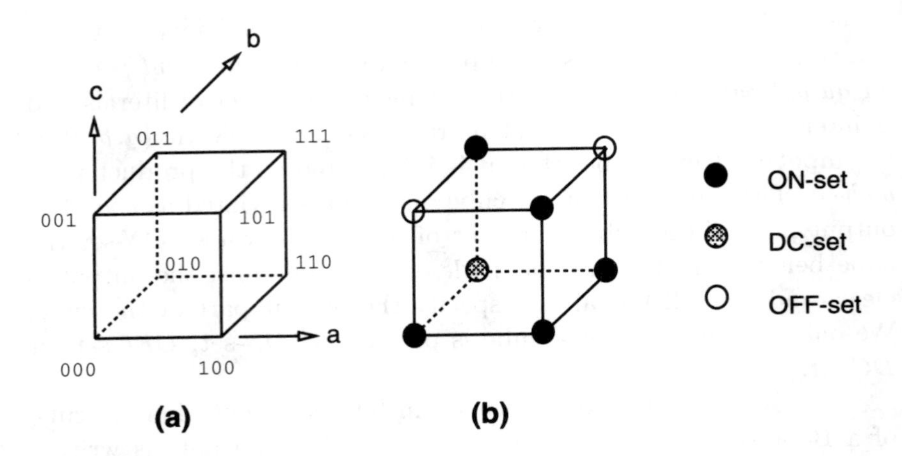

Figure 3.2: Boolean N-cube for function f

$r)\}$. For example, the distance between abc and $ab\bar{c}$ is 1; the distance between ab and abc is 0; and the distance between $b\bar{c}def$ and $abcf$ is 1. For multiple-output functions, if the intersection of the output parts of cubes q and r is empty, then the distance between q and r is the distance between the input parts plus 1, else it is the distance between the input parts.

The *universal cube* or *universe* is the cube with all $-$ entries, or no literals.

A *cover* is defined as a set of cubes and is interpreted as a sum-of-products expression for a function, e.g., $\{\{a, b, c\}, \{d, e, f\}\}$ is a single-output cover, interpreted as $abc + def$.

A cube q *contains* a cube (or vertex) r if the literals in the input part of cube q are a subset of the literals in the input part of cube r and the outputs in the output part of q are a superset of the outputs in the output part of cube r. In bit-vector notation, the cube $0-\ \ 1$ of a two-input, single-output function contains the cube $00\ \ 1$. Similarly, the cube $0-\ \ 11$ of a two-input, two-output function contains the cube $0-\ \ 10$. A cube is said to be contained by a cover if every minterm contained by the cube is contained by some cube in the cover. For example, the cover $\{00--, -1-1\}$ contains the cube $0--1$.

If a cube q contains only *ON*-set and *DC*-set vertices of a Boolean function f, then q is called an *implicant* of f. A *prime implicant* or prime of f is an implicant which is not contained by any other implicant of f and which is not entirely contained in the *DC*-

set of f. An *essential prime implicant* or essential prime is a prime implicant which includes one or more *ON-set* vertices which are not included in any other prime implicant. These vertices are termed *essential vertices.* An *optional prime implicant* is a prime implicant for which all vertices are included in other prime implicants.

A minimal cover for a function f is generated by selecting all of the essential prime implicants and a minimal set of optional prime implicants such that all vertices in the *ON*-set of f are included in the cover. For the example in Figure 3.2(b), there are 3 essential prime implicants and no optional prime implicants. The minimal cover would be $f = \overline{c} + \overline{a}\, b + a\, \overline{b}$.

A *relatively essential vertex* of a cube q in a cover C is a vertex in the *ON*-set that is contained by q and is not contained in any other cube in C. In the example of Figure 3.2(b), $a\,\overline{b}\,c$ is a relatively essential vertex of the cube $a\,\overline{b}$, while the other vertex in this cube, $a\,\overline{b}\,\overline{c}$, is not a relatively essential vertex since it is also contained in the cube $\overline{c}$.

A two-input, two-output function can also be represented as a multiple-output cover, with cubes that have input as well as output parts. For example, the two-output function $\mathcal{F} = \{11\ \ 01,\ 00\ \ 10,\ 10\ \ 11\}$ has two cubes in each of its components $\mathcal{F}_1$ and $\mathcal{F}_2$. If the inputs are a and b, $\mathcal{F}_1$ can be represented algebraically as $\overline{a} \cdot \overline{b} + a \cdot \overline{b}$, and $\mathcal{F}_2$ is $a \cdot b + a \cdot \overline{b}$. The cube $a \cdot \overline{b}$ is shared by $\mathcal{F}_1$ and $\mathcal{F}_2$, because its output part indicates that it belongs to both their *ON*-sets.

Given a Boolean function, the function resulting when some argument x_i of the function f is replaced by a constant b is called a *cofactor* of the function and is denoted $f|_{x_i=b}$. That is, for any arguments $x_1, \cdots, x_n$,

$$f|_{x_i=b}(x_1, \cdots, x_n) \;=\; f(x_1, \cdots, x_{i-1}, b, x_{i+1}, \cdots, x_n)$$

Using this notation the *Shannon expansion* of a function around variable x_i is given by:

$$f \;=\; x_i \cdot f|_{x_i=1} \;+\; \overline{x_i} \cdot f|_{x_i=0}$$

We can also cofactor a Boolean function with respect to a cube c by sequentially cofactoring the function with each literal in the cube.

We denote the value at the output of a circuit C on the application of an input vector v by $C(v)$. For example, the circuit $abc + def$ evaluates to 1 on the vector $abcdef$. We denote the Boolean value of a cube q given a vector v of inputs by $q(v)$. For example,

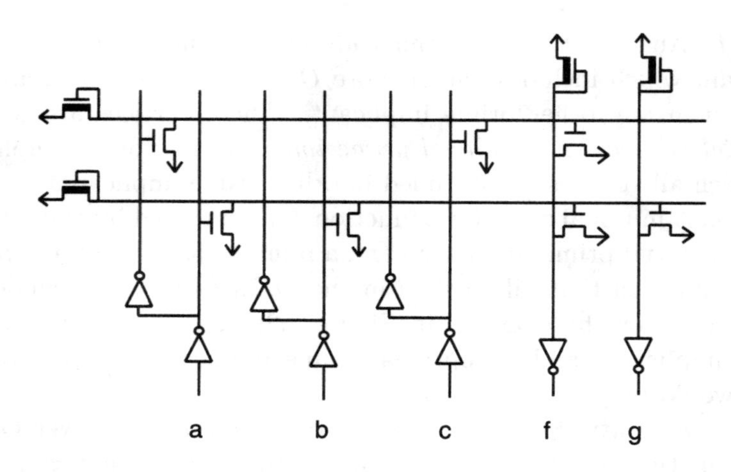

Figure 3.3: Programmable logic array

given the cube $q = abc$, $q(abcdef) = 1$. Similarly, we denote the Boolean value of a literal l in a cube q for an input vector v by $q_l(v)$. For example, given the cube $q = abc$, $q_b(abcedef) = 1$.

3.3 Programmable Logic Arrays

Two-level circuits can be implemented as *programmable logic arrays* (PLAs) [1]. A CMOS PLA is shown in Figure 3.3, whose output marked f implements the logic function of Figure 3.1. Note that while the input plane and output plane are both NOR planes, we have inverters at the outputs. A sum-of-products representation can be directly mapped to a NOR-NOR PLA with output inverters by complementing each literal in the input plane. The function $f = a \cdot b + \overline{a} \cdot c$ has been implemented as $\overline{(\overline{a} + \overline{b}) \cdot (a + \overline{c})}$.

It is of great interest to minimize the number of product terms, i.e., rows in the PLA, as well as tō minimize the number of literals, i.e., transistors in the PLA. PLAs can implement multiple-output functions that share product terms across outputs as shown in Figure 3.3. The multiple-output cover is represented as shown in Figure 3.4. The two outputs share the first cube in their *ON*-sets, i.e., $a \cdot b$. Similarly, in the PLA of Figure 3.3 the first row from the bottom feeds transistors in both columns in the output plane. The number of columns in a PLA equals two times the number of inputs plus the number of outputs, the number of rows equals the number of product terms in the cover, the number of transistors in the input plane equals

```
11-  11
0-1  10
```

Figure 3.4: Multiple-output cover

the number of 1 or 0 literals in the input part of the multiple-output cover, and the number of transistors in the output plane equals the number of 1s in the output part of the multiple-output cover.

3.4 Primality and Irredundancy Properties

We will define the fundamental notions of primality and irredundancy of single-output and multiple-output functions in this section. The properties of primality and irredundancy have an intimate relationship to stuck-at fault testability as will be shown in Chapter 5.

As defined in Section 3.2, an implicant c of a function f is prime if it is not contained in any other implicant d or is not entirely contained in the *DC*-set of f. An alternate definition is as follows: an implicant c is prime if no 0 or 1 literal can be *raised* to a $-$ (don't-care) without the resulting implicant intersecting the *OFF*-set of any component of the multiple-output function. (Changing a literal to a don't-care is often called *raising the literal* because a don't-care is sometimes represented by the value 2, and therefore changing a literal from a value of 0 or 1 to a 2 corresponds to raising its value.) For instance, a cube 111 of a three-input, single-output function would be a prime cube if each of $11-$, $1-1$ and -11 intersected the *OFF*-set. A literal in a cube is said to be prime if raising that particular literal to a $-$ results in a cube that intersects the *OFF*-set. Thus, 110 may not be a prime cube of a function f because $11-$ is an implicant of f, but the first two literals may be prime in the implicant 110 because -10 and $1-0$ intersect the *OFF*-set of f. It is clear that all the literals contained in a cube have to be prime in order for the cube to be prime.

An implicant in a cover is *irredundant* if it contains an essential or a relatively essential vertex. Else it is *redundant*. A redundant cube can be removed from a cover without changing the functionality of the cover.

A cover is prime if each of the implicants in the cover is prime. A cover is irredundant if each of the implicants is irredundant.

The above definitions apply to both completely specified

and incompletely specified functions. A prime cube can intersect the *DC*-set of a function (but cannot be contained in it), and a relatively essential vertex has to be in the *ON*-set of the function.

3.5 Boolean Operations on Logic Functions

Operations on Boolean functions can be defined on single output as well as on multiple-output functions. In the case of multiple-output functions, the same Boolean operations are performed component-wise on the outputs.

The *intersection* or *product* of two logic functions f and g, $h = f \cdot g$ is defined to be the logic function h whose components h_i have *ON*-sets equal to the intersection of the *ON*-sets of the corresponding components of f and g, and have *OFF*-sets equal to the union of the *OFF*-sets of the corresponding components of f and g. The *DC*-sets of the h_i can be derived using the fact that the union of the *ON*-, *OFF*-, and *DC*-sets should be the universal set.

The *difference* between two logic functions f and g, $h = f - g$ is a logic function given by the intersection of f with the complement of g.

The *union* or *sum* of two completely specified logic functions f and g, $h = f \cup g$ or $h = f + g$ is a completely specified logic function such that the *ON*-sets of the components of h, i.e., h_i, are the union of the *ON*-sets of f_i and g_i. We can generalize the definition of sum to incompletely specified functions, as we did for intersection.

A completely specified logic function is a *tautology* written as $f \equiv 1$ if the *OFF*-sets of all its components are empty. In other words, the outputs of f are 1 for all inputs.

3.6 Operations on Cubes and Covers

For the purposes of minimizing two-level logic functions and generating tests under various fault models for two-level circuits, efficient procedures for performing Boolean operations on sum-of-products representations or covers are desirable. A package for performing various Boolean operations such as intersection, union, difference and complementation is part of the ESPRESSO program [3].

In this section, we will describe the disjoint SHARP operation, which is useful for computing the difference of two covers. The disjoint SHARP operation finds many applications in two-level logic testing and two-level logic minimization.

3.6.1 Cube Intersection

The intersection of two cubes can be computed in time linear in the number of inputs. Having defined cubes as sets, we compute the intersection of cubes c and d, denoted $q = c \cap d$, by actually taking the *union* of the sets of literals in c and d. If $q = c \cap d$ computed in this fashion contains both a literal l and its complement $\bar{l}$, then the intersection is empty. In bit-vector notation, $q_i = c_i$ if $c_i = d_i$, is c_i if $d_i = X$ and is d_i if $c_i = X$. If $c_i = 1$ and $d_i = 0$ or $c_i = 0$ and $d_i = 1$, then q is empty, i.e., $q = \phi$. For example, the intersection of $11X$ and -10 is 110, and the intersection of $11X$ and $X0X$ is ϕ.

Two covers can be intersected by intersecting each pair of cubes in the covers.

3.6.2 Disjoint SHARP

The disjoint SHARP [2] operation, denoted as $\ominus$, can be used to compute $C - D$ or $C \cap \overline{D}$ where C and D are covers. $C \ominus D$ gives the same result as $C - D$ and all the cubes in $C \ominus D$ have empty pair-wise intersections.

We first give a method of subtracting a cube c from another cube d, i.e., computing $c \ominus d$. We first check that $c \cap d = \phi$. If so, then $c \ominus d = c$. Else,

1. $Q = \phi$. $q = c$.
2. i = 0.
3. i = i + 1. If $i > N$ (number of literals) go to Step 7.
4. If q_i equals d_i or if d_i is $-$, then go to Step 3.
5. If $q_i = -$, $p = q$, except $p_i = \overline{d_i}$. $Q = Q \cup p$.
6. Set q_i to d_i. Go to Step 3.
7. $c \ominus d$ has been computed in Q.

We give an example of subtracting $- - 11$ from the universal cube $- - - -$. We enter Step 5 with $i = 3$. p is set to $- - 0-$ and added to Q. q is set to $- - 1-$. We enter Step 5 again with $i = 4$. p is set to $--10$ and added to Q. q is set to $--11$, and we end at Step 7 with Q being $\{--0-,\ --10\}$.

In order to compute the difference of a cover C and a cube d, $C \ominus d$, we do the following:

1. $Q = \phi$. $C_1 = C$.

2. Pick a (new) cube $c \in C_1$.

3. Compute $I = c \ominus d$.

4. $Q = Q \cup I$. Remove c from C_1.

5. If $C_1 = \phi$, then $C \ominus d$ has been computed in Q. Else go to Step 2.

Consider an example of subtracting -001 from $\{--0-,\ --10\}$. We subtract -001 from $--0-$ to obtain $\{-10-,\ -000\}$. The intersection of -001 and $--10$ is ϕ, and hence the result is $\{-10-,\ -000,\ --10\}$.

In order to compute the difference of a cover C and another cover D, $C \ominus D$ we do the following:

1. $Q = C$. $D_1 = D$.

2. Pick a (new) cube $d \in D_1$.

3. Compute $I = Q \ominus d$ using the above procedure.

4. $Q = I$. Remove d from D_1.

5. If $D_1 = \phi$ then $C \ominus D$ has been computed in Q. Else go to Step 2.

Note that we can compute the *OFF*-set of a completely specified logic function by subtracting its *ON*-set from the universal cube using the disjoint SHARP operation. However, as we show by example in the next section, the complement of a function can be much larger (exponential in terms of the number of inputs to the function) than the function in sum-of-products representation.

3.6.3 Single Cube Containment

In order to keep cover sizes small, it is desirable to ensure some form of minimality for the cover. An easily satisfiable property is that no cube c of a cover contains another cube d of the cover. The above property is called *single cube containment minimality*.

3.7 Complexity of Two-Level Circuits

Two-level circuits can be unmanageable for certain functions. For instance, the parity function over n inputs, i.e., $a_1 \oplus a_2 \cdots \oplus a_n$, has 2^{n-1} product terms in a sum-of-products representation and is equally large in a product-of-sums representation. The Achilles heel function over $2n$ inputs, i.e., $(a_1 + b_1) \cdot (a_2 + b_2) \cdots (a_n + b_n)$, has 2^n product terms in sum-of-products representation but has a linear-sized product-of-sums representation. Thus, complementing the function $a_1 \cdot b_1 + a_2 \cdot b_2 \cdots + a_n \cdot b_n$ in sum-of-products form will result in 2^n product terms in the representation of the complement.

Integer multiplication over n-bit operands, comparison of two n-bit operands, and addition and subtraction of n-bit operands all have sum-of-product and product-of-sums realizations that grow exponentially with n. Multilevel representations are therefore of great practical interest.

Problems

1. Compute the intersection, union, and difference of the two covers f and g below. Ensure that the resulting cover is single cube containment minimal.

$$\begin{aligned} f &= a \cdot b + \overline{a} \cdot \overline{b} + \overline{c} \\ g &= \overline{a} \cdot b + \overline{c} + d \end{aligned}$$

2. Given

$$f = a \cdot \overline{b} + \overline{a} \cdot \overline{b} \cdot c + c \cdot d \cdot \overline{e} + \overline{b} \cdot \overline{c} \cdot d \cdot \overline{e}$$

   (a) Generate all the primes of f.

   (b) List all the essential primes.

   (c) For each essential prime, find a vertex which is essential, i.e., a vertex in that prime but in no other prime.

3. Compute the complement of the function below using the disjoint SHARP operation.

$$f = a \cdot b + c \cdot d + e \cdot f$$

4. Give an example of an irredundant cover that is not prime and a prime cover that is redundant.

5. Show that each cube in a sum-of-products expansion of

$$(a_1 + b_1) \cdot (a_2 + b_2) \cdots (a_n + b_n)$$

is an essential prime implicant.

6. Draw a NOR-NOR PLA that implements the multiple-output function below and uses a minimum number of rows.

$$\begin{aligned} f_1 &= a \cdot b + \overline{a} \cdot \overline{b} + \overline{c} \\ f_2 &= \overline{a} \cdot b + \overline{c} + d \end{aligned}$$

7. Use the Shannon expansion to prove the two identities below.

$$(f \cdot g)_{x_i} = f_{x_i} \cdot g_{x_i}$$

$$(\overline{f})_{x_i} = \overline{f_{x_i}}$$

8. You are given a cover C of a function f such that C has all the primes of f. Let c be any cube (not necessarily in C). Show that C_c, the cofactor of C with respect to the cube c, has all the primes of f_c.

9. Prove the correctness of the disjoint SHARP operation $C \ominus D$, where C and D are covers, i.e., prove that every pair of cubes generated by $C \ominus D$ is disjoint, that every cube generated is in $C - D$, and that every vertex in $C - D$ is covered by at least one of the cubes.

10. Give an upper bound on the maximum number of cubes that are generated when a n-input cube with k literals is subtracted from another with l literals using the disjoint SHARP operation.

11. Let f be a completely specified function. Suppose C_1 and C_0 are prime covers of f_x and $f_{\overline{x}}$. Prove that a prime cover C of f is obtained by the following operations:

    (a) for each $c \in C_1$ if $c \subseteq C_0$, then $c \in C$, else $x \cdot c \in C$

    (b) for each $c \in C_0$ if $c \subseteq C_1$, then $c \in C$, else $\overline{x} \cdot c \in C$

REFERENCES

[1] H. Fleisher and L. I. Maissel. An Introduction to Array Logic. IBM *Journal of Research and Development*, 19(3):98–109, March 1975.

[2] S. J. Hong, R. G. Cain, and D. L. Ostapko. MINI: A Heuristic Approach for Logic Minimization. IBM *Journal of Research and Development*, 18(4):443–458, September 1974.

[3] R. Rudell and A. Sangiovanni-Vincentelli. Multiple-Valued Minimization for PLA Optimization. *IEEE Transactions on Computer-Aided Design of Integrated Circuits*, CAD-6(5):727–751, September 1987.

Chapter 4

Synthesis of Two-Level Circuits

4.1 Two-Level Boolean Minimization

Two-level Boolean minimization is used to find a sum-of-products representation for a Boolean function that is optimum according to a given cost function. The typical cost functions used are the number of product terms in a two-level realization, the number of literals, or a combination of both.

With any of these cost functions, the problem of two-level Boolean minimization contains the subproblem of finding the solution of a minimum covering problem which has been shown to be *nondeterministic polynomial-time* (NP)-complete [3]. Neverthless, sophisticated exact minimizers (e.g., [2, 6]) have been developed whose average-case behavior for most commonly encountered functions is acceptable. Furthermore, heuristic minimization methods exist (e.g., [1, 4]) which have been shown to produce results that are close to the minimum[1] within reasonable amounts of *central processing unit* (CPU) time, even for large Boolean fuctions.

The two steps in two-level Boolean minimization are:

1. Generation of the set of prime implicants for a given function.

2. Selection of a minimum set of prime implicants to implement the function.

[1]In some cases the minimum result is known because exact minimization is possible; in other cases the result produced by the heuristic minimizer is compared against an established lower bound on the minimum solution.

We will briefly describe the Quine-McCluskey method [5] in Section 4.2, which was the first algorithmic method proposed for two-level minimization and which follows the two steps outlined above. State-of-the-art exact minimization algorithms are all based on the Quine-McCluskey method and also follow the two steps above, but are able to outperform the Quine-McCluskey method significantly due to superior prime generation, implicant table generation, and covering techniques. We describe the techniques used in ESPRESSO-EXACT [6] in Section 4.5. Before describing the algorithm of ESPRESSO-EXACT, we describe the algorithm used to determine if a two-level cover is a tautology in Section 4.3. We briefly describe heuristic methods for logic minimization in Section 4.6.

4.2 The Quine-McCluskey Method

4.2.1 Prime Implicant Generation

The Quine-McCluskey method is best illustrated with an example. Consider the completely specified Boolean function shown in Figure 4.1(a). It has been represented as a list of minterms. Each minterm has an associated decimal value obtained by converting the binary number represented by the minterm into a decimal number — for instance the value of 0000 is 0 and that of 1100 is 14. Each pair of minterms is checked to see if they can be merged into a cube. Two cubes can be merged if they differ only in one position and they both have their X's in the same positions. The cubes generated in our example by merging the pairs of minterms are shown in Figure 4.1(b). These cubes are termed 1-cubes because they have exactly one X or $-$ entry. A minterm is thus a 0-cube. Next, these 1-cubes are examined in pairs to see if they can be merged into 2-cubes. Two 2-cubes can be formed as shown in Figure 4.1(c). If a k-cube is formed by merging two $k-1$-cubes, then the two $k-1$-cubes are not primes and are marked so that they can be discarded later. The process ends when no merging is possible in the final set of L-cubes. The unmarked k-cubes, where $0 \leq k \leq L$, are the complete set of prime implicants of the function. We have five prime implicants for the function in this example.

In the case of incompletely specified functions the initial list of minterms includes those minterms in the ON-set and the DC-set. All prime implicants consisting of minterms only in the DC-set are discarded.

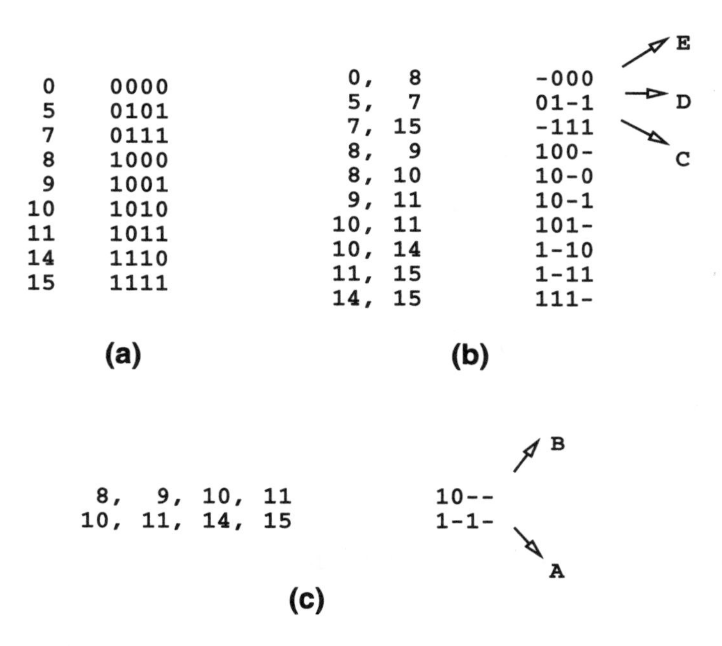

Figure 4.1: Prime implicant generation

4.2.2 Prime Implicant Table

The prime implicants of the function in Figure 4.1 have been marked as **A**, **B**, **C**, **D**, and **E**. We now construct the prime implicant table as shown in Figure 4.2. The rows of the prime implicant table are the minterms of the *ON*-set of the function (we do not need to choose minterms only in the *DC*-set), and the columns are the prime implicants. An "X" in the prime implicant table in row i and column j signifies that the minterm corresponding to row i is contained by the prime corresponding to column j. For instance, the minterm 0000 is contained only by prime **E**, -000.

Given the prime implicant table, we have to select a minimum set of primes (columns) such that there is at least one X in every row. This is the classical minimum unate covering problem which has been shown to be NP-complete [3].

4.2.3 Essential Prime Implicants

A row with a single X signifies an essential prime implicant. Any prime cover for the function will have to contain the prime that contains the minterm corresponding to this row, because this minterm is

	A	B	C	D	E
0000					X
0101				X	
0111			X	X	
1000		X			X
1001		X			
1010	X	X			
1011	X	X			
1110	X				
1111	X		X		

Figure 4.2: Prime implicant table

not contained by any other prime (column). In the prime implicant table of Figure 4.2 **A**, **B**, **D**, and **E** are essential prime implicants.

We select the essential prime implicants since they have to be contained in any prime cover. This results in a cover for the function, since selecting **A**, **B**, **D**, and **E** results in a column of X's.

4.2.4 Dominated Columns

Some functions may not have essential prime implicants. Consider the hypothetical prime implicant table of Figure 4.3(a). There is no row with a single X. It is necessary to make an arbitrary selection of a prime to begin with. Assume that prime **A** is selected. We obtain the reduced table of Figure 4.3(b) after deleting column **A** and the two rows contained by **A**, namely 0 and 1. A column **U** of a prime implicant table is said to dominate another column **V** if **U** contains every row contained by **V**. In the reduced table of Figure 4.3(b) column **B** is dominated by column **C** and column **H** is dominated by column **G**. We can delete the dominated columns, since selecting the dominating column will result in covering more uncontained minterms than the dominated column. Note that the dominating column might not exist in a minimum solution. Further note that if minimizing the literal count was our objective, then we can only delete dominated columns that correspond to primes with

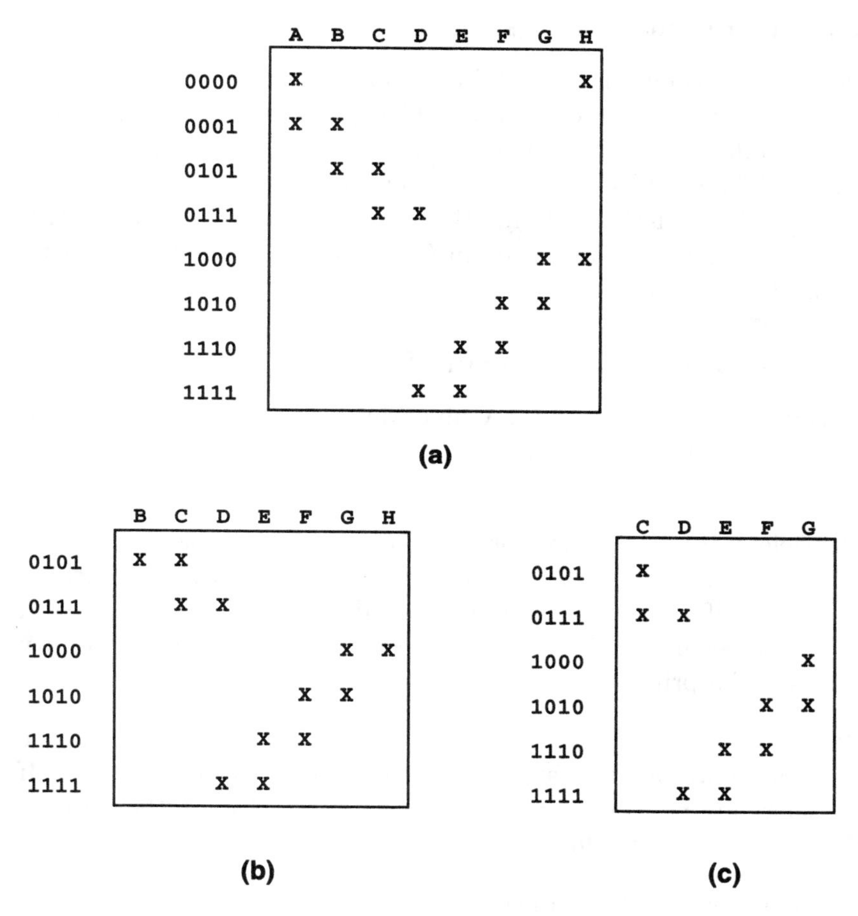

	A	B	C	D	E	F	G	H
0000	X							X
0001	X	X						
0101		X	X					
0111			X	X				
1000							X	X
1010						X	X	
1110					X	X		
1111				X	X			

(a)

	B	C	D	E	F	G	H
0101	X	X					
0111		X	X				
1000						X	X
1010					X	X	
1110				X	X		
1111			X	X			

(b)

	C	D	E	F	G
0101	X				
0111	X	X			
1000					X
1010				X	X
1110			X	X	
1111		X	X		

(c)

Figure 4.3: Cyclic prime implicant table

equal or more literals than the dominating prime.

Reducing the table of Figure 4.3(b) gives the table of Figure 4.3(c). In this table **C** and **G** are relatively essential prime implicants. Choosing **C** and **G** results in the selection of **E**, which completes the cover $f = \{ \mathbf{A}, \mathbf{C}, \mathbf{E}, \mathbf{G} \}$. We are not guaranteed that this cover is minimum; we have to backtrack to our arbitrary choice of selecting prime **A** and delete prime **A** from the table, i.e., explore the possibility of constructing a cover that does not have **A** in it. This results in $f = \{ \mathbf{B}, \mathbf{D}, \mathbf{F}, \mathbf{H} \}$.

4.2.5 Dominating Rows

A row i of a prime implicant table is said to dominate another row j if i has an X in every column in which j has an X. Any minimum expression derived from a table which contains both rows i and j can be derived from a table which only contains the dominated row.

After deleting column **B** from the table of Figure 4.3(b), because it is dominated by column **C**, row 0111 dominates row 0101 and can be deleted.

4.2.6 A Branching Covering Strategy

The covering procedure of the Quine-McCluskey method is summarized below. The input to the procedure is the prime implicant table T.

1. Delete the dominated primes (columns) and the dominating minterms (rows) in T. Detect essential primes in T[2] by checking to see if any minterm is contained by a single prime. Add these essential primes to the selected set. Repeat until no new essential primes are detected.

2. If the size of the selected set of primes equals or exceeds the best solution thus far, return from this level of recursion. If there are no elements left to be contained, declare the selected set as the best solution recorded thus far.

3. Heuristically select a prime.

4. Add this prime to the selected set and recur for the subtable resulting from deleting the prime and all minterms that are contained by this prime. Then, recur for the subtable resulting from deleting this prime without adding it to the selected set.

4.3 Two-Level Tautology

In this section we will describe the tautology checking algorithms used in ESPRESSO [1]. Tautology is a much used procedure in current exact and heuristic logic minimizers. During the prime implicant generation and prime implicant table generation steps of two-level

[2]These primes are not necessarily essential primes of the original function or table.

Boolean minimization, repeated checks of the containment of a cube by a cover are performed. In order to check whether $c \subseteq T$ (i.e., cube c is contained by cube T) we can check whether T_c, the cofactor of T with respect to the cube c, is a tautology. (See Problem 11.)

Consider a Boolean function f with inputs x_1, x_2, x_3, and x_4. $c = 0--1$ and $T = \{000-,\ -1-1,\ 0-11\}$. We compute T_c by cofactoring T with respect to $\overline{x_1}$ to obtain $T_{\overline{x_1}} = \{00-,\ 1-1,\ -11\}$. $T_c = T_{\overline{x_1}x_4} = \{00,\ 1-,\ -1\}$. T_c is a tautology, and therefore c is contained by T.

In order to explain the tautology algorithm we will need some definitions.

4.3.1 Unate Functions

A logic function f is monotone increasing (decreasing) in a variable x_j if changes in x_j from $0 \rightarrow 1$ cause f to change from $0 \rightarrow 1$ or stay constant ($1 \rightarrow 0$ or stay constant).

A function is unate in x_j if it is either monotone increasing in x_j or monotone decreasing. A function is unate if it is unate in all its variables. For example consider $f = x_1 \cdot \overline{x_2} + \overline{x_2} \cdot x_3$. It is monotone increasing in x_1 and x_3 and monotone decreasing in x_2, and thus f is unate. $f = \overline{x_1} \cdot x_2 + x_1 \cdot \overline{x_2}$ is nonunate or binate in x_1 and x_2.

Given a cover for a function if a variable x_j is a $-$ or a 1 ($-$ or a 0) in each cube, then f is unate in x_j. As an example consider $f = x_1 \cdot \overline{x_2} + \overline{x_2} \cdot x_3$. It satisfies the above definition for all variables. Note that this is a sufficiency check for unateness and is not necessary. To see this consider $f = x_1 \cdot x_2 \cdot \overline{x_3} + \overline{x_2}$. For variable x_2, f does not satisfy the above definition, but f *is* unate in x_2. f is equivalent to $x_1 \cdot \overline{x_3} + \overline{x_2}$ which does satisfy the above check for the unateness of a cover. The unateness of a particular cover representation of a function f has the simple check above, but checking whether a function is unate is much harder. We will be using the unateness of a cover check extensively in the tautology checking algorithm.

Unate functions are interesting from a tautology check standpoint because of the following theorems taken from [1].

Theorem 4.3.1 *A unate cover is a tautology if and only if the cover contains a row of all X's (or $-$'s).*

Proof. Necessity: A unate cover has to have a 1 or – (0 or –) in each cube for any variable x_j. Assume that a cover is monotone increasing in x_1 and x_3 and monotone decreasing in x_2 and x_4. Then the cover cannot contain 0101, unless it contains – – – –.
Sufficiency: Trivial. □

Theorem 4.3.2 *Let C be a cover for f. Suppose the rows and columns of f can be rearranged so that:*

$$C = \begin{pmatrix} A & T \\ T & D \end{pmatrix}$$

where T is a block of all –'s.

Suppose A is not a tautology. Then $f \equiv 1$ if and only if $D \equiv 1$.

Proof. (⇒): Since A is not a tautology, $\overline{A}$ is nonempty. Further, we know that $\overline{A} \cap \overline{D}$ is not contained in C. Therefore, if $f \equiv 1$, then $\overline{D} \equiv 0$.
(⇐): Trivial. □

Corollary 4.3.1 *Let U be the set of unate columns in a cover C. Rearrange C so that the U columns come first as shown below where B are the binate columns:*

$$C = (U \;\; B)$$

If U has no rows of all –'s, then $f \not\equiv 1$.

Proof. Since the function corresponding to U, namely $f(U)$, is unate, then $f(U) \equiv 1$ if and only if U has a row of all –'s by Theorem 4.3.1. If $f(U) \not\equiv 1$, then $f \not\equiv 1$. □

4.3.2 Tautology Procedure

A tautology procedure that uses the concepts developed in the previous section is given in Figure 4.4.

If $T \equiv 1$, then F is a tautology; if $T \equiv 0$, then F is not a tautology; and if $T \equiv -1$, then the question is as yet unanswered.

```
TAUTOLOGY(F):
{
    T = SPECIAL_CASES(F) ;
    if (T ≡ −1) (T, F) = UNATE_REDUCTION(F) ;
    if (T ≡ −1) (T, F) = COMP_REDUCTION(F) ;
    if (T ≡ −1) return(T) ;
    j = BINATE_INPUT_SELECT(F) ;
    if (TAUTOLOGY(F_{x_j}) ≡ 0) return(T = 0) ;
    if (TAUTOLOGY(F_{x̄_j}) ≡ 0) return(T = 0) ;

    return(T = 1) ;
}
```

Figure 4.4: Procedure to determine tautology

The procedure **SPECIAL_CASES** looks for a row of all −'s in F in which case the function is a tautology, or the procedure looks for a column of all 1s or all 0s in which case the function is not a tautology. For each cube the number of minterms in the cube can be computed as 2^k, where k is the number of −'s in the cube. The procedure sums these numbers over all the cubes and checks if the sum is $\geq 2^N$, where N is the number of inputs to F. If not, then F is not a tautology. If the procedure is provided with a list of minterms, then it checks if the function is a tautology by checking if the computed number is exactly equal to 2^N.

The procedure **UNATE_REDUCTION** corresponds to Theorem 4.3.2 and Corollary 4.3.1.

The procedure **COMP_REDUCTION** is based on the theorem below.

Theorem 4.3.3 *Suppose f has the matrix form as follows:*

$$M_f = \begin{pmatrix} A_1 & D & . & . & D \\ D & A_2 & . & . & D \\ & . & . & . & \\ D & . & . & . & A_P \end{pmatrix}$$

where D represents a block of all −'s. Then M_f is a tautology if and only if for some k, $1 \leq k \leq P$, $A_k \equiv 1$.

Proof. $\overline{A_1} \cap \overline{A_2} \cap \cdots \overline{A_P}$ is not contained in M_f and has to be empty for f to be a tautology. Since the A_k's all depend on different sets of inputs, one of the A_k's has to be a tautology, if f is a tautology. If none of the A_k's are a tautology, then $\overline{A_1} \cap \overline{A_2} \cap \cdots \overline{A_P}$ is not empty and f is not a tautology. □

Procedure **BINATE_INPUT_SELECT** heuristically selects a binate variable of the cover. A highly binate variable is selected, which is such that a large number of cubes in the cover have a 0 in the position corresponding to this variable, and a large number of cubes have a 1 in the position corresponding to this variable. The reasoning behind this choice is that the number of cubes in each of the cofactors of F with respect to a highly binate variable will be significantly smaller than the number of cubes in F. The procedure recurs on the two different cofactors of the function with respect to the chosen variable.

4.3.3 Example

Consider the function below with inputs x_1, x_2, x_3, and x_4. We reorder the inputs so the unate inputs x_1 and x_3 are first.

x_1	x_2	x_3	x_4
1	0	–	1
–	1	–	1
–	1	1	0
–	–	–	0
1	1	1	–
–	–	–	1

x_1	x_3	x_2	x_4
1	–	0	1
–	–	1	1
–	1	1	0
–	–	–	0
1	1	1	–
–	–	–	1

At this point if the subfunction corresponding to the first two columns did not have a row of all –'s, then we could declare the function to be nontautologous by Corollary 4.3.1. However, there are several rows of all –'s, and we reorder them to place the matrix in the form so that Theorem 4.3.2 is applicable. The matrix after reordering the cubes and after applying Theorem 4.3.2 is shown below.

x_1	x_3	x_2	x_4
1	–	0	1
–	1	1	0
1	1	1	–
–	–	1	1
–	–	–	0
–	–	–	1

x_2	x_4
1	1
–	0
–	1

After applying Theorem 4.3.2 the resulting two-input function is in the form such that Theorem 4.3.2 can be applied again. The function below:

x_4
0
1

results after the second application, which is determined to be a tautology by procedure **SPECIAL_CASES**.

In general, the function has to be cofactored with respect to chosen binate variables because at some stage we may have a nontrivial function for which none of the reduction theorems are applicable.

4.4 Complementation

Complementing a sum-of-products representation of a logic function f involves interchanging the *ON*-set and the *OFF*-set of f to produce $\overline{f}$. This assumes that the *OFF*-set of the function f is available in sum-of-products form. However, this may not always be the case. We may be given only the *ON*-set of a completely specified function in sum-of-products form and asked to generate a sum-of-products representation for the *OFF*-set. In the case of incompletely specified functions we may be given the *ON*-set and the *DC*-set and be asked to generate the *OFF*-set. Clearly, one can use the disjoint SHARP operation to compute the *OFF*-set R of a function as $U \ominus F$ where F is the *ON*-set of the function. (U is the universe cube with all $-$ entries.) Using disjoint SHARP may result in a sum-of-products representation for R that has a needlessly large number of cubes. Ideally, we would like to directly generate a near-minimum sum-of-products form for R, given F.

4.4.1 Basic Procedure

The complementation procedure used in ESPRESSO is shown in Figure 4.5. In this procedure N refers to the total number of variables in the function.

The basis of the complementation procedure is the identity

$$\overline{F} \;=\; x_j \cdot \overline{F_{x_j}} \;+\; \overline{x_j} \cdot \overline{F_{\overline{x_j}}}$$

which can be easily derived by complementing both sides in the Shan-

```
COMPLEMENTATION(F):
{
    if (row of all 2's) return(φ) ;
    if (F is a unate cover in all variables)
        return(UNATE_COMPLEMENT(F)) ;

    c = first cube in F ;
    for (i = 1; i ≤ N; i = i + 1) {
        if (c[i] not equal to d[i] for any cube d ∈ F)
            c[i] = 2 ;
    }
    R = UNATE_COMPLEMENT(c) ;
    F = F_c ;

    j = BINATE_INPUT_SELECT(F) ;
    R = R ∪ (x_j · COMPLEMENT(F_{x_j})) ;
    R = R ∪ (x̄_j · COMPLEMENT(F_{x̄_j})) ;
    return(R) ;
}
```

Figure 4.5: Procedure to compute complement

non expansion for F as shown below.

$$
\begin{aligned}
\overline{F} &= \overline{x_j \cdot F_{x_j} \; + \; \overline{x_j} \cdot F_{\overline{x_j}}} \\
&= (\overline{x_j} + \overline{F_{x_j}}) \;\cdot\; (x_j + \overline{F_{\overline{x_j}}}) \\
&= \overline{x_j} \cdot \overline{F_{\overline{x_j}}} \; + \; x_j \cdot \overline{F_{x_j}} \; + \; \overline{F_{x_j}} \cdot \overline{F_{\overline{x_j}}}
\end{aligned}
$$

The third term in the last equation is contained by the first two terms and can be discarded.

We can thus select a variable x_j and evaluate the complement of the two cofactors of F with respect to x_j to evaluate $\overline{F}$. The covers F_{x_j} and $F_{\overline{x_j}}$ have one fewer input than F and also have fewer cubes. The selection of the binate variable can use the same heuristic as in the **TAUTOLOGY** procedure; a highly binate variable is selected which appears in a large number of cubes in true form and in a large number of cubes in complemented form.

The result R is computed by taking the union of the complements of F_{x_j} and $F_{\overline{x_j}}$ and deleting any cube that appears twice in F_{x_j} and $F_{\overline{x_j}}$. Optionally, R can be made single cube containment minimal to minimize its cardinality.

There are several ways of increasing the efficiency of the basic procedure by exploiting special cases and detecting unate covers. We will describe these methods in the following sections.

4.4.2 Special Cases

The first special case in the procedure of Figure 4.5 occurs if F has a cube of all 2s (the universe in the given subspace). This will never happen at the top level of the complementation procedure, but may occur in any of the recursive calls when we attempt to complement a cofactor.

Another case is when a particular input to F is always 0 or always 1, i.e., a column of the cover F has all 0s or all 1s. In this case, we extract a cube c which represents all the columns where this case occurs. The cover can be written as

$$F = c \bigcap F_c$$

and therefore

$$\overline{F} = \overline{c} \bigcup \overline{F_c}$$

4.4.3 Unate Complementation

We can also improve the efficiency of the procedure by detecting unate covers that have to be complemented. If a cover F is monotone increasing in x_j then

$$\overline{F} = \overline{x_j} \cdot \overline{F_{\overline{x_j}}} + \overline{F_{x_j}} \tag{4.1}$$

and if F is monotone decreasing in x_j then

$$\overline{F} = x_j \cdot \overline{F_{x_j}} + \overline{F_{\overline{x_j}}} \tag{4.2}$$

The proof of the above identities is left as an exercise (Problem 10).

If a cover F is unate in all the variables, we call a modified version of the **COMPLEMENT** procedure, namely the procedure **UNATE_COMPLEMENT** which has the same recursive structure as **COMPLEMENT**. The main difference is that we select a unate variable x_j, and depending on whether F is monotone increasing or monotone decreasing in x_j we use Equation 4.1 or Equation 4.2 to compute $\overline{F}$.

The selection of the unate variable follows a simple heuristic where the variable with the fewest – entries in the given cover F is selected. This has the effect of eliminating the most cubes in one branch of the recursion.

4.4.4 Example

We will consider the complementation of the cover below which has five inputs.

x_1	x_2	x_3	x_4	x_5
–	1	1	1	1
–	1	1	1	0
–	1	0	1	–
–	1	0	–	1

There is no row of all 2s, and F is not a unate cover. However, we find that x_2 is a column containing all 1s. Therefore, we have $R = -0---$, and we have to compute the complement of F_{x_2} shown below.

x_1	x_3	x_4	x_5
–	1	1	1
–	1	1	0
–	0	1	–
–	0	–	1

F_{x_2} above does not fall into any of the special cases. We select a highly binate variable for cofactoring, namely x_3. This gives us the two cofactors $F_{x_2 \cdot x_3}$ and $F_{x_2 \cdot \overline{x_3}}$ shown below.

x_1	x_4	x_5
–	1	1
–	1	0

x_1	x_4	x_5
–	1	–
–	–	1

Focusing on $F_{x_2 \cdot x_3}$, the selection of the binate input x_5 will produce $F_{x_2 \cdot x_3 \cdot x_5}$ and $F_{x_2 \cdot x_3 \cdot \overline{x_5}}$ shown below.

x_1	x_4
–	1

x_1	x_4
–	1

Complementing the above covers and merging them produces $\overline{F_{x_2 \cdot x_3}}$ shown below.

x_1	x_4	x_5
–	0	1
–	0	0

The unate complementer produces $\overline{F_{x_2 \cdot \overline{x_3}}}$ which is

x_1	x_4	x_5
–	0	0

Merging $\overline{F_{x_2 \cdot x_3}}$ and $\overline{F_{x_2 \cdot \overline{x_3}}}$ produces $\overline{F_{x_2}}$ shown below.

x_1	x_3	x_4	x_5
–	1	0	1
–	1	0	0
–	0	0	0

Finally, $\overline{F}$ is

x_1	x_2	x_3	x_4	x_5
–	0	–	–	–
–	–	1	0	1
–	–	1	0	0
–	–	0	0	0

which is obtained by merging $\overline{F_{x_2}}$ with $\overline{x_2}$.

4.5 Exact Minimization Methods

There are many problems with the Quine-McCluskey procedure presented in Section 4.2. First, prime generation begins from an exhaustive listing of minterms in the *ON*-set and *DC*-set of the function. Secondly, the minterms are checked for pair-wise merges, most of which may not be possible. A prime may be generated multiple times during the merging steps. The prime implicant table has a set of rows corresponding to the minterms in the *ON*-set, making the prime implicant table very large. Finally, the covering method presented does not use any bounding strategy.

The above problems are alleviated in minimization programs like ESPRESSO-EXACT [6]. ESPRESSO-EXACT generates all the prime implicants of a function from a sum-of-products representation of the function, which is a list of cubes rather than minterms. It generates a reduced prime implicant table where each row is a collection of minterms that are all contained by the same set of primes. A bounding strategy based on a maximal independent set heuristic is used in the covering step. These steps will be discussed in the following sections.

4.5.1 Prime Implicant Generation

Expanding a cube means raising 0 or 1 literals in the input part of a cube to $-$, or a 0 literal in the output part of a cube to a 1. We will represent a function f in terms of its *ON*-set F, *DC*-set D, and *OFF*-set R. For any cube c, in order for cube c to expand into an implicant of F, it is necessary that after the expansion c is distance-1 or more from each cube $r^i \in R$. We can express this condition by writing a Boolean expression, focusing on the single output case.

We let c_j^0 be a Boolean variable representing the condition that variable j of c can be raised from a 0 to a $-$, c_j^1 be a Boolean variable representing the condition that variable j of c can be raised from a 1 to a $-$, and the pair of variables $(r^i)_j^0$ and $(r^i)_j^1$ have values of 0 and 1 if variable j in cube r^i is a 0, 1 and 0 if the variable is a 1, and 1 and 1 if the variable is a $-$. The reason for encoding the $(r^i)_j^0$ will become clear later.

For any variable x_j we can express the condition that r^i and an expansion of c be disjoint in x_j as:

$$\overline{G_{ij}} \;=\; (r^i)_j^0 c_j^0 + (r^i)_j^1 c_j^1 \;=\; 0$$

or equivalently:

$$G_{ij} \;=\; (\overline{(r^i)_j^0} + \overline{c_j^0})(\overline{(r^i)_j^1} + \overline{c_j^1}) \;=\; 1$$

The values of $(r^i)_j^0$ and $(r^i)_j^1$ are known to be either 0 or 1, and the variables of above equation are c_j.

Continuing, note that r^i and c are disjoint if they are disjoint for some variable j. This is written as:

$$H_i \;=\; \bigcup_{j=1}^{N} G_{ij} \;=\; 1$$

Finally the expansion of c is disjoint from R only if it is disjoint from all cubes $r^i \in R$, and we express this as:

$$I \;=\; \bigcap_{i=1}^{|R|} H_i \;=\; 1$$

We therefore have a Boolean expression which expresses the condition that an assignment of $\{0,\ 1\}$ to the variables c_j^0 and c_j^1 results in an implicant of f. We write this in full as:

$$I \;=\; \bigcap_{i=1}^{|R|} \bigcup_{j=1}^{N} (\overline{(r^i)_j^0} + \overline{c_j^0})(\overline{(r^i)_j^1} + \overline{c_j^1}) \;=\; 1$$

An implicant of the function I corresponds to an assignment of {0, 1} to the variables c_j^0 and c_j^1 which results in a corresponding implicant of f. Furthermore, a prime implicant of I corresponds to an assignment of {0, 1} to the variables which is maximal in the sense that no other variable which is a 0 can be made a 1; therefore a prime implicant of I corresponds to a prime implicant of f.

We note the following:

1. By construction we see that I contains only the complements of the variables and is therefore unate. Note that if we had not chosen to represent the raising of literals from 0 to a $-$ and 1 to a $-$ separately with the c_j^0 and c_j^1 (and the $(r^i)_j^0$ and $(r^i)_j^1$), this resulting function would not have been unate.

2. Any cover for a unate function contains all of the prime implicants of the function, i.e., every prime is essential (Proposition 3.3.6 of [1]).

3. The prime implicants of a unate function may be obtained by expanding the product-of-sum-of-product form into a sum-of-products form and then performing single cube containment on the resulting cover. Given a cover F for a unate function f if a cube q can be removed from F and F remains a cover for f, then it means that q is contained by some single cube in F (Propositions 3.3.5 and 3.3.7 in [1]).

Thus, if we repeatedly intersect the sum-of-product forms given by H_i, then we will obtain a sum-of-products cover which can then be made single cube containmnent minimal. The primes for f can be derived from those of I.

An example corresponding to the function of Figure 4.1 is given below. R for the function with inputs x_1, x_2, x_3, and x_4 is $\{001-,\ 00-1,\ 01-0,\ 110-\}$. The four sum-of-products expressions corresponding to the four cubes that have to be intersected are:

$$\overline{c_1^1} + \overline{c_2^1} + \overline{c_3^0}$$

$$\overline{c_1^1} + \overline{c_2^1} + \overline{c_4^0}$$

$$\overline{c_1^1} + \overline{c_2^0} + \overline{c_4^1}$$

$$\overline{c_1^0} + \overline{c_2^0} + \overline{c_3^1}$$

If a literal in R is a $-$, then it results in a $c_j^0 c_j^1$ term which is null given the definition of c_j^0 and c_j^1; we cannot raise a 0 to a $-$ and a 1 to a $-$ in the same input and the same cube.

Intersecting the first two expressions above gives:

$$\overline{c_1^1} + \overline{c_2^1} + \overline{c_3^0}\,\overline{c_4^0}$$

Intersecting the third and fourth expressions and making the result single cube containment minimal gives:

$$\overline{c_1^0}\,\overline{c_4^1} + \overline{c_1^1}\,\overline{c_3^1} + \overline{c_2^0} + \overline{c_3^1}\,\overline{c_4^1}$$

Intersecting the cubes resulting from the previous intersections gives:

$$\overline{c_1^1}\,\overline{c_3^1} + \overline{c_1^1}\,\overline{c_2^0} + \overline{c_1^0}\,\overline{c_2^1}\,\overline{c_4^1} + \overline{c_2^1}\,\overline{c_3^1}\,\overline{c_4^1} + \overline{c_2^0}\,\overline{c_3^0}\,\overline{c_4^0}$$

which correspond to the five primes of Figure 4.1(b).

The set of prime implicants for the function is $\{1-1-,\ 10--,\ 01-1,\ -111,\ -000\}$. We did *not* have to exhaustively list the minterms of either F or R for the function.

The above strategy can be generalized to the multiple output case. We can enumerate the cubes in the *OFF*-set for each output, and have additional c_j variables for the output part. The Boolean expression for I can be modified to specify that for each output, the expanded cube should not intersect any of the cubes in the *OFF*-set for that output. We can also generate primes that cover a given cube in a cover without generating all the primes of the function (see Section 4.6.4).

4.5.2 Reduced Prime Implicant Table Generation

We now have a set of cubes that correspond to the set of prime implicants for the function f that is to be minimized. Call this set Q. Q is first split into the relatively essential set E_r and the relatively redundant set R_r. A cube $c \in F$ belongs to E_r if $Q \cup D - c$ fails to contain c, or c belongs to R_r if $Q \cup D - c$ contains c. (Note that $-$ in this context corresponds to removing c from the list of cubes $Q \cup D$ and not the Boolean difference.) The set E_r has to be retained in any cover for the function f and is the set of essential prime implicants for f.

The prime implicants of R_r are further divided into the totally redundant subset R_t and the partially redundant subset R_p. A cube $c \in R_r$ belongs to R_t if $E_r \cup D$ contains c, or c belongs to R_p if

$E_r \cup D$ fails to contain c. The cubes of R_t are totally redundant in the sense that, because they are completely contained by the set of essential primes, they can never be in a minimum cover for f. The cubes of R_p are relatively redundant, because although any single cube of R_p can be removed, it is not possible to simultaneously remove all of the cubes of R_p while still maintaining a cover for f. The tautology algorithm described in Section 4.3 is used to split Q into E_r, R_r, and R_p. What remains in R_p causes the most difficulty in trying to extract a minimum cover of f.

We now have to generate the reduced prime implicant table where the cubes in R_p will correspond to the columns and the rows will correspond to collections of minterms all of which are contained by the same primes in R_p. The key to the generation is a simple modification of the tautology algorithm. Rather than testing whether a function is a tautology, we determine which subsets of cubes in a function would have to be removed in order to prevent a function from becoming a tautology.

Consider forming $H = E_r \cup R_p - c$, where $c \in R_p$, and using the tautology algorithm to determine if H_c is a tautology. H_c is a tautology because every cube of R_p is contained by the union of E_r and the remaining cubes of R_p. When we get to a leaf in the tautology algorithm (i.e., when we are able to determine that the function is a tautology) we examine the cubes which are in the cover at this leaf. If there is a cube from E_r (or D) which is the universe (in this leaf), then it is not possible to prevent the function from being a tautology in this leaf. Otherwise, all of the cubes of R_p which are the universe (in this leaf) must be removed in order to avoid this leaf from becoming a tautology. In terms of determining how a cover contains the cube, this is equivalent to saying that the cover will fail to contain the cube if and only if all of the cubes of R_p which are universal in this leaf are discarded.

In this way, the prime implicant table is formed with a cube of R_p associated with each column. At each leaf which is a tautology (and for which no cube from E_r is the universal cube) we add a row to our table with an "X" for each column where a cube from R_p is universal (in this leaf). A minimum cover of this table corresponds to a minimum subset of primes of R_p which must be retained in the cover for f.

The algorithm proceeds by forming H_c for each $c \in R_p$ and calling a modified version of the **TAUTOLOGY** procedure of Section 4.3. After determining how c can be contained, c can be moved

	0-1-	-01-	-101	1-01
001-	X	X		
1101			X	X

Figure 4.6: Reduced prime implicant table

to the set E_r since we now know how all the minterms of c can be contained by selecting primes from R_p.

An example prime implicant table is shown in Figure 4.6. This reduced table is generated from the set of primes, for a function $f(x_1, x_2, x_3, x_4)$ given by:

$$Q = \{0-1-,\ -01-,\ 01--,\ 10--,\ -101,\ 1-01\}$$

We have:

$$E_r = \{01--,\ 10--\}$$

$$R_p = \{0-1-,\ -01-,\ -101,\ 1-01\}$$

The table has four columns corresponding to the initial set of cubes in R_p. We construct H_c with $c = 0-1-$. After cofactoring the cubes of $H = E_r \cup R_p - c$ with respect to c, we obtain:

$$H_c = \{0-,\ 1-\}$$

over the inputs x_2 and x_4, which correspond to cubes $-01-$ and $01--$ of R_p and E_r, respectively. At this point the tautology procedure selects the binate variable x_2 and cofactors H_c with respect to x_2 and $\overline{x_2}$. This results in two leaves, with cubes that are universal (in their corresponding leaves). The cube $1-$ corresponding to $01--$ is ignored since $01--$ belongs to E_r. We add a row to the prime implicant table corresponding to $c \cdot \overline{x_2}$, namely $001-$. This row is obviously contained by c, and is also contained by $-01-$. X's are added to the appropriate locations in the table.

Next we have:

$$E_r = \{01--,\ 10--,\ 0-1-\}$$

$$R_p = \{-01-,\ -101,\ 1-01\}$$

We set $c = -01-$.

$$H_c = \{0-,\ 1-\}$$

over the inputs x_1 and x_4. The first cube $0-$ corresponds to cube $0-1-$ in E_r and is ignored. The second cube $1-$ corresponds to cube $10--$ in E_r and is also ignored.

We now have:

$$E_r = \{01--,\ 10--,\ 0-1-,\ -01-\}$$

$$R_p = \{-101,\ 1-01\}$$

We set $c = -101$.

$$H_c = \{0,\ 1\}$$

over the input x_1. The first cube corresponds to cube $01--$ in E_r and is ignored. We add a row to the prime implicant table corresponding to $c \cdot x_1$, namely 1101, which is contained by $c = -101$ and $1-01$ in R_p. No more rows are added to the table when we set $c = 1-01$.

If we had generated a prime implicant table using the Quine-McCluskey method for this example, it would have had six columns corresponding to the six prime implicants, and eleven rows. Substantial savings are possible by using the more sophisticated strategy presented in this and previous sections.

4.5.3 Branch-and-Bound Covering

The branching covering step of the Quine-McCluskey method can be improved by adding a bounding strategy. The covering algorithm used in ESPRESSO-EXACT is described below.

The procedure below receives as input the prime implicant table, T, corresponding to the logic function.

1. Delete dominated columns and dominating rows. Detect essential primes by checking to see if any minterm is contained by a single prime. Add these essential primes to the selected set. Repeat until no new essential elements are detected.

2. If the size of the selected set *plus* the lower bound calculated using a maximal independent set heuristic (to be described later) equals or exceeds the best solution thus far, return from this level of recursion. If there are no elements left to be contained, declare the selected set as the best solution recorded thus far.

3. Heuristically select a prime.

4. Add this prime to the selected set and recur for the subtable resulting from deleting the prime and all minterms that are contained by this prime. Then, recur for the subtable resulting from deleting this prime without adding it to the selected set.

At Step 2, we find a maximal set of rows of T all of which are pair-wise disjoint, i.e., they do not share an X in any column. Assume that each row corresponds to a minterm. We have a set of minterms all of which are contained by different sets of primes. This means that we have to select a different prime for each minterm in the independent set. Thus, a lower bound on the number of primes we have to select to contain all the minterms is given by the cardinality of the maximal independent set. Note that this is just a heuristic as rows do not correspond to disjoint cubes.

For the prime implicant table of Figure 4.3, a maximal independent set of rows is (0000, 0101, 1010, 1111). This implies that any minimum solution will have at least four product terms. Therefore, after selecting prime **A** and finding the solution $f = \{ \mathbf{A}, \mathbf{C}, \mathbf{E}, \mathbf{G} \}$ we do *not* have to recur, since this is a minimum product term cover.

Finding a maximum independent set is itself an NP-hard problem [3]. However, a maximal independent set that is not maximum provides a correct lower bound on the size of the final solution. Of course, the maximum set will provide the best bound, and therefore it is of interest to develop a good heuristic that is efficient both in terms of time complexity and quality of the solution.

A greedy algorithm is used in ESPRESSO-EXACT based on mapping the problem into finding a maximal clique (a completely connected subgraph) in a graph. The graph is constructed where the nodes correspond to rows in the table, and an edge is placed between two nodes if the two corresponding rows are disjoint. The algorithm to find a maximal clique is as follows:

1. Initialize the clique to contain no nodes.

2. Pick the node of the largest degree (the node which has the most number of edges connected to it) and add this node to the clique. Break ties by choosing the node which is connected to the largest number of other nodes of maximum degrees.

3. Remove all nodes and their edges from the graph which are not connected to the current clique.

4. Repeat Steps 1 and 2 while there are still nodes in the graph that are not in the current clique.

The node of largest degree in Step 2 corresponds to the row which is disjoint from the maximum number of other rows in the matrix. The tie-breaker attempts to preserve as many of the remaining nodes of maximum degree as possible.

4.6 Heuristic Minimization Methods

Heuristic minimization methods fall into two broad categories:

1. Minimization methods that follow the strategy of exact minimizers but do not necessarily generate the entire set of prime implicants and solve the covering problem heuristically.

2. Minimization methods based on the iterative expansion, reduction, and reshaping of implicants in a cover.

We will briefly describe state-of-the-art heuristic minimization techniques in this section.

4.6.1 Heuristics Based on Exact Minimization

A set of prime implicants that contain a *particular* cube c can be generated using a simple modification of the strategy presented in Section 4.5.1. For instance, if a cube c has a 1 in position j, then we set the variable c_j^0 to 0. If a cube c has a $-$ in position j, then we set both c_j^0 and c_j^1 to 0 in the expression for I. The prime implicants of I will correspond to the prime implicants of f that contain c and not the entire set. Prime implicants can be generated that contain a heuristically chosen set of cubes in the given cover of the function f that is to be minimized. A heuristic covering algorithm that does not backtrack can be used on the prime implicant table generated from this subset of primes.

4.6.2 Heuristics Based on Iterative Improvement

The program MINI [4] pioneered heuristic approaches based on iterative improvement. The MINI process starts with an initial solution and iteratively improves it. There are three basic modifications that are performed on the implicants of the function. First, each implicant is reduced to the smallest possible size while still maintaining

```
--11  11
1001  01
1101  10
-001  10                 0000  11
-1-0  01                 1110  10
1010  01                 00-0  01
0110  10                 -000  01
-010  10                 0-01  01
-100  10                 0101  11
1000  10                 -101  01

 (a)                      (b)
```

Figure 4.7: Function and complement of function

```
--11  11                 --11  11
1001  01                 1001  01
1101  10                 1101  10
-001  10                 -0-1  10
-1-0  01      ------>    -1-0  01
1010  01                 1010  01
0110  10                 0-10  10
-010  10
-100  10                 -100  10
1000  10                 10-0  10
```

Figure 4.8: First EXPAND operation carried out on the ON-set

proper containment of minterms. Second, the implicants are examined in pairs to see if they can be reshaped by reducing one and enlarging the other by the same set of minterms. Third, each implicant is enlarged to its maximal size and any other implicants that are contained are removed. The first process, which may reduce an implicant to nothing, and the third process, which removes contained implicants, may reduce the number of implicants in the solution. The second process facilitates the reduction of the solution size that occurs in the other two processes. The order in which the implicants are reduced, reshaped, and expanded is crucial to the success of the procedure. We will not describe the details of these processes here but instead will provide an example that illustrates the procedures.

In Figure 4.7(a) we have a function f, and its complement $\overline{f}$ is shown in Figure 4.7(b). In Figure 4.8, we illustrate the expansion step in which the objective is to expand cubes into prime cubes. However, in this figure, not every cube has been expanded. The cube -010 10 is contained by the preceding expanded cubes and is

```
--11  11              --11  11
1001  01              1001  01
1101  10              1101  10
-0-1  10              -001  10
-1-0  01    ----->    -1-0  01
1010  01              1010  01
0-10  10              0-10  10
-100  10              -100  10
10-0  10              10-0  10
```

Figure 4.9: REDUCE operation carried out on the expanded ON-set

```
--11  11              --11  11
1001  01              1001  11
1101  10              1101  10
-001  10              0001  10
-1-0  01    ----->    -110  01
1010  01              1010  11
0-10  10              0-10  10
-100  10              -100  11
10-0  10              1000  10
```

Figure 4.10: RESHAPE operation carried out on the reduced ON-set

removed from the cover for F. In Figure 4.9, the reduction step is illustrated. Only one cube $-0-1$ 10 is reduced to -001 10. The reshaping step attempts to move the solution from the current local minimum. The cubes are reshaped in pairs. If we number the cubes from the top as 1, 2, 3, etc., then the pairs (2, 4), (5, 8), and (6, 9) have been reshaped in Figure 4.10. Note that the containment of each pair remains the same. For instance, in the pair (2, 4), i.e., (1001 01, -001 10), the minterm 1001 10 is added to the first cube and removed from the second to result in 1001 11, 0001 10. After the reshape operation, the expand operation is carried out again and results in the removal of one more cube. The result of the final expand is shown in Figure 4.11. A cover with 8 cubes is the final result.

The heuristic version of the program ESPRESSO also follows an iterative improvement approach. The three basic procedures of ESPRESSO are **REDUCE**, **IRREDUNDANT**, and **EXPAND**. The procedure **EXPAND** replaces each cube in the cover F with a prime cube which contains the cube. **REDUCE** replaces each cube in the cover F with the smallest cube contained in the cube which is necessary to still represent the same function. **EXPAND**

```
--11  11                --11  11
1001  11                10-1  11
1101  10                1-0-  10
0001  10                -0-1  10
-110  01    ------>     -1-0  01
1010  11                101-  11
0-10  10                0-1-  10
-100  11                -100  11
1000  10
```

Figure 4.11: Second EXPAND operation carried out on the reshaped ON-set

and **REDUCE** are similar to the corresponding operations in MINI. **IRREDUNDANT** extracts from the cover F a minimal subcover which is still sufficient to represent the function. **IRREDUNDANT** contains a heuristic covering algorithm.

The heuristic algorithms of ESPRESSO and MINI have been shown to produce results close to the minimum within reasonable amounts of CPU time for large commonly designed functions. We will describe the methods used in ESPRESSO in greater detail in the following sections.

4.6.3 ESPRESSO Minimization Loop

We will describe the basic minimization loop of ESPRESSO in this section. The strategy used is shown in Figure 4.12.

The minimization loop of ESPRESSO requires cover representations for the *ON*-set, *DC*-set, and *OFF*-set of the function. Given the *ON*-set F and the *don't-care* set D, a sum-of-products representation (denoted R) for the *OFF*-set can be generated using the **COMPLEMENT** procedure described in Section 4.4.

Given F, D, and R the minimization loop consists of iteratively reducing the implicants in F to nonprime cubes (**REDUCE**), expanding the cubes to prime implicants (**EXPAND**), and extracting a minimal subset of the prime implicants (**IRREDUNDANT**). The iteration continues until there is no further reduction in the number of cubes in the function. A procedure **MAKE_SPARSE** is used in the case of multiple-output functions to lower 1s in the output part of F to 0s so as to minimize the number of transistors in a *programmable logic array* (PLA) implementation. We will describe the three main procedures of ESPRESSO in the following sections.

```
MINIMIZE(F, D):
{
    R = COMPLEMENT(F ∪ D) ;

    do {
        φ1 = |F| ;
        F = REDUCE(F, D) ;
        F = EXPAND(F, R) ;
        F = IRREDUNDANT(F, D) ;
    } while (|F| < φ1) ;

    F = MAKE_SPARSE(F, D, R) ;
    return(F) ;
}
```

Figure 4.12: The ESPRESSO minimization loop

4.6.4 EXPAND

The **EXPAND** procedure examines each cube $c \in F$ (where F is a cover of the *ON*-set of the function f that is to be minimized) and replaces c with a prime implicant d such that $c \subseteq d$. If c is not prime, then d covers more minterms of F than c does, and hence it is said that c has expanded into a larger cube. If c is known to be prime from a previous expansion, then there is no reason to attempt to expand c. Note that each c is replaced with a single prime implicant d (out of all the possible prime implicants which cover c) so that the number of cubes in the cover can never increase during the **EXPAND** step.

The goal for the minimization program is to minimize the number of cubes in F. There are several criteria that can be used in the **EXPAND** procedure to achieve this goal. We can define an optimally expanded prime as a prime d for which:

1. d covers the largest number of cubes of F.

2. Among all cubes d which cover the same number of cubes of F, d covers the largest number of minterms of F.

Condition 1 is a local statement of the minimization objective, and Condition 2 states that ties should be broken by covering as many minterms of F as possible.

By enumerating all primes $d \supseteq c$ it is easy to choose an optimally expanded prime to replace c. The method described in Section 4.5.1 can be modified to generate only the primes that cover a given cube c.

We let c_j^0 be a Boolean variable representing the condition that variable j of c can be raised from a 0 to a $-$, c_j^1 be a Boolean variable representing the condition that variable j of c can be raised from a 1 to a $-$, and the pair of variables $(r^i)_j^0$ and $(r^i)_j^1$ have values of 0 and 1 if variable j in cube r^i is a 0, 1 and 0 if the variable is a 1, and 1 and 1 if the variable is a $-$. Given this encoding and the *OFF*-set R of the given function f we can write an expression for a unate function I whose prime implicants are the prime implicants of f. (See Section 4.5.1 for details.)

$$I = \bigcap_{i=1}^{|R|} \bigcup_{j=1}^{N} (\overline{(r^i)_j^0} + \overline{c_j^0})(\overline{(r^i)_j^1} + \overline{c_j^1}) = 1$$

Since I is unate, we can obtain the prime implicants for I by repeatedly intersecting the sum-of-products expressions and performing single cube containment on the resulting cover.

In order to only generate the primes of f that cover cube c, we merely set certain c_j^0's and c_j^1's to fixed values. If the position corresponding to variable j in c is a 1, we set c_j^0 to 0, since clearly the position cannot be expanded from a 0 to a 1. Similarly, if the position corresponding to variable j in c is a 0, we set c_j^1 to 0. If the position corresponding to variable j in c is a $-$, we set both c_j^0 and c_j^1 to 0. Fixing values of the c_j^0 and c_j^1 variables results in greater efficiency in obtaining a sum-of-products representation for I.

In some cases, it may not be possible to generate all of the primes that cover a given cube c within a reasonable amount of time. This occurs if c is a cube with very few or no $-$ entries. In such cases, nonprime literals in c are heuristically selected and raised to $-$ prior to generating all the primes.

4.6.5 IRREDUNDANT

The **IRREDUNDANT** procedure extracts from a cover a minimal subset which is still sufficient to cover the same function. While the procedure **EXPAND** ensures that no single cube in F covers any other cube, it is possible that a cube in F is covered by the disjunction of multiple cubes in F.

IRREDUNDANT is given the *ON*-set for the function F and the *DC*-set D. A straightforward means of achieving an irredundant cover is to pick each cube c in F and check if $F \cup D - c$ covers c. (Note that $-$ in this context corresponds to removing c from the list of cubes $F \cup D$ and is not the Boolean difference operator.) If so, c is deleted from F, and the process continues. However, this simple method may produce results that are far from optimal.

The **IRREDUNDANT** procedure of ESPRESSO is virtually identical to the reduced prime implicant table generation procedure described in Section 4.5.2. The procedure of Section 4.5.2 began with the entire set of prime implicants of the given function. In **IRREDUNDANT** we begin with the set of cubes in F (which are all prime cubes since **IRREDUNDANT** is called after **EXPAND**). This set is typically is a much smaller set than the total set of primes.

Given the cubes in F a reduced prime implicant table is generated and the covering problem corresponding to this table is solved. An exact strategy is used if the table is small else a greedy heuristic strategy where no backtracking is performed is used if the table is large.

4.6.6 REDUCE

The **REDUCE** procedure transforms an irredundant cover of prime implicants into a new cover by replacing each prime implicant, where possible, with a smaller nonprime implicant contained in the prime implicant. An irredundant, prime cover is a local minimum for the cover cardinality cost function, and **REDUCE** moves the solution away from the local minimum. The hope is that the subsequent **EXPAND** will determine a better set of prime implicants.

The **maximal reduction** of a cube c with respect to a cover F is the smallest cube contained in c that can replace c in F without changing the function realized. The maximal reduction of a cube c is denoted $\underline{c}$. Each cube c is replaced by $\underline{c}$ in the **REDUCE** step.

The **supercube** of a set of cubes Q is the smallest cube containing all the cubes in Q. The supercube of a set of cubes can be computed simply by using bitwise OR on the different cubes in Q. For example, the supercube of the cubes 101, 110, and 100, is $1--$. The supercube of cubes 00 and 11 is $--$.

The identity

$$\underline{c} \;=\; c \bigcap \mathit{supercube}(\overline{(F \cup D - c)_c}) \tag{4.3}$$

can be used to compute the maximal reduction of a cube. Hence, the operation of finding the maximal reduction of a cube can be reduced to finding the smallest cube which contains the complement of a cover.

As an example consider $F = \{1--,\ -1-,\ --1\}$ and $D = \phi$. Choose $c = 1--$. $F - c = \{-1-,\ --1\}$. $(F-c)_c = \{1-,\ -1\}$ and $\overline{(F-c)_c}$ is $\{00\}$. The supercube of $\{00\}$ is 00. $c \cap -00$ (the first variable of the function does not exist in the supercube and is therefore $-$) is 100. Therefore, the maximal reduction of $1--$ in the given cover is 100. Replacing $1--$ by 100 in F clearly does not change the functionality of the cover.

The reduction of a single cube depends on the form of the cover for the function. In particular, the order in which the cubes are processed for reduction affects the results of the **REDUCE** operation. The cubes which are reduced first will tend to reduce to smaller cubes, thus possibly preventing cubes which follow from reducing as much as they might have otherwise. A heuristic which has worked well in practice is to order the cubes in decreasing order of size.

The operation corresponding to Equation 4.3 can be performed by explicitly forming the cover for $\overline{(F \cup D - c)_c}$ and using bitwise OR on all the cubes in the cover. Alternately, given a cover the Shannon cofactor can be used to recursively compute the supercube of the complement, without explicitly generating the complement. In the **COMPLEMENT** procedure of Figure 4.5, rather than computing the complement of a function, we compute the supercube of the complement.

Given an F we cofactor F with respect to x_j to obtain F_{x_j} and $F_{\overline{x_j}}$. We recursively compute the supercubes of $\overline{F_{x_j}}$ and $\overline{F_{\overline{x_j}}}$, call them s_{x_j} and $s_{\overline{x_j}}$. We compute the supercube of $\overline{F}$ as the supercube of $x_j \cdot s_{x_j}$ and $\overline{x_j} \cdot s_{\overline{x_j}}$.

At the leaves of the recursion, if we encounter a single cube c and we have to compute the supercube for its complement, then we can use the identity below.

$$supercube(\overline{c}) = \begin{cases} \phi & if\ c\ has\ no\ literals \\ \overline{c} & if\ c\ has\ 1\ literal \\ universe & if\ c\ has\ 2\ or\ more\ literals \end{cases} \quad (4.4)$$

As an example consider $c = 1--$. $\overline{c} = \{0--\}$. The supercube of $\overline{c}$ is $0--$ since $\overline{c}$ corresponds to a single cube. Now consider $c = 11-$.

$\overline{c} = \{0--,\ -0-\}$. The supercube of this set of cubes is the universe $---$.

Problems

1. Given

$$\begin{aligned} f_1 &= \overline{a} \cdot \overline{b} \cdot c + \overline{a} \cdot b \cdot \overline{c} + \overline{a} \cdot \overline{b} \cdot \overline{c} \\ f_2 &= \overline{a} \cdot b \cdot c + \overline{a} \cdot \overline{b} \cdot c + a \cdot b \cdot c + \overline{a} \cdot b \cdot \overline{c} \end{aligned}$$

   (a) Generate all the primes of f_1 using the Quine-McCluskey method.

   (b) Generate all the primes of f_2 using the Quine-McCluskey method.

   (c) Generate a reduced prime implicant table for f_1.

   (d) Generate a reduced prime implicant table for f_2.

2. Extend the reduced prime implicant generation algorithm to multiple-output functions and apply it to f_1 and f_2 in Problem 1.

3. Give an example of a function where the Quine-McCluskey procedure generates the same k-cube multiple times.

4. Using tautology, verify that f is irredundant.

$$f = b \cdot d + b \cdot \overline{e} \cdot f + \overline{a} \cdot \overline{b} \cdot c \cdot e + a \cdot \overline{b} \cdot \overline{d} + a \cdot c \cdot f + \overline{c} \cdot \overline{d}$$

5. Consider the function

$$f = a \cdot b + a \cdot c \cdot e + b \cdot c \cdot d + a \cdot b \cdot c \cdot e + c \cdot d \cdot e + b \cdot \overline{c} \cdot e + \overline{a} \cdot b \cdot c \cdot e$$

   (a) Show that f is positive unate in all its variables.

   (b) Give one essential minterm (vertex) for each essential prime of f.

6. Give a polynomial-time algorithm to determine if a disjoint two-level cover, i.e., a cover where each pair of cubes is disjoint, is a tautology.

7. Two two-level covers C_1 and C_2 can be checked for equivalence by checking $C_1 \cdot C_2 + \overline{C_1} \cdot \overline{C_2}$ for tautology. Give a method for checking the equivalence of C_1 and C_2 which uses repeated tautology checking on covers that are no larger than the larger of C_1 and C_2.

8. Prove that every prime of a unate function is an essential prime implicant.

9. Prove that given an arbitrary cover F for a unate function f, making F single cube containment minimal results in a minimum cover for f.

10. Prove that the complement of a unate cover F can be expressed as

$$\overline{F} = x_i \cdot \overline{F_{x_i}} + \overline{F_{\overline{x_i}}}$$

if F is monotone decreasing in x_i.

11. Prove that a set of cubes C covers a cube c if and only if C_c, the cofactor of C with respect to c, is a tautology.

12. Let $h(x, y) = f(x) + g(y)$ where x and y are disjoint sets of variables. Prove or disprove that a minimum cover of h can be obtained by finding minimum covers of f and g separately.

13. Consider a completely specified function f with a cover F in the form of

$$F = \begin{pmatrix} A & T \\ T & B \end{pmatrix}$$

where T is a block of all 2's. Thus, the function is the sum of two functions which have no variables in common. One might conjecture that the minimum cover of $\overline{f}$ is simply the cross product of the minimum covers of the complements taken separately. Give an example of such a function f with the property that the minimum cover of $\overline{f}$ is strictly less than the size of the minimum cover of the complement of A times the size of the minimum cover of the complement of B.

14. Give an example of an incompletely specified function which has a unate cover but whose minimum cover is not unate.

15. Prove Equation 4.3.

16. Prove Equation 4.4.

REFERENCES

[1] R. K. Brayton, G. D. Hachtel, C. McMullen, and A. Sangiovanni-Vincentelli. *Logic Minimization Algorithms for VLSI Synthesis.* Kluwer Academic Publishers, Norwell, MA, 1984.

[2] M. Dagenais, V. K. Agarwal, and N. Rumin. McBOOLE: A Procedure for Exact Boolean Minimization. *IEEE Transactions on Computer-Aided Design of Integrated Circuits*, CAD-5(1):229–237, January 1986.

[3] M. R. Garey and D. S. Johnson. *Computers and Intractability: A Guide to the Theory of NP-Completeness.* W. H. Freeman and Company, New York, NY, 1979.

[4] S. J. Hong, R. G. Cain, and D. L. Ostapko. MINI: A Heuristic Approach for Logic Minimization. IBM *Journal of Research and Development*, 18(4):443–458, September 1974.

[5] E. J. McCluskey. Minimization of Boolean Functions. *Bell Systems Technical Journal*, 35(6):1417–1444, November 1956.

[6] R. Rudell and A. Sangiovanni-Vincentelli. Multiple-Valued Minimization for PLA Optimization. *IEEE Transactions on Computer-Aided Design of Integrated Circuits*, CAD-6(5):727–751, September 1987.

Chapter 5

Testability of Two-Level Circuits

5.1 Introduction

Just as synthesis and optimization techniques for area are easier to understand in terms of two-level circuits, similarly synthesis for testability techniques are also easier to understand on two-level circuits. In this chapter we will derive conditions for the testability of two-level circuits under various fault models. The fault models we will consider are described in Section 5.2. In the case of dynamic fault models we will need timing analysis terminology which we present in Section 5.5. Under the delay fault model, robust or nonrobust testing (to be defined in this chapter) can be performed. Different forms of robust and nonrobust tests have been investigated and are described in Section 5.6. For each considered fault model we will derive necessary and sufficient conditions for a fault in a given circuit to be testable. We will provide test generation methods and synthesis for testability approaches based on the necessary and sufficient testability conditions.

5.2 Fault Models

There are a number of fault models which are used to describe the most common types of physical faults which might occur in a fabricated circuit. These models can be broadly classified into two categories – static (logic) fault models and dynamic (parametric) fault models. We describe these models below.

Static fault models are models for those types of physical faults which have an effect on the logic functionality of the circuit and can be detected independent of delays in the circuit. Static fault models have had widespread use in industry since they can model a large number of typical fabrication failures. They are simple to test since the circuits can be run at a low speed, and usually a small test vector set can be constructed to test a majority of the static faults in a circuit. Static faults are detected by applying a single vector to the primary inputs of a combinational logic circuit and observing the logical values at the circuit outputs. Typically, a stream of vectors is applied to the circuit, and the circuit is allowed to stabilize upon the application of each vector by waiting an arbitrarily long period of time. The static fault models considered in this book are the single stuck-at, multiple stuck-at, and the bridging fault models.

Dynamic fault models are models for those types of physical faults which do not necessarily affect the static operation of the circuit and thus can only be detected by tests which either measure delay through a circuit or which are applied at the same speed at which the circuit is specified to operate. Typically, a vector pair $\langle v_1, v_2 \rangle$ is applied to the primary inputs of a combinational logic circuit. The circuit is allowed to stabilize after the application of vector v_1, and the primary outputs of the circuit are sampled at time τ (the clock period of the circuit) after the application of vector v_2. The three dynamic fault models considered in this book are the gate delay fault, transistor stuck-open fault, and path delay fault models.

5.2.1 Single-Stuck-At Fault Model

The single stuck-at fault model is the most widely used fault model. This model assumes that a single fault existing at a given wire in the circuit causes that wire to appear to be permanently at a high voltage level (stuck-at-1), or to appear to be permanently at a low voltage level (stuck-at-0). One example of the type of manufacturing error modeled by a single stuck-at fault would be shorts to either the power supply or to ground.

Specifically, a circuit has a stuck-at-1 fault if there exists some wire in the circuit such that the logical value associated with that wire is 1 and is independent of the values applied at the inputs of the circuit. A stuck-at-0 fault is similarly defined. It is also assumed that only one such fault can be present in a circuit. A circuit C

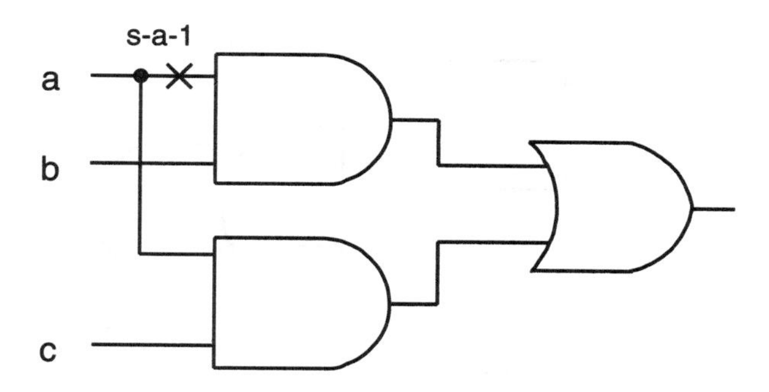

Figure 5.1: Single stuck-at fault in a combinational circuit

is completely single stuck-at fault testable if there exists for each stuck-at fault f in C, an input vector v such that the output of the faulty circuit containing f is different from the output of the fault-free circuit when v is applied to both of them.

Consider the circuit of Figure 5.1 which contains a stuck-at-1 fault which can be detected by a vector $a = 0$, $b = 1$, and $c = X$ (X means that the input is a don't-care, i.e., it can assume any value). The response of the faulty circuit is a 1, but the response of the fault-free (or true) circuit is a 0.

5.2.2 Multiple-Stuck-At Fault Model

The multiple stuck-at fault model is a more comprehensive stuck-at fault model in which multiple stuck-at faults are assumed to exist in the circuit simultaneously. Specifically, a circuit has a multiple stuck-at fault if there exists one or more stuck-at faults in the circuit. A multiple stuck-at fault is often referred to as a multifault.

A circuit C is completely multiple stuck-at fault testable if there exists for each set of stuck-at faults S in C, an input vector v such that the output of the faulty circuit containing S is different from the output of the fault-free circuit when v is applied to both of them. Multifault testability is obviously a much more stringent form of testability than single stuck-at fault testability. However, both model the same type of physical faults.

A multiple stuck-at fault is shown in the circuit of Figure 5.2 which can be detected by the vector $a = 1$, $b = 0$, and $c = 1$.

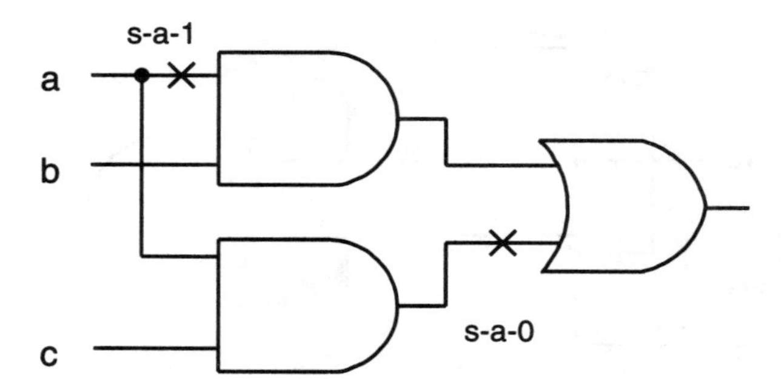

Figure 5.2: Multiple stuck-at fault in a combinational circuit

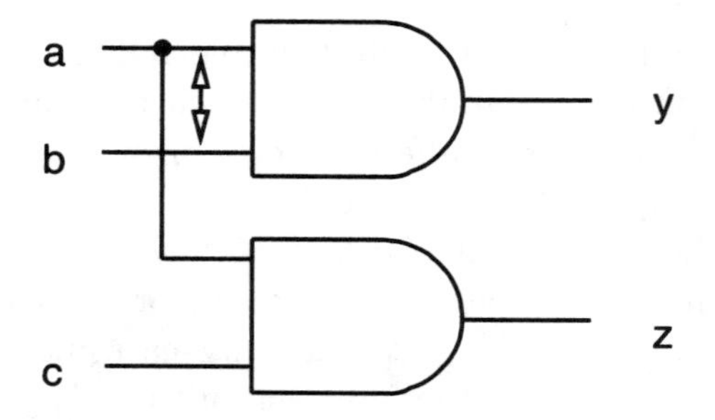

Figure 5.3: Bridging fault in a combinational circuit

5.2.3 Bridging Fault Model

A bridging fault in a circuit occurs when two or more lines are shorted together. There are two types of bridging faults – the OR bridging fault where if any of the lines is a 1, then all the lines become 1, and the AND bridging fault where if any of the lines is a 0, then all the lines become 0.

An example of an OR bridging fault is shown in Figure 5.3. The function of output y is changed from $a \cdot b$ to $a + b$, and the function of output z is changed from $a \cdot c$ to $(a + b) \cdot c$.

5.2.4 Gate Delay Fault Model

A circuit has a gate delay fault if there exists some gate in the circuit such that the output of the gate and the circuit output are both slow to make a transition ($0 \rightarrow 1$ or $1 \rightarrow 0$) when one or more inputs to the gate change value. This means that the rise time or the fall time of the output of the gate might change, or the time taken for the transition to propagate through the gate might change. It is assumed

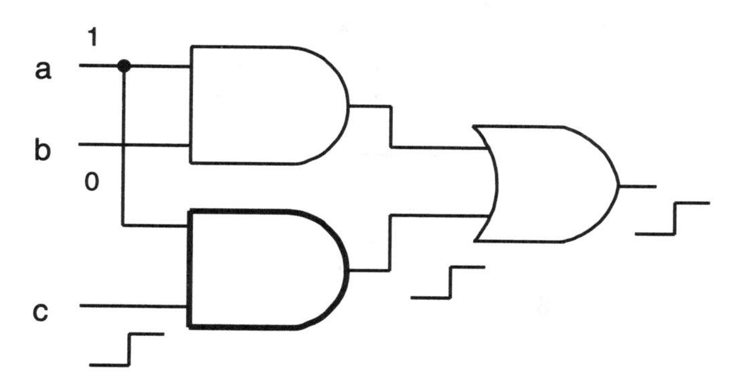

Figure 5.4: Gate delay fault in a combinational circuit

that a single gate delay fault is catastrophic enough to result in an extra delay along *any* path through that gate to a primary output.

Consider the circuit of Figure 5.4. If the bottom AND gate has a gate delay fault for the $0 \rightarrow 1$ transition, then the fault can be detected using the test vector pair shown.

5.2.5 Transistor Stuck-Open Fault Model

The transistor stuck-open fault model was first developed by Wadsack [11]. A circuit has a stuck-open fault if there exists some transistor in the circuit which is rendered nonconducting by a fault or if there exists some interconnect which has a break causing it to be non-conducting. In a physical system, often transistors or interconnect are not rendered entirely nonconducting by faults, but instead have a very high resistance resulting in slow delays through the circuit. These high resistance "stuck-open" faults are also detected using this model. The stuck-open fault model is more comprehensive than the gate delay fault model since it can model a fault, on an individual input or transistor of a gate, as causing an extra delay through a circuit, whereas a gate delay fault models a fault as causing all paths through a gate to be delayed. Thus, the stuck-open fault model uses a finer grain approach.

Tests for stuck-open faults are dynamic in nature for two reasons. First, as mentioned in the preceding paragraph stuck-open faults are often not totally nonconducting but will allow the circuit to reach the correct value given enough time. Second, even if a pull-down transistor is rendered completely nonconducting, eventually the output will go low as a result of leakage. Thus, stuck-open faults need

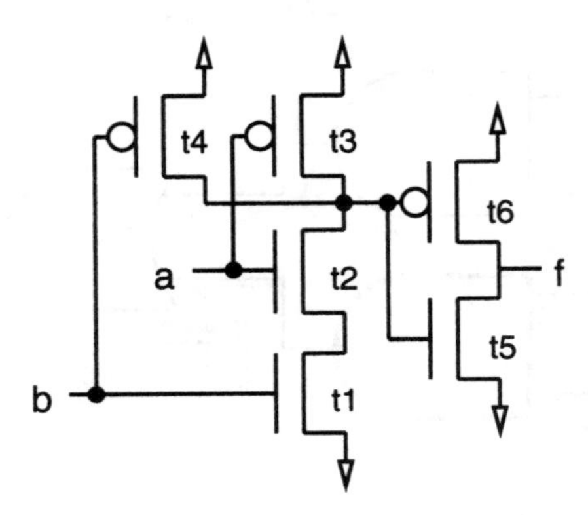

Figure 5.5: Static CMOS AND gate implementation

to be treated as dynamic faults during testing.

It is noteworthy that not all stuck-open faults are dynamic. If a pull-up transistor in an inverter is stuck-open, then the output, having reached 0 once, will stay at that value forever as if it had a stuck-at-0 fault.

Like gate delay faults, a pair of test vectors is required to detect stuck-open faults instead of a single test vector which suffices in the case of stuck-at faults. This is due to the fact that stuck-open faults require the circuit to be in a certain state in order for a fault to be detected. Consider the static *complementary metal oxide semiconductor* (CMOS) AND gate shown in Figure 5.5. If transistor $t3$ is faulty and nonconducting, this fault can only be detected if the output f is first made high by asserting both inputs a and b, and then input a is brought low to detect the fault. In the faulty circuit transistor $t3$ will not conduct and output f will remain high. Output f may eventually go low if $t3$ is partially conducting, so the output must be observed shortly (depending on the expected delay of the gate) after applying the second vector to determine if the fault exists.

It can be noted that there is no distinguishable difference between a given transistor having a stuck-open fault or the interconnect to the gate of that same transistor having a stuck-open fault. If the circuit of Figure 5.5 is considered again, the fault in transistor $t3$ which was described previously would be indistinguishable from a stuck-open fault on the section of interconnect from input a to the gate of transistor $t3$. Both faults would require the same test vector pair in order to be detected. Thus, test vectors which test all pos-

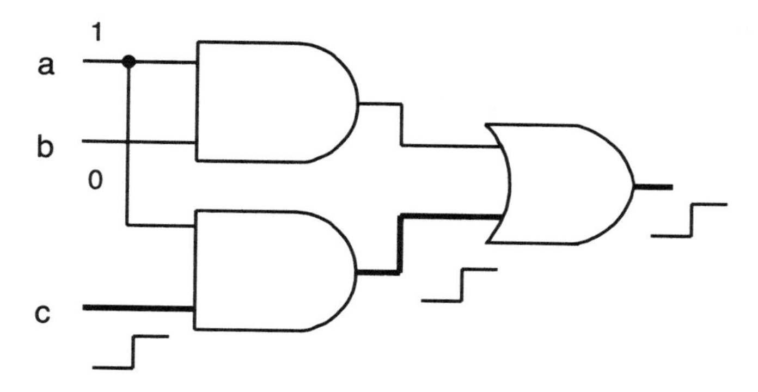

Figure 5.6: Path delay fault in a combinational circuit

sible stuck-open faults on the inputs of a gate also test all internal transistors of the gate for stuck-open faults.

5.2.6 Path Delay Fault Model

A circuit has a path delay fault if there exists some path from a primary input to a primary output through a set of gates and interconnect such that a transition on the primary input is slow to propagate along the path to the primary output. A path delay fault can result from several different physical phenomena. One cause can be a cumulative effect due to each device in a path being slower than specified due to a variation in temperature, voltage, or processing parameters. Another cause can be a single defective gate such as that detected by the gate delay fault model. An additional cause could be additional delay resulting from a stuck-open fault which results in a device which is either nonconducting or which has a very high resistance. Thus, the path delay fault model is more comprehensive than either the gate delay fault model or the stuck-open fault model.

Consider the circuit of Figure 5.6. If the path shown in bold has a path delay fault for the $0 \rightarrow 1$ transition, then the fault can be detected using the test vector pair shown.

5.2.7 Complexity of Test Generation

A *decision problem* [4] is a problem that has a *yes* or a *no* answer. Some decision problems form an equivalence class called *nondeterministic polynomial-time* (NP)-complete. The characteristics of these problems are that a solution to any problem can be transformed into a solution to another problem in polynomial time and no known polynomial-time algorithm can solve any of the problems. In addition, these problems are not provably intractable. Many optimization problems can be transformed into decision problems, and the corresponding decision problem can be proven to be NP-complete (e.g., the traveling salesman problem). There is another class of decision problems that are not NP-complete but are at least as hard as NP-complete problems. Such problems are called NP-hard.

Test generation for combinational circuits is an NP-complete problem and test generation for sequential circuits is an NP-hard problem under the single stuck-at fault model [7] as well as under other commonly used fault models. There is no known polynomial-time algorithm that can be used to generate tests for circuits. For some classes of circuits it is possible to generate tests efficiently, using various heuristic search techniques. However, in some cases design for testability or synthesis for testability methods must be used to obtain high fault coverage.

5.3 Single Stuck-At Faults

In this section we will provide necessary and sufficient conditions for a two-level circuit to be fully single stuck-at fault testable. This will in turn provide synthesis for testability procedures as well as a test generation mechanism.

5.3.1 Conditions for Testability

We can relate primality and irredundancy of a two-level cover of a function to the testability of stuck-at faults in the corresponding circuit.

Theorem 5.3.1 *A single-output two-level circuit C is fully testable for all single stuck-at faults if and only if the corresponding cover E is prime and irredundant.*

Proof. Necessity: If any implicant c in the cover E is not prime in literal l, it means that there is an AND gate in the two-level circuit C such that an input to the AND gate from the input l can be removed without changing the functionality of the circuit. This implies that there is a stuck-at-1 redundancy on that input to the AND gate in C.

If any implicant c in the cover E is redundant, then the corresponding AND gate in the circuit can be removed without changing the functionality of the circuit. This implies that there is a stuck-at-0 redundancy on the output of the AND gate in C.

Sufficiency: Since C is a single-output two-level circuit, the set of faults we have to consider are the stuck-at-1 faults on the inputs to the AND gates in C, stuck-at-0 faults on the outputs of the AND gates, and the stuck-at-1 and stuck-at-0 faults on the output of C. Stuck-at-0 faults at the inputs to an AND gate are equivalent to stuck-at-0 faults on the output. Similarly, stuck-at-1 faults at the output of any AND gate are equivalent to the stuck-at-1 fault at the output of C.

A stuck-at-1 fault on input l of an AND gate in the circuit can be tested by applying a primality test for the literal l in the corresponding cube c in the cover E. Since l is prime in c, we know that $c - \{l\} \cup \{\bar{l}\}$ intersects the *OFF*-set of C. Call this *OFF*-set vertex v. Then v is a primality test for literal l, and when the test is applied to C, it will detect the stuck-at-1 fault on input l of the AND gate. In the fault-free circuit the output of the AND gate and C are at 0, but all inputs to c other than l are at 1, since $v \subseteq c - \{l\} \cup \{\bar{l}\}$. In the faulty circuit the AND gate will have the value 1, and therefore C will have the value 1 at the output.

A stuck-at-0 fault on the output of an AND gate in the circuit can be tested by applying an irredundancy test for the corresponding cube c in the cover E. The irredundancy test corresponds to a relatively essential vertex of cube c. In the fault-free circuit the output of the AND gate and C are 1, but all cubes other than c are at 0 values. The faulty circuit will have the AND gate at 0, and therefore C will also be at 0.

E has a nonempty *ON*-set and *OFF*-set, and therefore the stuck-at-0 and stuck-at-1 faults on the output of C are easily testable. □

Unfortunately, the above theorem cannot be generalized to multiple-output circuits as we will illustrate in the next section.

5.3.2 Synthesis for Full Testability

The Quine-McCluskey procedure described in Chapter 4 will produce prime and irredundant covers that are fully testable for single stuck-at faults. Prime and irredundant multiple-output functions are, however, not fully single stuck-at fault testable.

Recall that a necessary condition for an implicant c to be prime is that there is no other implicant d that contains c. In the case

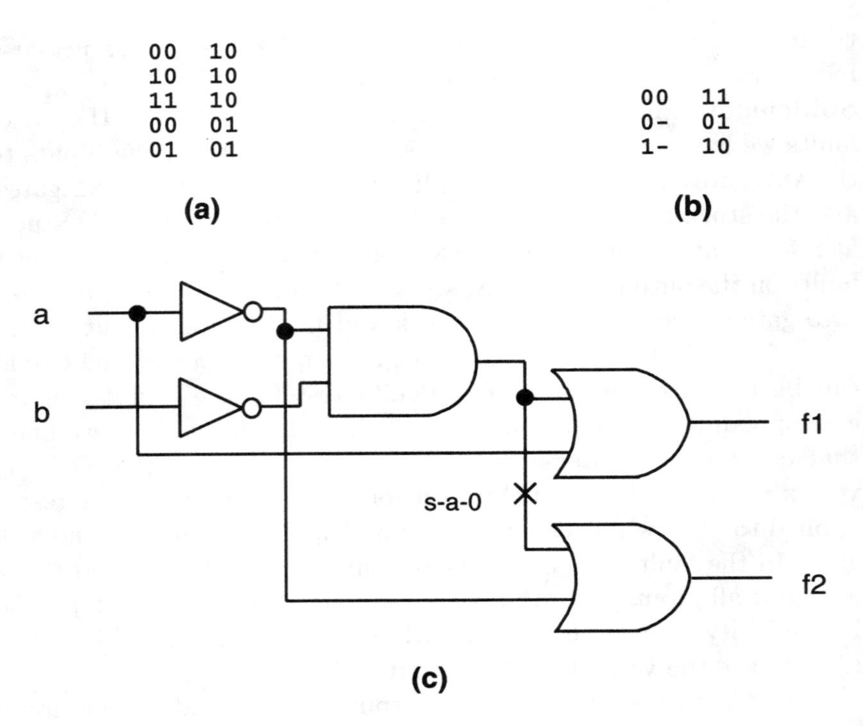

Figure 5.7: Multiple-output prime and irredundant cover is not fully single stuck-at fault testable

of multiple-output functions c and d may have identical input parts, but d may contain vertices of additional outputs than the output part of c. For instance, we may have a two-input, two-output function, with an implicant 00 11 that contains the implicant 00 10. Therefore, 00 10 cannot be a prime implicant of this function.

Consider the example of Figure 5.7. The *ON*-sets of the two outputs are given separately in Figure 5.7(a). A multiple-output prime and irredundant cover is shown in Figure 5.7(b). The corresponding circuit is shown in Figure 5.7(c). In this circuit, the stuck-at-0 fault on the input to the bottom OR gate is redundant.

In the case of multiple-output functions, the covers produced by exact or heuristic logic minimization algorithms have to be "processed" to achieve a minimal number of literals and connections which corresponds to a fully testable circuit. In this processing, each output is made prime and irredundant separately. The **MAKE_SPARSE** procedure in ESPRESSO (see Section 4.6.3) performs this processing and thereby guarantees full testability of the final cover. Literals that

```
00  11        00  10        -0  10
0-  01        0-  01        0-  01
1-  10        1-  10        1-  10

  (a)           (b)           (c)
```

Figure 5.8: Transformations to a multiple-output cover for full testability

are 1 in the output parts of cubes are lowered, and 0 or 1 literals in the input parts of cubes are raised to $-$ iteratively until convergence is reached.

The transformations that take place in a multiple-output cover during the removal of stuck-at fault redundancies are illustrated in Figure 5.8. A prime and irredundant realization of the function is given in Figure 5.8(a). Since the prime 00 11 unnecessarily contains the vertex 00 01 which is already contained by the prime 0 $-$ 01, we replace the prime 00 11 by the implicant 00 10 to arrive at the cover of Figure 5.8(b). Now, the implicant 00 10 is not prime in its input part, and we raise it to -0 10. The final cover, shown in Figure 5.8(c) will produce an implementation that is fully single stuck-at fault testable.

5.3.3 Test Generation Methods

Stuck-at fault tests can be generated for two-level circuits by manipulating the corresponding two-level cover. In order to generate a primality test for input l in cube c of cover E, we compute $r_c = c - \{l\} \cup \{\bar{l}\}$, and intersect r_c with the complement of E. Alternatively, we can compute the difference $r_c - E$ (for example, using disjoint SHARP operation). If the difference is empty, the literal is not prime.

To generate an irredundancy test for cube c in cover E, we compute $c - D$, where D is the set of cubes in E other than c.

We have not explicitly considered functions with don't-care sets so far. However, the notion of primality and irredundancy was defined for incompletely specified functions as well. The following theorem will relate testability to don't-care sets and is also useful in the synthesis of sequential circuits for stuck-at fault testability.

Theorem 5.3.2 *Given a single-output two-level circuit C that is prime and irredundant under a don't-care set D, no minterm in D is* *required to detect any single stuck-at faults in C.*

Proof. First, note that minterms in D may detect single stuck-at faults in C. However, the theorem merely states the existence of some stuck-at fault test for any given fault outside D.

The primality test for a literal l in a cube q corresponds to a test for stuck-at-1 fault at the input of the corresponding AND gate. Cube q is prime under the don't-care set D. This means that there is no other implicant that is contained in the union of the *ON*-set and *DC*-set that contains q. Therefore, removing l from the cube q will result in a cube that intersects the *OFF*-set. Thus, we can find a test in the *OFF*-set for any stuck-at-1 fault to any AND gate in C.

The irredundancy test for a cube tests for stuck-at-0 faults at the output of the AND gates. The irredundancy test for a cube q has to lie within the *ON*-set since it is a relatively essential vertex of q.

The stuck-at faults at the primary output of C are trivially detected by vectors in the *ON*-set and *OFF*-set. □

Consider the example of Figure 5.9. In Figure 5.9(a), an XOR gate that is fully testable for single stuck-at faults is shown. If, however, a don't-care set $D = \{11\}$ is specified, then the implementation is not prime and irredundant under the don't-care set. This implies that the two marked single stuck-at faults are redundant. Both these faults require the vector 11 to be detected. A prime and irredundant realization under $D = \{11\}$ is shown in Figure 5.9(b) which is simply an OR gate. Note that the vector 11 does detect the stuck-at-0 fault on the output of the gate, but it is not *required* to detect any faults on the inputs or the output of the gate.

5.4 Multiple Stuck-At Faults

In this section we give conditions for full multiple stuck-at fault (multifault) testability in two-level circuits, derive synthesis procedures that result in fully multifault testable circuits, and give test generation methods to obtain complete multifault test sets.

5.4.1 Conditions for Testability

Primality and irredundancy of a single-output circuit are obviously necessary conditions for multifault testability. The theorem below,

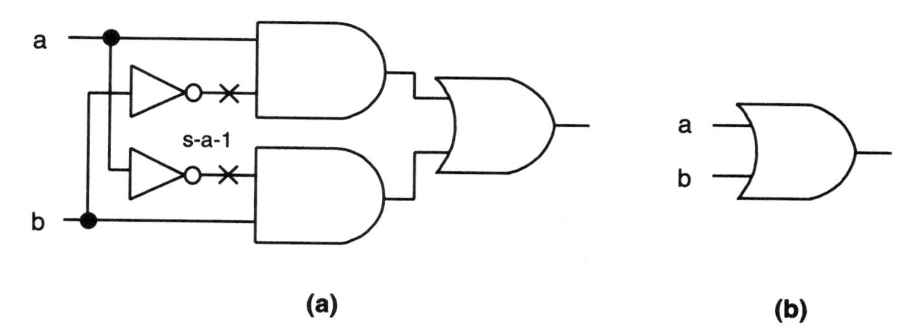

Figure 5.9: Prime and irredundant realizations without and with a don't-care set

originally proved in [6], shows that it is a sufficient condition too.

Theorem 5.4.1 *Given a single-output prime and irredundant two-level circuit C, the set of tests, S, comprising the primality tests for each literal in each cube and the irredundancy test for each cube in the cover E corresponding to C detects all multifaults in C.*

Proof. Consider the effect of a multifault m on E. For each single fault f in the multifault m some wire e in C gets stuck to 1 or 0. The effect of the fault m falls into three classes: Literals are uniformly stuck-at-0, and as a result cubes are uniformly removed from E; literals are uniformly stuck-at-1, and as a result cubes are uniformly expanded (i.e., literals are removed from the cubes) in E; some tagged literals are stuck-at-1 and some are stuck-at-0, and as a result cubes are simultaneously removed from E and expanded in E.

In the first case, if m strictly removes cubes from E, then by the hypothesis of the theorem, there exists an irredundancy test in S, call it v, for a removed cube c. The vector v is a member of the ON-set that is uniquely contained by c, and so the removal of other cubes in E cannot possibly result in containing v. Hence, the faulty network gives a value of 0 on the vector v. Thus, v detects the multifault m.

In the second or third case, if m strictly expands literals in E or expands literals in cubes that are not removed, by the hypothesis to the theorem, there is a primality test v for an expanded literal l in some cube c that is not removed from E. This means that $E(v) = 0$. However, if l is expanded in c in the faulty cover, call it E_f, then $E_f(v) = 1$. Therefore, v detects m.

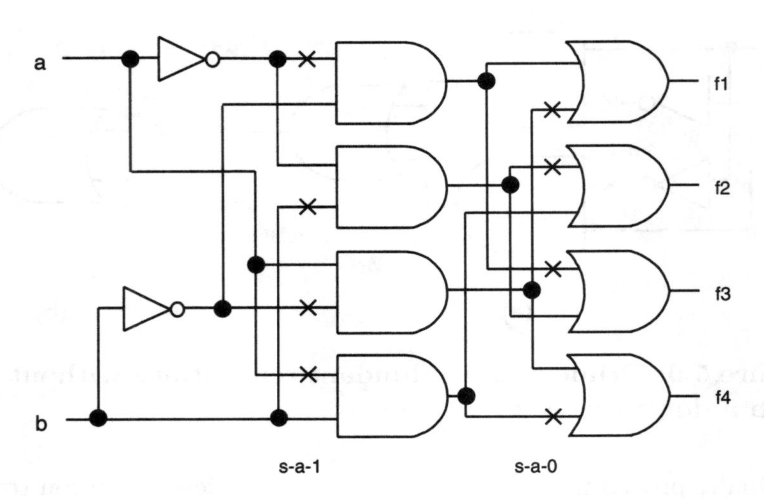

Figure 5.10: A fully single stuck-at fault testable but multifault redundant two-level circuit

Thus, in all three cases we can find a $v \in S$ that detects m, and therefore S detects all multifaults of C. □

5.4.2 Synthesis for Full Testability

Single-output two-level circuits can be synthesized to be fully multifault testable simply by obtaining prime and irredundant realizations. As described in Section 5.3.2 multiple-output prime and irredundant two-level circuits are not necessarily fully single stuck-at fault testable. However, they can be made fully single stuck-at fault testable by a postprocessing step.

It has been conjectured that a fully single stuck-at fault testable multiple-output two-level circuit is fully multifault testable as well. Consider the circuit of Figure 5.10, taken from [1]. It is fully single stuck-at fault testable, but the multifault corresponding to the eight marked stuck-at faults is redundant.

In order to obtain fully multifault testable multiple-output two-level circuits, we have to ensure that the *cone* corresponding to each output is prime and irredundant with respect to that output so that Theorem 5.4.1 is applicable for that output. Note that this does not preclude the sharing of product terms across the different outputs. An example of a shared realization is shown in Figure 5.11(b); it requires fewer product terms than the realization of Figure 5.11(a)

```
1-  10
-1  10          1-  11
-0  01          -1  10
10  01          -0  01

  (a)             (b)
```

Figure 5.11: Fully multifault testable multiple-output two-level circuits

which does not have any sharing of product terms across the outputs. Note that the use of the MAKE_SPARSE procedure of ESPRESSO guarantees that this condition is satisfied.

5.4.3 Test Generation Methods

The test generation methods for the case of single stuck-at faults (see Section 5.3.3) are applicable here as well.

5.5 Timing Analysis Terminology

In this section we introduce terminology that will allow us to discuss timing issues as well as fault models for temporal behavior. A combinational circuit can be viewed as a directed graph where vertices or nodes in the graph correspond to gates in the circuit and edges correspond to connections in the circuit. Primary inputs are source vertices while primary outputs are sinks. A *path* in a combinational circuit is an alternating sequence of vertices and edges, $\{v_0, e_0, ..., v_n, e_n, v_{n+1}\}$, where edge e_i, $1 \leq i \leq n$, connects the output of vertex v_i to an input of vertex v_{i+1}. For $1 \leq i \leq n$, v_i is a gate g_i, v_0 is a primary input, and v_{n+1} is a primary output. Each e_i is a wire in the actual circuit (or a net).

Each gate g_i (wire e_i) is assumed to have a delay $d(g_i)$ ($d(e_i)$) which can be a fixed quantity under a fixed delay model or can vary in a given range under the monotone speedup delay model (originally coined in [9]). The *length* of a path $P = \{e_0, v_0, e_1, ..., e_n, v_n, e_{n+1}\}$ is defined as $D(P) = \sum_{i=0}^{n} d(v_i) + \sum_{i=0}^{n+1} d(e_i)$.

An *event* is a transition, either $0 \rightarrow 1$ or $1 \rightarrow 0$ at a gate. Consider a sequence of events $\{r_0,\ r_1,\ ...,\ r_n\}$ occurring at gates $\{g_0,\ g_1,\ ...,\ g_n\}$ along a path, such that r_i occurs as a result of event r_{i-1}. The event r_0 is said to propagate along the path. If there exists a vector pair such that under *appropriate* delays in the circuit, i.e., some choice of numbers for the $d(g_i)$'s and $d(e_i)$'s, an event could propagate along a path, then the path is said to be *event sensitizable.* If there exists an input vector pair such that under *arbitrary* delays in the circuit, i.e., all possible choices of numbers for the $d(g_i)$'s and $d(e_i)$'s, an event propagates along a path, then the path is said to be *single event sensitizable.*

A *controlling value* at a gate input is the value that determines the value at the output of the gate independent of the other inputs. For example, 0 is a controlling value for an AND gate. A *noncontrolling value* at a gate input is the value which is not a controlling value for the gate. For example, 1 is a noncontrolling value for an AND gate. We say that a gate g has the *controlled value* if one of its inputs has a controlling value; otherwise, we say that g has the *noncontrolled value.*

Let $\pi = \{v_0, e_0, ..., v_n, e_n, v_{n+1}\}$ be a path. The inputs of v_i other than e_{i-1} are referred to as the *side-inputs* to π. The input e_{i-1} is referred to as the π-input to v_i. We say that π is *statically sensitizable* if there exists an input vector such that all the side-inputs along π settle to noncontrolling values.

We say that an input vector w *sensitizes to a 1* path π in C if and only if the value of v_{n+1} is 1, and for each v_i, $1 \leq i \leq n+1$, if v_i has a controlled value, then the edge e_{i-1} presents a controlling value (note that other inputs to v_i may also present controlling values). Similarly, we say that an input vector w *sensitizes to a 0* path π in C if and only if the value of v_{n+1} is 0, and for each v_i, $1 \leq i \leq n+1$, if v_i has a controlled value, then the edge e_{i-1} presents a controlling value.

We say that an input vector w *statically sensitizes to a 1* path π in C if and only if the value of v_{n+1} is 1, and for each v_i, $1 \leq i \leq n+1$, if v_i has a controlled value, then the edge e_{i-1} is the only input of v_i that presents a controlling value. Similarly, we say that an input vector w *statically sensitizes to a 0* path π in C if and only if the value of v_{n+1} is 0, and for each v_i, $1 \leq i \leq n+1$, if v_i has a controlled value, then the edge e_{i-1} is the only input of v_i that presents a controlling value. Note that by definition, if a vector w statically sensitizes a path, then it either statically sensitizes the path to a 1 or to a 0. Also, if a path can be statically sensitized to a 1, it does not mean that it can be statically sensitized to a 0.

We may now observe that for a path π in a circuit C to be *single event sensitizable* (to a 1) it is necessary that there exist a vector pair $\langle w_1, w_2 \rangle$ such that w_1 statically sensitizes the path π to a 0 and w_2 statically sensitizes the path π to a 1. Moreover, all side-inputs along π must have (the same) noncontrolling values for both w_1 and w_2. In addition, at each gate v_i along π if the value of v_i on w_1 was the controlled value, then the value of v_i on w_2 is the noncontrolled value, and if the value of v_i on w_1 was the noncontrolled value, then the value of v_i on w_2 is the controlled value.

5.6 Robust and Nonrobust Testing

5.6.1 Introduction

Delay faults and transistor stuck-open faults require the application of a vector pair $\langle v_1,\ v_2 \rangle$ in order to be detected. The response of a circuit to an applied vector pair depends on the delays of the circuit elements, i.e., gates and interconnects (also referred to as links).

Consider the circuit of Figure 5.12(a). A vector pair has been applied to the primary inputs. A $1 \rightarrow 0$ transition propagates along a single path to the circuit output. A $1 \rightarrow 0$ transition will occur at the circuit output regardless of the delays of the circuit elements. If gate 1 has a gate delay fault, it will be detected regardless of the delays of the circuit elements.

Now consider a different vector pair applied to the same circuit as illustrated in Figure 5.12(b). In this case, two $1 \rightarrow 0$ transitions race to set the AND gate output to a 0. Even if the path corresponding to $\{a, 1, 3\}$ has a delay fault, this vector pair may not detect the fault. For instance, if the path $\{b, 1, 3\}$ is not faulty, the correct transition is seen at the output.

Finally, consider the application of a third vector pair to the same circuit as shown in Figure 5.12(c). The circuit will make a $0 \rightarrow 1$ transition which may or may not be preceded by a $0 \rightarrow 1 \rightarrow 0$ glitch, depending on the relative time it takes for the $0 \rightarrow 1$ and $1 \rightarrow 0$ transitions to reach gate 3. It is possible under certain circuit delays that even if gate 2 is faulty, at the sample time τ the value of 1 is seen at the circuit output. We may erroneously conclude that the circuit is fault-free.

5.6.2 Hazard-Free Robust Path Delay Faults

A two-pattern test $\langle v_1,\ v_2 \rangle$ is termed a *hazard-free robust test* for a path delay fault if the test can be proven to detect the fault under

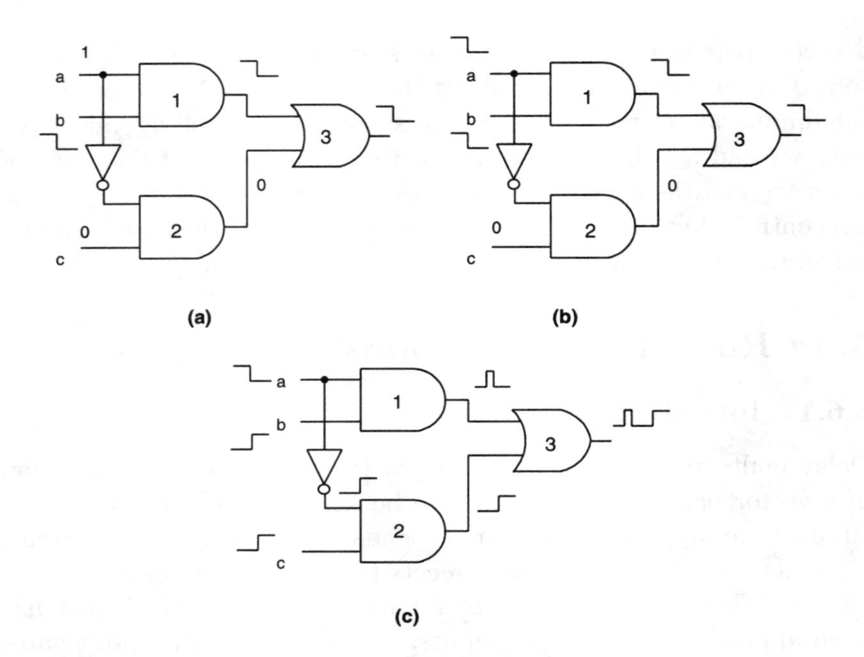

Figure 5.12: Robust and nonrobust testing

arbitrary delays in the circuit and is not invalidated by hazards or races.

Throughout this book we will assume that a hazard-free robust test for a path delay fault in a circuit C is a vector pair $\langle v_1, v_2 \rangle$ such that $C(v_1) = 0$ and $C(v_2) = 1$. We are also assuming single-output circuits for this discussion, but the results can be generalized to multiple-output circuits. Let the expected transition time on the vector pair be τ. The application of the vector pair is as follows: Vector v_1 is applied to C, and the values on nets are allowed to settle for an arbitrary amount of time. Vector v_2 is then applied to C. At time τ the output of C is sampled; if the value is 1 then no fault is detected, otherwise a fault is detected. Next, the vector pair $\langle v_2, v_1 \rangle$ is applied to propagate the opposite event along the path and detect faults corresponding to the $1 \rightarrow 0$ event. From the above discussion, the following definition can be obtained.

Definition 5.6.1 *A vector pair is a hazard-free robust path delay fault test for a path π if and only if it single event sensitizes π.*

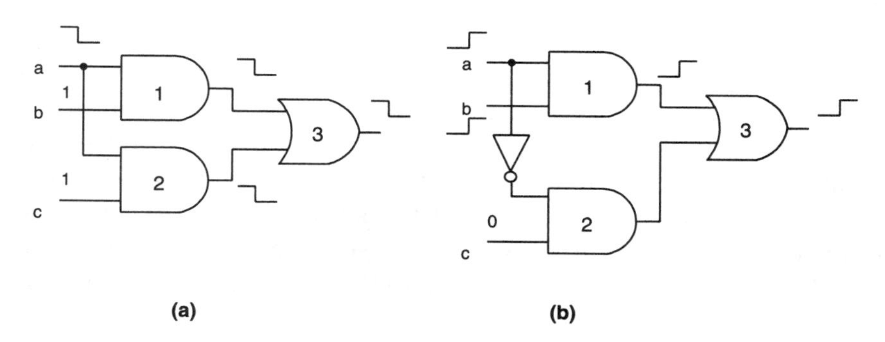

Figure 5.13: General robust testing in two-level circuits

5.6.3 General Robust Path Delay Faults

In *general robust* testing we can exploit a mechanism that eases the burden on both test generation as well as synthesis for testability approaches.

While testing for a particular transition along a path π general robust testing allows the convergence of multiple transitions of the same parity at a gate, provided the transitions correspond to a change in value from controlling to noncontrolling. The output of the gate changes only after the slowest transition. For example, in Figure 5.13(a), we have two $1 \to 0$ transitions, corresponding to paths passing through AND gates **1** and **2** and the OR gate. The output of the OR gate goes low only after the slower $1 \to 0$ transition. We are, in fact, validating the delays of both paths beginning from input **a** and passing through gates **1** and **2** if the test is successfully applied. Note that if the test fails, we cannot diagnose the faulty path. We thus lose diagnosability as we move from the hazard-free robust to the general robust world. In Figure 5.13(b) we present another example that allows converging $0 \to 1$ transitions at an AND gate.

Definition 5.6.2 *Let* $\pi = \{g_0,\ e_0,\ \ldots,\ g_n,\ e_n,\ g_{n+1}\}$ *be a path in a circuit* C. *A two-pattern test* $\langle v_1,\ v_2 \rangle$ *is a general robust test for* π *on a* $0 \to 1$ $(1 \to 0)$ *transition if:*

1. v_1 *sensitizes* π *to a 0* (1) *and* v_2 *sensitizes* π *to a* 1 (0).
2. *For each side-input* f_{ij} *at a gate* g_i *in* π, *if* v_2 *produces a noncontrolling value on* f_{ij} *and if* v_1 *produces a noncontrolling value at the* π *input to* g_i, *then* $\langle v_1,\ v_2 \rangle$ *gives a steady noncontrolling value on* f_{ij}.

In general robust testing if a vector pair $\langle v_1, v_2 \rangle$ is a test for a $0 \to 1$ transition on a path, it is not necessary that the vector pair $\langle v_2, v_1 \rangle$ is a general robust test for the $1 \to 0$ transition. This is because if the mechanism of allowing converging transitions is allowed, while the slower of the transitions is propagated in the first case (say two converging $0 \to 1$ transitions at an AND gate), the faster of the two is propagated in the second case (two converging $1 \to 0$ transitions at the AND gate). Therefore, the $\langle v_2, v_1 \rangle$ test is invalidated by a race.

The general robust test may involve hazards, because the above definition does not specify the value on side-inputs for v_1 when a path transition from a controlling value is being tested. This is because even under the presence of hazards on side-inputs, a delay on π will appear as a delay on the transition at the output of the gate. However, for the opposite transition from noncontrolling to controlling values on the path input to the gate, no transitions are allowed on the side-inputs to this gate.

The possible advantages of using delay tests with hazards are a reduction in the size of the test set and a higher degree of fault coverage (for multilevel circuits). A circuit is shown in Figure 5.14 (taken from [10]) for which no hazard-free robust path delay fault test exists for the path shown in bold but a general robust test exists. For the $0 \to 1$ transition on a, we apply $\langle b = 0, c = 1, d = 0, f = 0, g = 1 \rangle$ This results in converging $0 \to 1$ transitions at the 3-input AND gate. For the $1 \to 0$ transition on a, we apply $\langle b = 0, d = 1, e = 0, f = 0, g = 1 \rangle$ This could potentially result in a $0 \to 1 \to 0$ hazard at the bottom input of the final OR gate; however, this will not invalidate the test for this path. This circuit is a multilevel circuit. It can be shown that if a path in a two-level circuit is testable for both the $0 \to 1$ and $1 \to 0$ transitions under the general robust model, it is testable under the hazard-free robust model as well.

5.6.4 Hazard-Free Robust Gate Delay Faults

In the hazard-free robust gate delay fault model, it is not necessary that a path through the gate be tested for single event sensitizability. Transitions are allowed on multiple paths, but *all* the paths have to pass through the gate.

As in the hazard-free robust path delay fault case a hazard-free robust gate delay fault test in a circuit C is a vector pair $\langle v_1, v_2 \rangle$ such that $C(v_1) = 0$ and $C(v_2) = 1$. Vector v_1 is applied to C, and

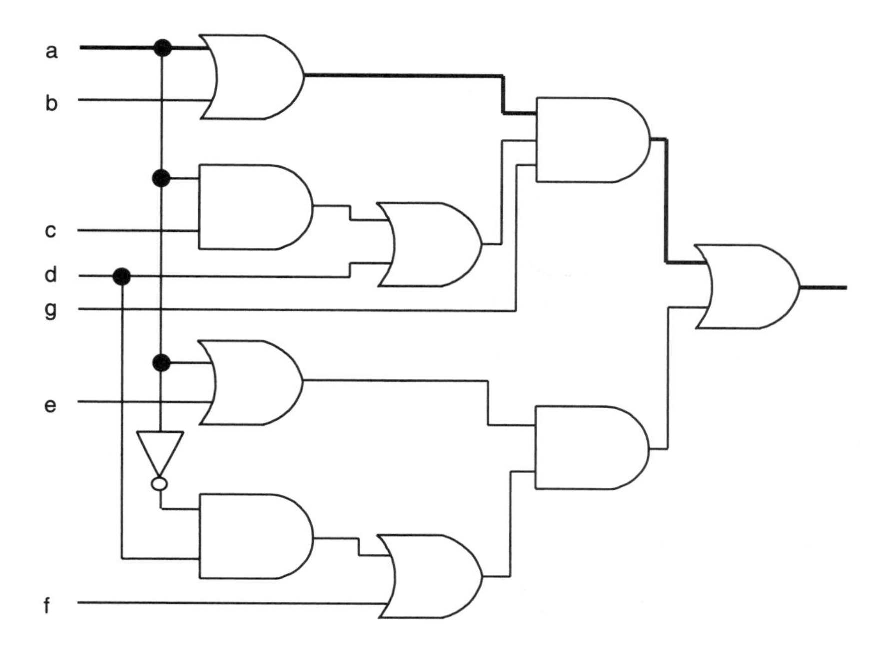

Figure 5.14: A circuit with a general robust test but no hazard-free test

the values on nets are allowed to settle. Vector v_2 is then applied to C. At time τ the output of C is sampled, if the value is 1, then no fault is detected, otherwise a fault is detected. Next, the vector pair $\langle v_2, v_1 \rangle$ is applied to propagate the opposite event on the gate and detect the fault corresponding to the $1 \rightarrow 0$ event.

An example is shown in Figure 5.15. The gate drawn in bold has a gate delay fault. We allow converging $1 \rightarrow 0$ transitions through the gate, since both transitions will be delayed if the gate is faulty. Toggling a from $1 \rightarrow 0$ will robustly test the gate for the $1 \rightarrow 0$ transition.

5.6.5 General Robust Gate Delay Faults

This is the analog of the general robust path delay fault: Controlling to noncontrolling transitions are allowed on paths (which do not pass through the gate under test) that converge as side-inputs to paths that pass through the gate under test, provided the converging transitions are of the same polarity. An example is shown in Figure 5.16. The gate shown in bold is being tested for the $0 \rightarrow 1$ transition.

We will not consider this fault model further, since we can

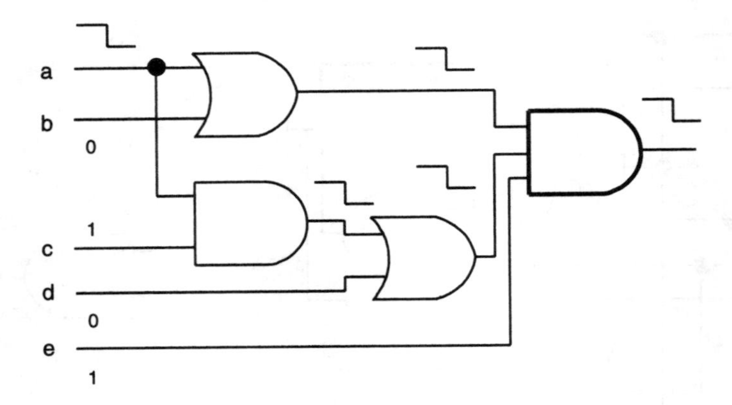

Figure 5.15: Hazard-free robust gate delay fault testing

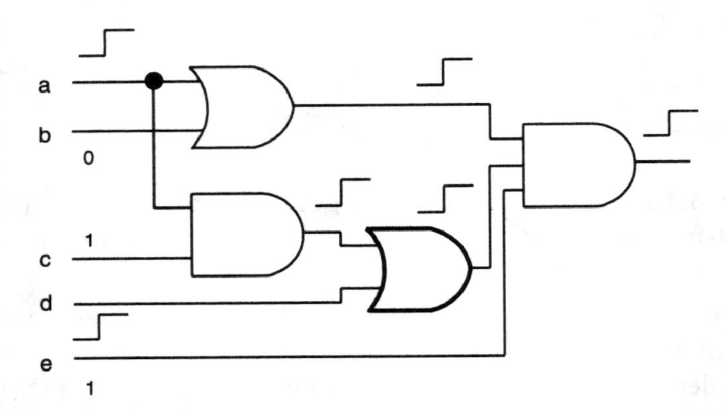

Figure 5.16: General robust gate delay fault testing

easily obtain efficient synthesis procedures for the hazard-free robust gate delay fault model.

5.6.6 Hazard-Free Robust Stuck-Open Faults

A comprehensive treatment of the testing for stuck-open faults in CMOS circuits can be found in [5]. The test generation and design for testability procedures described in [5] assume a test methodology where transistor stuck-open faults are treated as stuck-at faults, i.e., a vector pair $\langle v_1, v_2 \rangle$ is applied, and the circuit is allowed to stabilize, waiting an arbitrarily long period of time, after the application of *both* v_1 and v_2. The outputs of the circuit are sampled after stabilization. Recall that in the case of delay faults, the circuit is allowed to stabilize after the application of v_1, and its output is sampled upon the application of vector v_2 after waiting for a period of τ (typically the clock period of the circuit).

A transistor stuck-open fault can be treated as a link delay fault if the delay fault test methodology is adopted. Consider the AND gate of Figure 5.5. Under a hazard-free robust delay test methodology, propagating $0 \rightarrow 1$ and $1 \rightarrow 0$ transitions through each input of the AND gate to the output of the circuit will robustly detect all transistor stuck-open faults within the gate. The term *hazard-free robust transistor stuck-open faults*, coined in [2], corresponds to the situation where transistor stuck-open faults are treated as link delay faults.

For instance, the top link of the gate shown in bold in Figure 5.15 may be faulty. Note that the test shown in Figure 5.15 does *not* detect the link delay fault on the top link. This is because if the middle link to the same gate is not faulty, then the circuit will switch at the appropriate time. Setting $d = 1$ will force the detection of the fault. Thus, the hazard-free robust link delay fault or hazard-free robust transistor stuck-open fault model is more stringent than the gate delay fault model.

Having discussed the various fault models and their testing strategies, we move on to strategies for synthesizing two-level circuits so that they are fully testable under these fault models.

5.7 Hazard-Free Robust Path Delay Faults

In this section we will consider the testing of two-level circuits under the hazard-free robust path delay fault model. We will develop necessary and sufficient conditions for a path in a two-level circuit to be hazard-free robustly testable in Section 5.7.1, give a logic minimization procedure that maximizes robust path delay fault testability in Section 5.7.2, and describe an efficient test generation method tailored to two-level circuits in Section 5.7.3.

5.7.1 Conditions for Testability

Paths in two-level circuits are very simple. They begin at a primary input, may pass through an inverter, pass through an AND gate, and finally pass through an OR gate. Because of the correspondence between a two-level circuit and a cover, a primary input of a circuit and a literal in a cube, and a gate in the circuit and a cube in the cover, we will use phrases such as: "Let C be a two-level single-output circuit, and let π be a path in C that starts with literal l in cube q" as shorthand for: "Let C be a two-level single-output circuit implementing cover E, and let π be a path in C that starts with primary input i and passes through gate g such that i corresponds to literal l in cover E and g corresponds to cube q in cover E."

We now proceed with a lemma that gives a necessary condition for a path in a two-level single-output circuit C to be robustly testable. For $\langle v_1, v_2 \rangle$ to be a test for the path delay fault, every side-input to the OR gate must be zero for both v_1 and v_2. This means that the path is statically sensitizable for both v_1 and v_2. Therefore, for v_1 there must be literals in v_1 which for every cube in D will force the cube to evaluate to 0. The same should be true for v_2. The following theorem formally states this fact. Recall that $d_m(v_1)$ refers to the value of the literal in the mth position in cube d given by the value of the literal in the vector v_1.

Lemma 5.7.1 *Let C be a two-level single-output circuit, and let π be a path in C that starts with literal l in cube q. Let $D = C - q$. If $\langle v_1, v_2 \rangle$ is a hazard-free robust delay fault test for π in C, then for every cube d in D there exists some literal m in both v_1 and v_2 such that $d_m(v_1) = 0$ and $d_m(v_2) = 0$.*

Proof. Suppose there does not exist any such literal m both in v_1 and v_2. Then there exists at least one cube e in D such that for every literal l in v_1 and v_2 either $e_l(v_1) = 1$ or $e_l(v_2) = 1$. Thus, under the appropriate input delays, the cube e can evaluate to a 1, causing a glitch and invalidating a delay fault test for π. □

We now give a theorem describing necessary and sufficient conditions for a two-level single-output circuit to be hazard-free robustly path delay fault testable.

Theorem 5.7.1 *Let C be a two-level single-output circuit. Let π be a path in C that starts with l in cube q. Let $D = C - q$. There exists a hazard-free robust delay fault test for π in C if and only if there exists a vertex v_2 that is a relatively essential vertex of q and a vertex $v_1 = v_2 - \{l\} \cup \{\bar{l}\}$ such that v_1 is in the OFF-set of C.*

Proof. Necessity: Suppose $\langle w_1, v_2 \rangle$ is a delay fault test for π. Suppose v_2 is not a relatively essential vertex of q. Then there exists a cube d in D such that $d(v_2) = 1$. This implies that the path π is not the only path sensitized to a 1 by v_2. This would imply that π is not the only gate that supplies a controlling value to the OR gate at the output of the circuit, and this contradicts the definition of robust path delay fault testability. Thus, v_2 is a relatively essential vertex of q.

We now wish to show that while the vector w_1 need not be distance-1 (in literal l) from v_2 to be a valid test, if such a vector exists, then we can always construct a vector v_1 such that v_1 is in the *OFF*-set and distance-1 from v_2 in literal l and $\langle v_1, v_2 \rangle$ is a delay fault test for π. By Lemma 5.7.1 for every cube d in D there exists some literal m in both w_1 and v_2 such that $d_m(w_1) = 0$ and $d_m(v_2) = 0$. Let the set of all such literals be M. By definition, all literals in M are the same in w_1 and v_2 and $D(w_1) = D(v_2) = 0$. Let $v_1 = v_2 - \{l\} \cup \{\bar{l}\}$. The literal l is not in M because it cannot be the same in w_1 and v_2 as the literal l must change values to propagate an event along π. Since $M \subseteq v_2$, therefore $M \subseteq v_1$. Therefore, $D(v_1) = 0$, and because v_1 contains $\bar{l}$ and $q_{\bar{l}}(v_1) = 0$, so v_1 is in the *OFF*-set of C. Thus, the existence of vertices v_1 and v_2 is necessary for a hazard-free robust delay fault test to exist for π.

Sufficiency: To show that the existence of these vertices is sufficient for a delay fault test to exist for π, we simply show that these vertices, when applied as a vector pair, constitute a delay fault test for π. After applying v_1 and waiting an arbitrarily long amount of time the circuit output settles to a 0. When v_2 is applied, it first causes the input net associated with l to rise. The other inputs to cube q are set to noncontrolling values by both v_1 and v_2 because only l changes in q across v_1 and v_2. So the event propagates through q to the output of C. Thus, these vertices may be used as a delay fault test for π. □

5.7.2 Synthesis for Maximal Testability

In the previous section, a necessary and sufficient condition for robust path delay fault testability of two-level circuits was given. While the conditions named implied primality and irredundancy, we shall show that primality and irredundancy are not sufficient conditions for robust path delay fault testability of two-level circuits. Thus, a two-level minimization procedure particularly tuned to optimize for robust path delay fault testability is required. We begin with the observation that different prime and irredundant two-level im-

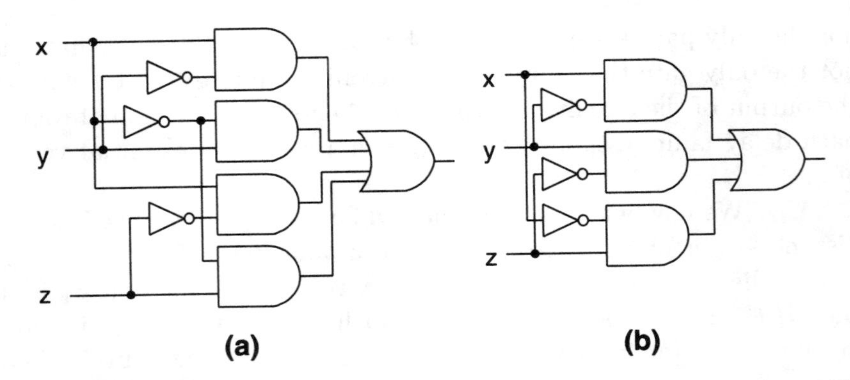

Figure 5.17: Different prime and irredundant implementations of a logic function

plementations of logic functions can have different path delay fault testabilities. Consider the prime and irredundant function shown in Figure 5.17(a). The literal x in the cube $x \cdot \overline{y}$ is not robustly testable to the output f. A different prime and irredundant implementation of the same logic function is shown in Figure 5.17(b). This implementation is completely robustly path delay fault testable.

The prime and irredundant network of Figure 5.17(b) satisfies the conditions given by Theorem 5.7.1, whereas the prime and irredundant network of Figure 5.17(a) does not. Thus, selecting an appropriate set of primes can enhance the path delay fault testability of a logic function. Note however that some functions do not have a fully path delay fault testable two-level realization. An example of such a function is given in Figure 5.18. This same function is given as Figure 2-4 in [3]. The paths through gate 5 are not robustly testable, since the corresponding cube does not contain a relatively essential vertex distance-1 from the *OFF*-set.

In the sequel, we will describe exact and heuristic covering algorithms that find a two-level implementation of logic maximizing path delay fault testability.

In the well-known Quine-McCluskey procedure (described in Chapter 4) for two-level Boolean minimization primes are generated from the specification of a Boolean function, and a covering step is performed wherein a minimum set of primes that form a cover for the function is selected. In the synthesis for testability procedure the prime implicant generation step is identical, and we use a variant of the branch-and-bound strategy for the covering step. During covering we select a minimum set of primes, attempting to maximize the num-

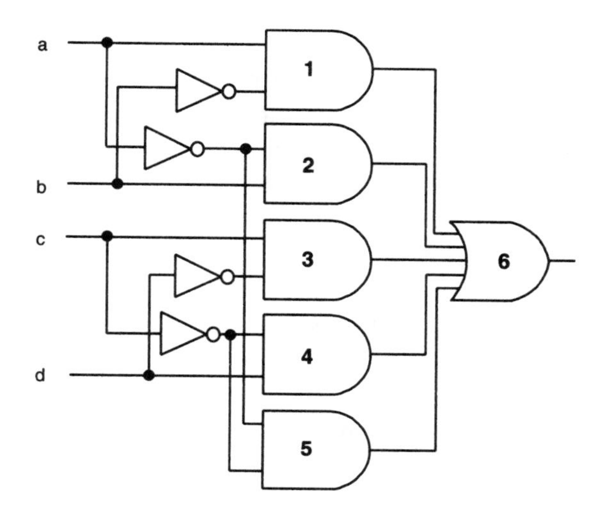

Figure 5.18: A prime and irredundant two-level cover that is not robustly path or gate delay fault testable

ber of relatively essential vertices in the primes that are distance-1 from the *OFF*-set of the function.

The procedure below receives as input the prime implicant table, T, corresponding to a single-output logic function. It searches for the most testable, minimum-product-term realization of a Boolean function.

1. Delete dominated columns and dominating rows in the prime implicant table. Detect essential primes by checking to see if any minterm is contained by a single prime. Add these essential primes to the selected set. Repeat until no new essential primes are detected.

2. If the size of the selected set *plus* the lower bound calculated using the maximal independent set heuristic described in Section 4.5.3 exceeds the best solution thus far, return from this level of recursion. Else if the size of the selected set *plus* the lower bound equals the best solution thus far, then if the number of untestable literals or paths[1] is greater than that of the best solution, return from this level of recursion. If there are no minterms left to be contained, declare the selected set as the best solution recorded thus far.

[1]A literal in a cover has to satisfy the conditions of Theorem 5.7.1 to be path delay fault testable.

3. Heuristically select a prime. Find the relatively essential vertices of all the primes in the selected set. This is done by using the disjoint SHARP operation described in Section 3.6. Search for an unselected prime that minimally affects (i.e., does not destroy) these relatively essential vertices and which itself contains relatively essential vertices that are distance-1 from the *OFF*-set in a maximum number of constituent literals. The information regarding the relatively essential vertices of the next prime we select is contained in the current implicant table T (the table with the selected primes and contained minterms removed). The goal is to maintain relatively essential vertices of the already selected set as well as add a new prime with more relatively essential vertices.

4. Add this prime to the selected set and recur for the subtable resulting from deleting the prime and all minterms that are contained by this prime. Then, recur for the subtable resulting from deleting this prime without adding it to the selected set.

In the above procedure all possible minimum covers of a function can be explored, and the minimum cover with the least number of untestable literals or paths can be selected. The procedure can easily be generalized to the multiple-output case. In the multiple-output case, we have two possible alternatives for generating primes.

1. Generate primes for the multiple-output function and perform covering as described above. In this case a minimum-product-term solution will result, but since some literals in the selected multiple-output primes may not be prime for *every* output, a large number of literals or paths may be untestable.

2. Generate primes for each of the single-output functions and merge primes with identical input parts. Thus, the primality and irredundancy requirements for robust path delay fault testability of a multiple-output function (see Theorem 5.7.1) will be satisfied for any selected output. The covering procedure can be used to find a minimum-product-term implementation for each output *under this smaller set of primes* that is maximally path delay fault testable. The testability of the implementation produced by this procedure is typically significantly higher than by the previous procedure, but the implementation may require a larger number of product terms.

The exact covering algorithm described above is viable for medium-sized functions with up to 15 inputs. However, for larger functions the number of primes in the function may be too large, or more often the branch-and-bound step which explores all minimum covers may require too much computation. Hence, an efficient heuristic strategy is desirable.

We describe a heuristic strategy that (unlike the exact covering algorithm) targets the delay testability of the selected prime and irredundant cover as a primary objective and the number of product terms in the cover as a secondary objective. The delay testability of the cover is, as before, measured as the inverse number of untestable literals or paths in the cover. The number of primes considered are a subset of the total number of primes, if the latter are too numerous. The input to the procedure is the prime implicant table, T, corresponding to the logic function.

1. Same as Step 1 of the exact procedure, except dominated primes in T are *not* deleted, since we are primarily concerned with testability, rather than product term count.

2. If there are no minterms left to be contained, stop. Else, select a prime which contains at least one uncontained minterm and which when added to the current selected set, results in the minimal number of untestable literals or paths in the selected set (including the prime). In the case of two primes resulting in the same number of untestable literals or paths select the prime which contains more uncontained minterms.

The above procedure represents a fast, greedy strategy to maximize testability without considering the product term count.

5.7.3 Test Generation Methods

We describe a test generation method for determining a vector pair that is a hazard-free robust path delay fault test for a path in a two-level circuit C based on a sum-of-products or cover representation E.

In order to determine whether a given path π is testable in a circuit C, we first determine the relatively essential vertices of the cube c that the path passes through by computing $c - D$, where D is the set of cubes in E other than c. If $c - D$ is empty, the path is untestable. Note that $c - D$ may be a collection of cubes or minterms.

If $c - D$ is not empty, we try to identify a distance-1 *OFF*-set vertex from any vertex in $c - D$.

For each cube $q \in c - D$ we compute $r_q = q - \{l\} \cup \{\bar{l}\}$ where l is the literal in c from which π begins. Obviously, l will be contained in all cubes in $c - D$. If r_q intersects the *OFF*-set of C, we have found a test for the fault. This is easy to do if we have an explicit representation of the *OFF*-set of C. An explicit representation can be obtained by complementing the *ON*-set since C is a completely specified logic implementation. We can walk through each cube g in the *OFF*-set and check if $g \cap r_q$ is nonempty. Alternatively, we can compute $r_q - E$ if we do not wish to generate the *OFF*-set. The computation of $r_q - E$ can be carried out using the disjoint SHARP operation described in Section 3.6 or by using the tautology algorithm described in Section 4.3. In the latter case we check if $r_q \subseteq E$. If not, we obtain the minterms that are in r_q (and therefore the *OFF*-set) but not in E.

If r_q intersects an *OFF*-set vertex v_2, then we have a hazard-free robust path delay fault test for π, $\langle v_1, v_2 \rangle$ where $v_1 = v_2 \cup \{l\} - \{\bar{l}\}$.

5.8 General Robust Path Delay Faults

The necessary and sufficient conditions for the general robust path delay fault testing of two-level circuits are the same as the hazard-free case. To understand this consider testing the $0 \rightarrow 1$ transition along a path π in a two-level circuit C with a vector pair $\langle v_1, v_2 \rangle$ If the path π passes through cube q, then v_2 has to be a relatively essential vertex of q. Else, we will have a race to set the output to a 1. The vector v_1 has to be in the *OFF*-set of C. Furthermore, when a $0 \rightarrow 1$ transition is observed at the output of C all the cubes other than q have steady 0 values. If they glitch, the glitch is propagated to the output and the test is invalidated. Note however that testing for $1 \rightarrow 0$ transitions along the paths is more relaxed than in the hazard-free case. We require a vertex (not a relatively essential vertex) distance-1 from the *OFF*-set in the literal corresponding to the chosen path. Each AND gate that makes a transition has a single $1 \rightarrow 0$ transition at its inputs. However, there can be converging $1 \rightarrow 0$ transitions at the inputs to the OR gate, the slowest of which is passed through to the output.

Thus, the synthesis and test generation procedures for general robust testing of two-level circuits are exactly the same as those for the hazard-free robust case.

5.9 Hazard-Free Robust Gate Delay Faults

5.9.1 Conditions for Testability

We begin with a simple lemma that gives sufficient conditions for robust gate delay fault testability.

Lemma 5.9.1 *Let C be a two-level single-output circuit, and let g be a gate in C. If a path π through g is robustly path delay fault testable, then g is robustly gate delay fault testable.*

Proof. Suppose there is a path π through g that is path delay fault testable. If there is any delay fault associated with g, be it a $1 \rightarrow 0$ or a $0 \rightarrow 1$ transition, it is reflected in the delay for the appropriate event propagating along π. Thus, any path delay fault test for π is a gate delay fault test for g. □

We now derive necessary and sufficient conditions directly in terms of the two-level circuit structure. As before, given the direct correspondence between the $0 \rightarrow 1$ and $1 \rightarrow 0$ transition gate delay fault tests, we will concentrate on a single transition fault per gate.

Theorem 5.9.1 *Let C be a two-level single-output circuit, and let g be an* AND *gate in C. The gate g is robustly gate delay fault testable if and only if there exists a vector pair $\langle v_1, v_2 \rangle$ such that v_2 is a relatively essential vertex of the cube q associated with g, and there exists a literal l in q such that $v_1 = v_2 - \{l\} \cup \{\bar{l}\}$ is in the OFF-set of C.*

Proof. Necessity: We could apply a similar argument to the one used in the proof of Theorem 5.7.1, but we thought it would be enlightening to see a different argument for the same property. Suppose $\langle w_1, w_2 \rangle$ is a gate delay fault test for g corresponding to a cube q. The two vectors might be more than distance-1 apart but we will proceed to construct a distance-1 vector pair as in the theorem.

Let S be the set of variables that change from w_1 to w_2. Let $x = \{l | \ (l \in w_2) \wedge (l, \bar{l} \ \not\in \ s)\}$, i.e., let x be the cube resulting from the elimination of all literals in S from w_2. First, observe that by the arguments of Lemma 5.7.1 under the appropriate delays on inputs, any vertex contained by x could be reached by applying the pair $\langle w_1, w_2 \rangle$ This implies that the intersection of x with each d in $D = C - q$ is empty. If not, then under appropriate delays some vertex contained by another cube in D could be reached, and a glitch would occur that invalidates the gate-delay test for q. Hence, what is contained by x? The cube x contains exactly members of the ON-set contained only by q, i.e., relatively essential vertices of q and members of the OFF-set.

We wish to find two adjacent vertices in the subspace determined by x, such that one is in the ON-set of C and the other is in the OFF-set. How can these vertices be located? The circuit values appearing on the inputs of C as the vector pair $\langle w_1, w_2 \rangle$ is applied may be visualized as a sequence of vertices forming a path in the Boolean N-space from vertex w_1 and vertex w_2. (Under certain delays on inputs the path may actually be traversed.) Furthermore, this path is restricted to be in the subspace determined by the cube x. As vertex w_1 is in the OFF-set and vertex w_2 is in the ON-set, there exist adjacent vertices v_1 and v_2 at some point on the path such that v_1 is in the OFF-set and v_2 is in the ON-set. Therefore these vertices v_1 and v_2 can be used as a test instead of $\langle w_1, w_2 \rangle$.

Sufficiency: Suppose there exists a vector pair $\langle v_1,\ v_2 \rangle$ such that v_2 is a relatively essential vertex of the cube q associated with g, and there exists a literal l in q such that $v_1 = v_2 - \{l\} \cup \{\bar{l}\}$ is in the OFF-set. By Theorem 5.7.1 this constitutes a robust path delay fault test for the path from the input associated with l through g to the output of C. Thus by Lemma 5.9.1 this constitutes a robust gate delay fault test for g. □

The above theorem did not explicitly consider the single OR gate in the two-level circuit. All paths have to pass through the OR gate, and hence a gate delay fault at the OR gate is also robustly testable if at least one AND gate is robustly testable. This is however merely a sufficiency condition and not a necessary one. The more general results of Section 9.9.1 can be used to determine necessary and sufficient conditions for the OR gate to be robustly gate delay fault testable.

5.9.2 Synthesis for Maximal Testability

It has been conjectured that primality and irredundancy of a cube in a cover is sufficient for the corresponding gate to be robustly gate delay fault testable. This would imply that each prime in the cover contains an relatively essential vertex that is distance-1 from the OFF-set. However, a two-level circuit that is prime and irredundant which contains a prime whose only relatively essential vertex is, in fact, distance-2 from the OFF-set is given in Figure 5.18.

Gate 5 is not robustly testable since the corresponding cube does not contain a relatively essential vertex distance-1 from the OFF-set. Thus, as was the case for robust path delay fault testability, modified covering strategies that satisfy necessary conditions for gate delay fault testability are required.

The necessary and sufficient conditions given in Section 5.9.1 for hazard-free robust gate delay fault testability in a two-level network are similar to the hazard-free robust path delay fault testability criteria. Maximizing robust gate delay fault testability entails maximizing the number of cubes in a cover that have relatively essential vertices that are distance-1 in some literal from the OFF-set (rather than in each literal as in the case of robust path delay fault testability). The covering procedure described in Section 5.7.2 can be used for maximizing hazard-free robust gate delay fault testability as well.

5.9.3 Test Generation Methods

Test generation under the hazard-free robust gate delay fault model for AND gates in a two-level circuit is exactly the same as testing for a chosen path through the gate. If any of the paths through the AND gate are robustly testable, then the gate is robustly testable. The method described in Section 5.7.3 can be used without modification.

If *any* of the AND gates in a single-output two-level circuit can be robustly tested, it follows that the OR gate in the circuit is robustly testable. The converse is not true. However, it is very rarely the case that not a single gate in a two-level circuit can be robustly tested, and therefore we will not consider this situation further. The results of Section 9.9 can be applied to determine necessary and sufficient conditions for robust testability of OR gates in a two-level circuit.

5.10 Hazard-Free Robust Stuck-Open Faults

The conditions for hazard-free stuck-open fault testability in two-level networks are exactly the same as hazard-free robust path delay fault testability. This is because there is a one-to-one correspondence between each link delay fault in a two-level network, a path in the two-level network, and a literal in each cube in the cover corresponding to the network. A complete hazard-free robust path delay fault test set for a two-level circuit corresponds to a complete hazard-free transistor stuck-open fault test set or a hazard-free link delay fault test set.

Problems

1. Generate a complete multifault test set for the circuit of Figure 5.18. Compact the primality test set by using vectors that are primality tests for multiple literals in different cubes.

2. Prove that any essential prime implicant of a logic function contains a vertex that is distance-1 from the *OFF*-set of the function.

3. Prove that every unate function in prime and irredundant two-level sum-of-products form is fully hazard-free robustly path delay fault testable.

4. Prove or disprove that if a function $f(x_1, x_2, \cdots, x_n)$ is unate in variable x_i, then in any two-level prime and irredundant realization of f all paths from x_i are hazard-free robustly path delay fault testable.

5. Prove that if a two-level circuit C is prime and irredundant under a don't-care set, then for any hazard-free robust path delay fault testable path $\pi \in C$, some test pair $\langle v_1, v_2 \rangle$ for π lies outside the don't-care set.

6. Give necessary and sufficient conditions, based on prime implicants and relatively essential vertices, for the OR gate in a two-level circuit to be hazard-free robustly gate delay fault testable.

7. Prove that complete general robust path delay fault testability implies complete multifault testability for two-level circuits.

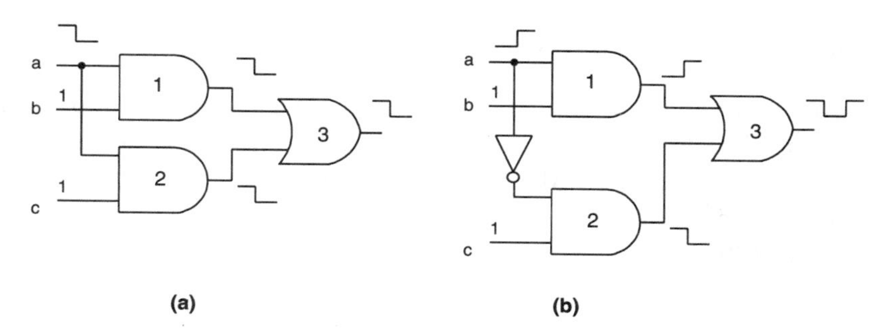

Figure 5.19: Validatable nonrobust testing in two-level combinational logic circuits

8. Give an example of a two-level circuit that is completely hazard-free robustly gate delay fault testable but is not completely hazard-free robustly path delay fault testable.

9. Prove or disprove that the primality of a two-level cover is a necessary condition for the corresponding two-level circuit to be completely hazard-free robust gate delay fault testable.

10. Consider a weakening of the general robust test methodology called *validatable nonrobust* (VNR) testing [8]. VNR testing corresponds to a methodology wherein multiple paths in a single-output two-level circuit corresponding to different cubes are validated for $1 \rightarrow 0$ transitions, because the slowest of these transitions is propagated to the output. Furthermore, for testing the $0 \rightarrow 1$ transitions on a path, other paths coverging at the OR gate are allowed to have $1 \rightarrow 0$ transitions (provided they have already been validated for $1 \rightarrow 0$ transitions). VNR testing is illustrated in Figure 5.19.

    Show that primality and irredundancy of a single-output two-level cover is a necessary and sufficient condition for complete VNR testability.

REFERENCES

[1] D. Bryan, F. Brglez, and R. Lisanke. Redundancy Identification and Removal. In *Proceedings of the International Workshop on*

Logic Synthesis, May 1989.

[2] M. J. Bryan, S. Devadas, and K. Keutzer. Necessary and Sufficient Conditions for Hazard-Free Robust Transistor Stuck-Open Fault Testability in Multilevel Circuits. *IEEE Transactions on Computer-Aided Design of Integrated Circuits*, 11(6):800–803, June 1992.

[3] G. L. Craig. *Stuck-Open Fault Testability in Combinational Networks.* PhD thesis, University of Wisconsin at Madison, 1987.

[4] M. R. Garey and D. S. Johnson. *Computers and Intractability: A Guide to the Theory of NP-Completeness.* W. H. Freeman and Company, New York, NY, 1979.

[5] N. K. Jha and S. Kundu. *Testing and Reliable Design of CMOS Circuits.* Kluwer Academic Publishers, Norwell, MA, 1990.

[6] I. Kohavi and Z. Kohavi. Detection of Multiple Faults in Combinational Logic Networks. *IEEE Transactions on Computers*, C-21(6):556–568, June 1972.

[7] T. Larrabee. *Efficient Generation of Test Patterns Using Boolean Satisfiability.* PhD thesis, Stanford University, February 1990.

[8] C. Lin, S. M. Reddy, and S. Patil. An Automatic Test Pattern Generator for the Detection of Path Delay Faults. In *Proceedings of the International Conference on Computer-Aided Design*, pages 284–287, 1987.

[9] P. McGeer and R. Brayton. Efficient Algorithms for Computing the Longest Viable Path in a Combinational Network. In *Proceedings of the* 26th *Design Automation Conference*, pages 561–567, June 1989.

[10] A. Pramanick and S. Reddy. On the Design of Path Delay Fault Testable Combinational Circuits. In *Proceedings of the* 20th *Fault Tolerant Computing Symposium*, pages 374–381, June 1990.

[11] R. L. Wadsack. Fault Modeling and Logic Simulation of CMOS and MOS Integrated Circuits. *Bell Systems Technical Journal*, 57(5):1449–1474, May-June 1978.

Chapter 6

Multilevel Combinational Circuits

In this chapter we introduce terminology regarding multilevel combinational logic circuits, describe special classes of multilevel circuits, and introduce the *binary decision diagram* (BDD) representation of logic circuits.

6.1 Boolean Networks

A *combinational logic circuit* is represented as a labeled, *directed, acyclic graph* (DAG) $G = (V, E)$ with each vertex v labeled with the name of a primitive gate such as AND, OR, or NOT, or with the name of a primary input or output. In general, a combinational logic circuit can be created from gates of arbitrary complexity, but to simplify the discussion we assume that the circuit is expressed in terms of primitive gates. There is an edge $\langle u,\ v\rangle$ between two vertices if the output of the gate associated with u is an input to gate v.

Each gate and edge in the circuit has an associated delay. For synthesis applications fixed gate and edge delays are assumed for the given technology.

The *fan-out* of a gate g (or a wire) is defined as the set of gates that use as an input the value generated by g. The *transitive fan-out* of g is defined recursively as follows. If g is a gate generating only a primary output, then its transitive fan-out is the null set. Else, the transitive fan-out of g is the union of the fan-outs of g and the transitive fan-out of every element in the fan-out of g. Similarly, the *fan-in* of a gate g (or a wire) is defined as the set of gates that provide inputs to g. The *transitive fan-in* of a gate is defined recursively like the transitive fan-out.

A *Boolean network* η is a DAG such that for each node i in η there is an associated cover F_i and a Boolean variable y_i representing the output of F_i. As there is a one-to-one correspondence between a node and its cover, we will use the terms interchangeably. There is a directed edge from i to j if and only if F_j explicitly depends on y_i or $\overline{y}_i$. Furthermore, some of the variables in η may be classified as primary inputs or primary outputs. We may directly implement a Boolean Network η by a combinational circuit $\mathcal{C}$ by replacing each cover F_i in η by a NAND-NAND network and by replacing each edge fanning out from a node in η by a *net*.

6.2 Special Classes of Circuits

Significant research has been done on testing problems associated with special classes of circuits. The study of such circuits sometimes lends insight into more general testing problems.

6.2.1 Fan-out-Free Circuits

A fan-out-free circuit is one in which the output of each gate and each circuit input is an input to at most one gate. A fan-out-free circuit is also called a tree. For example, the circuit $f = a \cdot b + c \cdot d$ is a tree. Many testability results have been proven about fan-out-free circuits which are not valid for general circuits. Among these are:

1. There exists a set of tests which detect all single and multiple stuck-at faults and is of minimal cardinality among all test sets for single faults [5].

2. The number of tests required to detect all stuck-at faults is bounded above by $n + 1$ and is bounded below by $2\sqrt{n}$ where n is the number of circuit inputs [5].

3. A set of tests which detect all stuck-at faults on circuit inputs will detect all single stuck-at faults.

6.2.2 Leaf-DAG Circuits

A leaf-DAG circuit is a generalization of a fan-out-free circuit where only the primary inputs are allowed to fan out to multiple gates, but the output of each gate in the circuit is an input to at most one gate. Note that any circuit can be converted into a leaf-DAG circuit by gate duplication. The leaf-DAG corresponding to a general circuit is a representation that will be used in Chapter 9 for analysis.

6.2.3 Algebraically Factored Circuits

An algebraically factored circuit is a circuit which is derived by a sequence of algebraic transformations from a two-level sum-of-products representation. For example, the circuit $(a+b)\cdot(c+d)$ is an algebraic factorization of the sum-of-products expression $a{\cdot}c + a{\cdot}d + b{\cdot}c + b{\cdot}d$.

A precise definition of algebraic factorization can be found in Chapter 7. We will present many interesting testability results about algebraically factored circuits in Chapter 9.

6.2.4 Multiplexor-Based Circuits

Arbitrary Boolean functions can be implemented using circuits whose only constituent gates are two-input multiplexors. An example of such a circuit is shown in Figure 9.23 in Chapter 9. We will present some interesting testability properties of multiplexor-based circuits in Chapter 9.

6.3 Binary Decision Diagrams

A variety of Boolean function representations have been developed. Classical representations like canonical sum-of-products, truth tables, and Karnaugh maps [9] are impractical because any function of n variables has a representation of size of 2^n. Representations like the set of prime and irredundant cubes [4] and the Boolean network [3] are generally used. However, such representations suffer from some critical drawbacks. First, certain common functions may require representations of exponential size. Second, simple operations like complementation may yield a function with exponential size. Finally, none of these representations have a *canonical* form, i.e., a function may have different representations. This makes it difficult to check for equivalency and tautology.

Binary decision diagrams (BDDs) were first proposed by Lee [10]. This approach was further developed by Akers [1]. As such, BDDs are not canonical. Bryant introduced restrictions on the ordering of variables and proposed a reduction algorithm which transformed BDDs to *reduced, ordered BDDs* (ROBDDs) [6] that have a canonical form.

A BDD is a rooted, directed graph with vertex set V containing two types of vertices. A *nonterminal* vertex v has as attributes an argument index $index(v) \in \{1, \cdots, n\}$ and two children $low(v)$, $high(v) \in V$. A *terminal* vertex v has as an attribute a value $value(v) \in \{0, 1\}$.

The correspondence between BDDs and Boolean functions is defined as follows. A BDD G having root vertex v denotes a function f_v defined recursively as:

1. If v is a terminal vertex:

   (a) If $value(v) = 1$, then $f_v = 1$.

   (b) If $value(v) = 0$, then $f_v = 0$.

2. If v is a nonterminal vertex with $index(v) = i$, then f_v is the function:

$$f_v(x_1, \cdots, x_n) = \overline{x_i} \cdot f_{low(v)}(x_1, \cdots, x_n) + x_i \cdot f_{high(v)}(x_1, \cdots, x_n)$$

   x_i is called the *decision variable* for vertex v.

BDDs can be categorized based on two additional properties:

1. *Freedom:* When traversing any path from a terminal vertex to the root vertex we encounter each decision variable at most once.

2. *Ordered:* We place the restriction that for any nonterminal vertex v, if $low(v)$ is also nonterminal, then we must have $index(v) < index(low(v))$. Similarly, if $high(v)$ is also nonterminal, then we must have $index(v) < index(high(v))$. Note that this implies the property of freedom.

6.4 Ordered Binary Decision Diagrams

From the conditions stated above, it is easy to see that a *ordered binary decision diagram* (OBDD) is an acyclic graph. The OBDDs for some simple functions are shown in Figure 6.1. Terminal vertices are represented as squares, while nonterminal vertices are represented

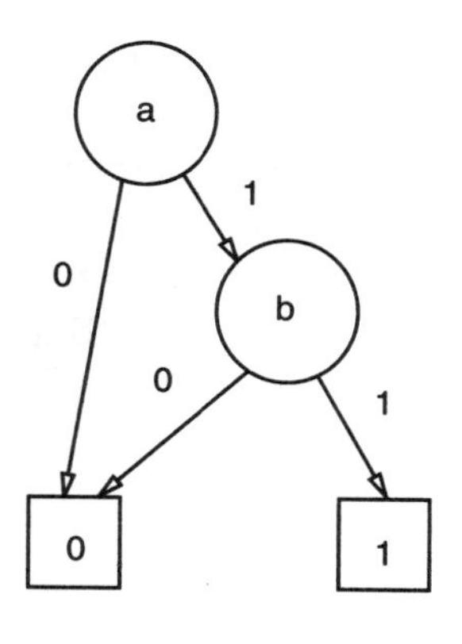

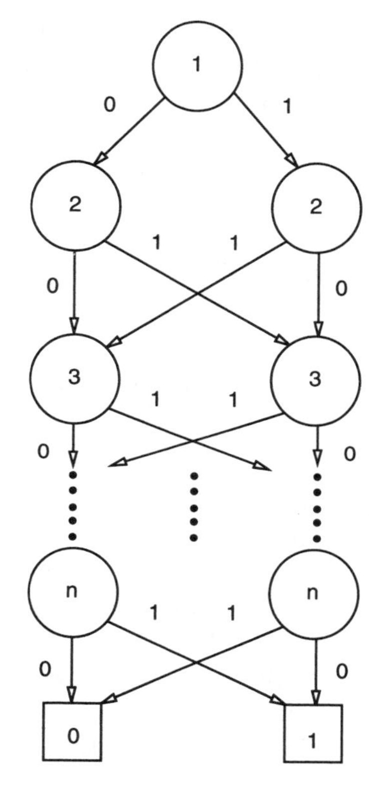

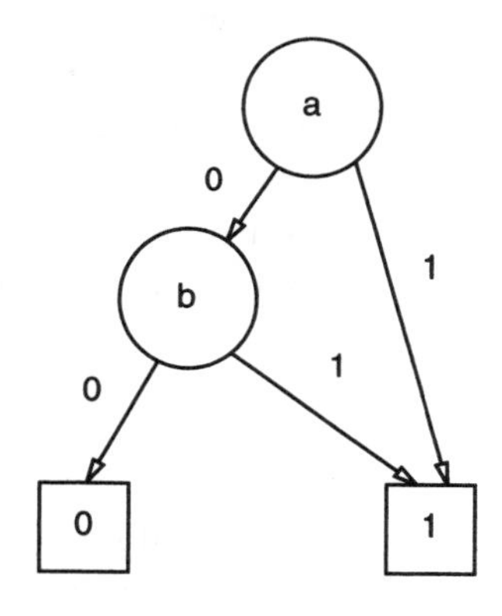

Figure 6.1: Example OBDDs

as circles. The low child is pointed to by the arrow marked 0, and the high child is pointed to by the arrow marked 1.

OBDDs represent the *OFF*-set and the *ON*-set of a function as disjoint covers, where each cube in the cover corresponds to a path from the root vertex to some terminal vertex. The OBDD at the top left of the figure represents the function $a \cdot b$. There are exactly two paths leading to the 0 terminal vertex. If a is a 0, then the function represented by the OBDD evaluates to a 0 since the 0 child of the vertex with index a is the 0 terminal vertex. If a is a 1 and b is a

0, the function evaluates to a 0. Thus, the *OFF*-set is represented as $\{\overline{a}, a \cdot \overline{b}\}$. The two cubes in the cover are disjoint. If a and b are both 1, the function evaluates to a 1. The *ON*-set is $\{a \cdot b\}$, where the single cube corresponds to the single path to the 1 terminal vertex.

The odd parity function of Figure 6.1 is an example of function which requires $2n - 1$ vertices in a OBDD representation but which has 2^{n-1} product terms in a minimum two-level representation. Thus, OBDD representations can be considerably more compact than two-level representations.

6.4.1 Reduced Ordered Binary Decision Diagrams

Consider the two vertices with index 3 in the OBDD of Figure 6.2(a). The functions rooted at these vertices are identical, namely x_3. We can delete one of the identical vertices and make the vertices that were pointing to the deleted vertex (those vertices whose low child or high child correspond to the deleted vertex) point instead to the other vertex. This does not change the Boolean function corresponding to the OBDD. The simplified OBDD is shown in Figure 6.2(b). In Figure 6.2(b) there is a vertex with index 2 whose low child and high child both point to the same vertex. This vertex is redundant because the function f rooted by the vertex corresponds to:

$$f \;=\; x_2 \cdot x_3 \;+\; \overline{x_2} \cdot x_3 \;=\; x_3$$

Thus, all the vertices that point to f can be made to point to its low or high child without changing the Boolean function corresponding to the OBDD as illustrated in Figure 6.2(c).

We will define the notion a reduced OBDD which is a minimal OBDD representation for a given function. In order to do so we first have to define isomorphism between OBDDs.

Definition 6.4.1 *Two OBDDs G_1 and G_2 are isomorphic if there exists a one-to-one function σ from the vertices of G_1 onto the vertices of G_2 such that for any vertex v if $\sigma(v) = w$, then either both v and w are terminal vertices with $value(v) = value(w)$, or both v and w are nonterminal vertices with $index(v) = index(w)$, $\sigma(low(v)) = low(w)$ and $\sigma(high(v)) = high(w)$.*

Since an OBDD only contains one root and the children of any nonterminal vertex are distinguished, the isomorphic mapping σ between OBDDs G_1 and G_2 is constrained and easily checked for.

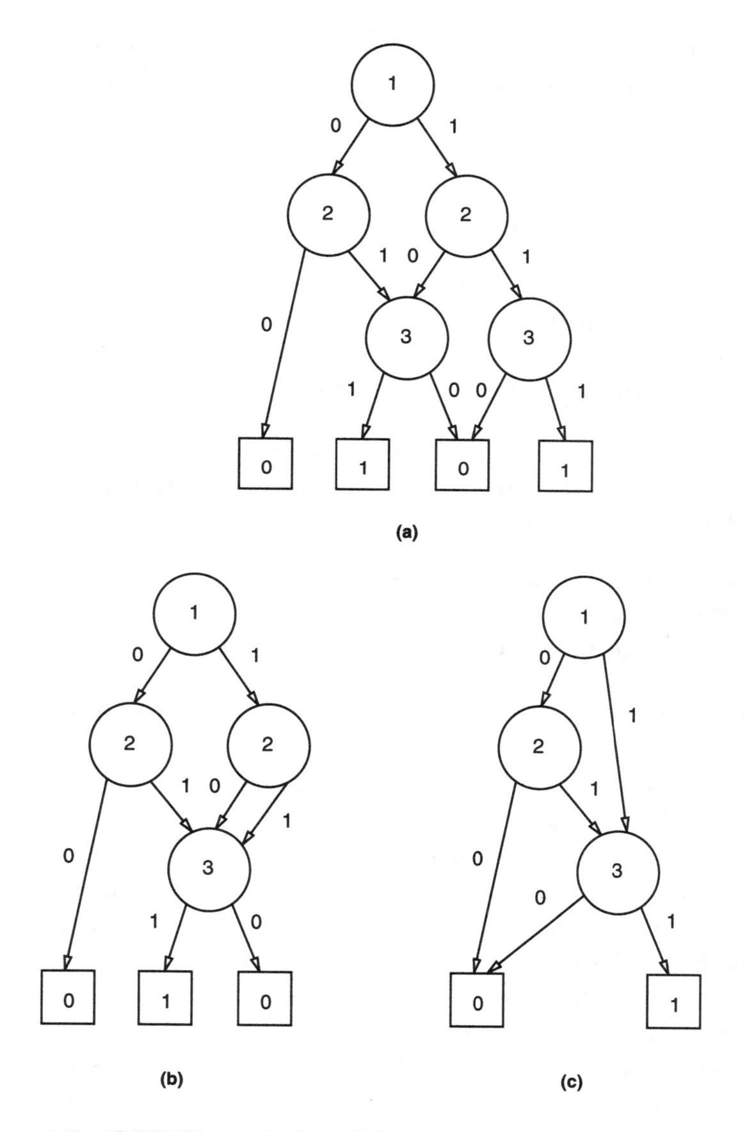

Figure 6.2: OBDD and simplified OBDDs

The root in G_1 must map to the root in G_2, the root's low child in G_1 must map to the root's low child in G_2, and so on all the way to the terminal vertices. Testing two OBDDs for isomorphism is thus a simple linear-time check.

Definition 6.4.2 [6] *A OBDD G is reduced if it contains no vertex v with $low(v) = high(v)$ nor does it contain distinct vertices v and w such that the subgraphs rooted by v and w are isomorphic.*

Henceforth, we will be dealing exclusively with ROBDDs. In the following sections we will describe the properties of ROBDDs, and describe in detail Boolean manipulation algorithms based on ROBDDs.

6.4.2 Canonicity Property

The following theorem, originally proven in [6], is a key property of ROBDDs, namely that they form a canonical representation for Boolean functions. Every function is represented by a unique ROBDD for a given ordering of inputs to the function.

Theorem 6.4.1 *For any Boolean function f, there is a unique (up to isomorphism) ROBDD denoting f, and any other OBDD denoting f contains more vertices.*

Proof. A sketch of the proof is given using using induction on the number of inputs.
Base case: If f has zero inputs, it can be either the unique 0 or 1 ROBDD.
Induction hypothesis: Any function g with a number of inputs $< k$ has a unique ROBDD.

Choose a function f with k inputs. Let G and G' be two ROBDDs for f under the same ordering. Let x_i be the input with the lowest index in the ROBDDs G and G'. Define the functions f_0 and f_1 as f_{x_i} and $f_{\overline{x_i}}$, respectively. Both f_0 and f_1 have less than k inputs, and by the induction hypothesis these are represented by unique ROBDDs G_0 and G_1.

We can have vertices in common between G_0 and G_1 or have no vertices in common between G_0 and G_1. If there are no vertices in common between G_0 and G_1 in G, and no vertices in common between G_0 and G_1 in G', then clearly G and G' are isomorphic.

Consider the case where there is a vertex u that is shared between G_0 and G_1 in G. There is a vertex u' in the G_0 of G' that corresponds to u. If u' is also in G_1 of G', then we have a correspondence between u in G and u' in G'. However, there could be another vertex u'' in the G_1 of G' that also corresponds to u. While the existence of this vertex implies that G and G' are not isomorphic, the existence of u' and u'' in G' is a contradiction to the statement of the theorem, since the two vertices root isomorphic subgraphs corresponding to u. (This would imply that G' is not reduced.) Therefore, u'' cannot exist, and G and G' are isomorphic. □

6.4.3 Reduction

The reduction algorithm transforms an arbitrary OBDD into a reduced OBDD. Proceeding from the terminal vertices up to the root, a unique integer identifier is assigned to each unique subgraph root. Thus, for each vertex v the algorithm assigns a label $id(v)$ such that for any two vertices v and w, $id(v) = id(w)$ if and only if the functions rooted at v and w are identical, i.e., $f_v = f_w$. Given this labeling, the algorithm constructs a graph with one vertex for each unique label.

The procedure works from the terminal vertices up to the root. First, two terminal vertices have the same label if and only if they have the same value attributes. Now assume that all terminal vertices and all nonterminal vertices with index greater than i have been labeled. As we proceed with the labeling of vertices with index i, a vertex v will have $id(v)$ equal to that of some vertex that has already been labeled if and only if one of the following two conditions is satisfied:

1. If $id(low(v)) = id(high(v))$, then vertex v is redundant, and $id(v)$ is set to $id(low(v))$.

2. If there is some labeled vertex w with $index(w) = index(v)$, $id(low(w)) = id(low(v))$, and $id(high(w)) = id(high(v))$, then the reduced subgraphs rooted by these two vertices are isomorphic, and $id(v)$ is set to $id(w)$.

The two conditions above correspond to the two requirements of the reduction property defined in Definition 6.4.2. Once all the vertices have been labeled, they are sorted according to their labels. Vertices with the same label are sorted in terms of decreasing indices (the attribute $index(v)$). This will place all vertices with the same label adjacent in the sorted list. Given a set of vertices with the same label, we can pick the first vertex in the set and make all vertices in the OBDD that point to the remaining vertices in the given set point instead to the first vertex.

The reduction algorithm is illustrated in Figure 6.3, using an ROBDD taken from [6]. We first assign the 0 and 1 terminal vertices a and b labels in Figure 6.3(a). Next, the right vertex with index 3 is assigned label c. Upon encountering the other vertex with index 3 we find that Condition 2 above is satisfied and we assign this vertex the label c as well. Proceeding upward we assign the label c to the right vertex with index 2 since Condition 1 is satisfied for

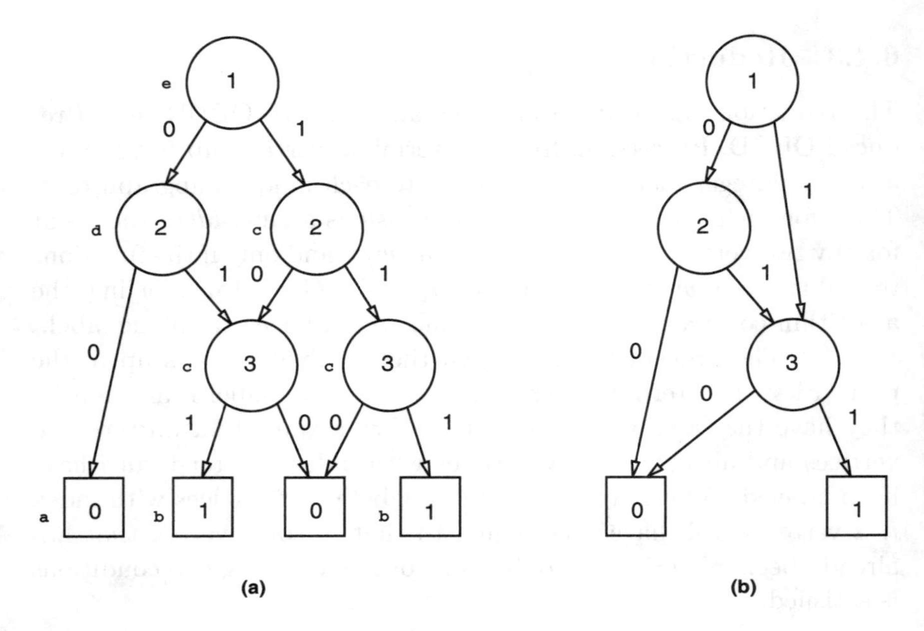

Figure 6.3: Reduction algorithm example

this vertex. (The low child and the high child of this vertex have the same label.) The left vertex with index 2 is assigned label d, and the root vertex is assigned the label e. Sorting and deleting redundant vertices results in the ROBDD of Figure 6.3(b).

6.4.4 Complementation

Complementing an ROBDD can be trivially accomplished by interchanging the 0 and 1 terminal vertices.

6.4.5 Cofactor

In order to cofactor an ROBDD with respect to a variable x_i, the variable is effectively set to 1 in the ROBDD. This is accomplished by determining all the vertices whose low child or high child corresponds to any vertex v with $index(v) = i$, and replacing their low child or high child by $high(v)$. This is illustrated in Figure 6.4 where the given ROBDD has been cofactored with respect to x_3. Similarly, an ROBDD can be cofactored with respect to $\overline{x_i}$ by using $low(v)$ to replace all vertices v with $index(v) = i$.

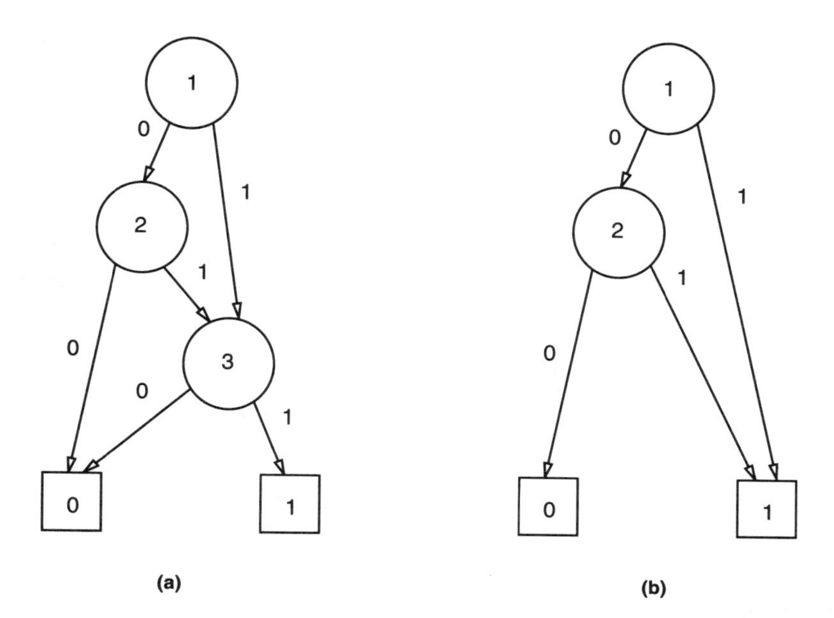

Figure 6.4: Cofactor example

6.4.6 APPLY

The most complex ROBDD operation is the **APPLY** operation where ROBDDs G_1 and G_2 are combined as $G_1 \langle op \rangle G_2$ where $\langle op \rangle$ is a Boolean function of 2 arguments. The result of the **APPLY** operation is another ROBDD. We will describe the **APPLY** operation for the case where $\langle op \rangle$ is the Boolean OR operator.

The algorithm proceeds from the roots of the two argument graphs downward, creating vertices in the resultant graph. The control structure of the algorithm is based on the following recursion:

$$f + g \;=\; x_i \cdot (f_{x_i} + g_{x_i}) \;+\; \overline{x_i} \cdot (f_{\overline{x_i}} + g_{\overline{x_i}})$$

From an ROBDD perspective we have:

$$f_v + g_w \;=\; x_i \cdot (f_{high(v)} + g_{high(w)}) \;+\; \overline{x_i} \cdot (f_{low(v)} + g_{low(w)}) \quad (6.1)$$

where f_v and g_w are the Boolean functions rooted at the vertices v and w.

We must consider several cases of the above recursion.

1. If v and w are terminal vertices, we simply generate a terminal vertex u with $value(u) = value(v) + value(w)$.

2. Else, if $index(v) = index(w) = i$, we follow Equation 6.1. Create vertex u with $index(u) = i$, and apply the algorithm recursively on $low(v)$ and $low(w)$ to generate $low(u)$ and on $high(v)$ and $high(w)$ to generate $high(u)$.

3. If $index(v) = i$ but $index(w) > i$, we create a vertex u having index i, and apply the algorithm recursively on $low(v)$ and w to generate $low(u)$ and on $high(v)$ and w to generate $high(u)$.

4. If $index(v) > i$ and $index(w) = i$, we create a vertex u having index i and apply the algorithm recursively on v and $low(w)$ to generate $low(u)$ and on v and $high(w)$ to generate $high(u)$.

Implementing the above algorithm directly results in an algorithm of exponential complexity in the number of input variables, since every call in which one of the arguments is a nonterminal vertex generates two recursive calls. Two refinements can be applied to reduce this complexity.

If the algorithm is applied to two vertices where one is a terminal vertex v with $value(v) = 1$, then we simply use the Boolean identity $f + 1 = 1$ or $f + 0 = f$ to return the result at this level of recursion. (If we were dealing with the AND operator, we would use the identities $f \cdot 0 = 0$ and $f \cdot 1 = f$.)

More importantly, the algorithm need not evaluate a given pair of vertices more than once. We can maintain a table containing entries of the form (v, w, u) indicating that the result of applying the algorithm to subgraphs with roots v and w was result u. Before applying the algorithm to a pair of vertices we first check whether the table contains an entry for these two vertices. If so, we can immediately return the result. Otherwise we make the two recursive calls, and upon returning, add a new entry to the table. This refinement drops the time complexity to $O(|G_1| \cdot |G_2|)$, where $|G_1|$ and $|G_2|$ are the number of vertices in the two given graphs.

We illustrate the algorithm with an example taken from [6]. The two ROBDDs to be operated on by an OR operator are shown in Figure 6.5. Each vertex in the two ROBDDs has been assigned a unique label. This label could correspond to the labels generated during ROBDD reduction. The labels are required to maintain the table entries described immediately above.

The OBDD resulting from the OR of the two ROBDDs is shown in Figure 6.6(a). First, we choose the pair of root vertices labeled $a1$ and $b1$. We create a vertex with index 1 and recursively

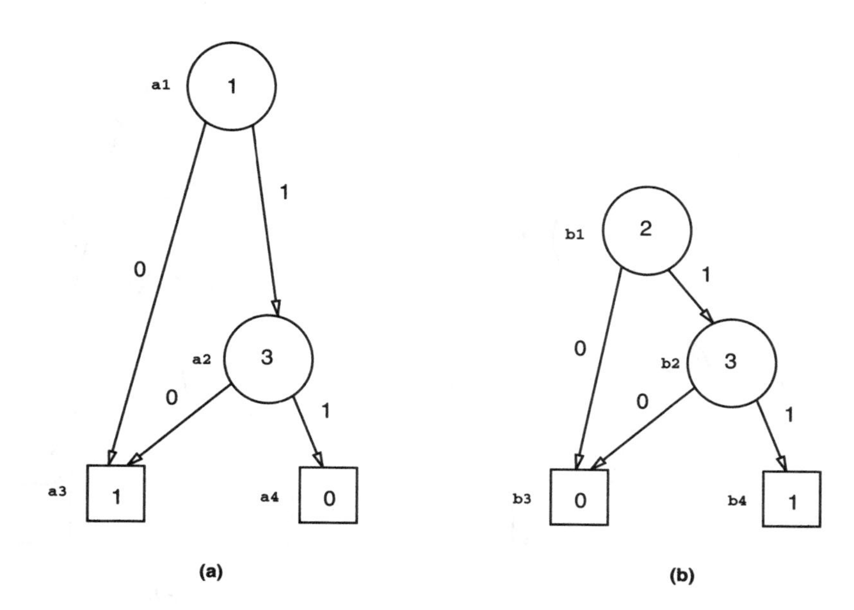

Figure 6.5: Example ROBDDs for the APPLY operation

apply the algorithm to the vertex pair $a3, b1$ and $a2, b1$. Since $a3$ corresponds to the 1 terminal vertex, we can immediately return the 1 terminal vertex as the result of the OR. We must still compute the OR of the $a2, b1$ vertex pair. This involves the computation of the OR of $a2, b3$ and $a2, b2$, and so on. Note that $a3, b3$ will appear as a vertex pair twice during the course of the algorithm.

Reducing the OBDD of Figure 6.6(a) results in the ROBDD of Figure 6.6(b).

6.4.7 Circuit Equivalence using ROBDDs

ROBDDs are a canonical representation of Boolean functions. Therefore, in order to check two different two-level or multilevel logic circuits for equivalence, we can use the following method.

1. Choose an ordering for the primary inputs of the circuits.

2. Create ROBDDs for the primary outputs of the two circuits.

3. Check if the ROBDDs are isomorphic. If so, the circuits are equivalent. If not, the circuits are not equivalent.

Any ordering will suffice, as long as the same ordering is chosen for both circuits. However, the size of the ROBDDs created

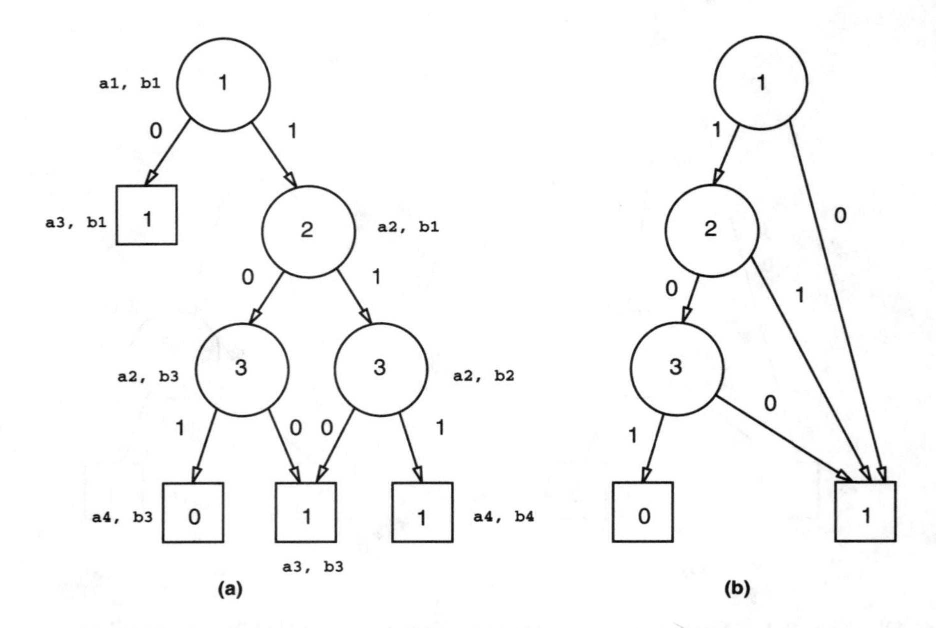

Figure 6.6: ROBDD resulting from the OR operation

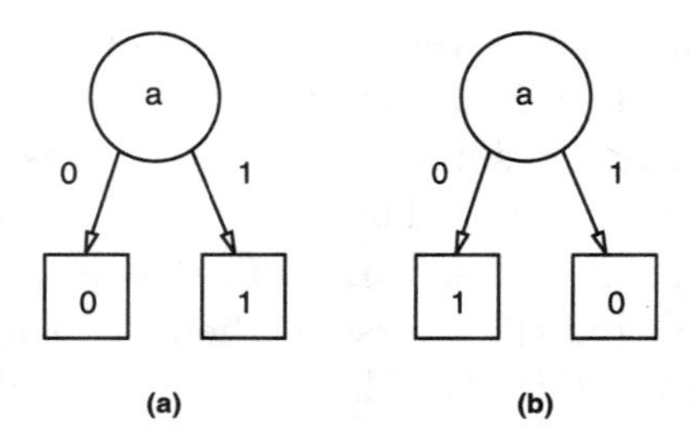

Figure 6.7: ROBDDs for primary inputs

is strongly dependent on the ordering chosen. We will touch upon methods to select orderings that result in small ROBDDs in Section 6.4.8.

The creation of ROBDDs of the outputs of a given combinational logic circuit requires the previously described algorithms for Boolean manipulation. ROBDDs for the primary inputs simply correspond to graphs with a single nonterminal vertex. For example, the ROBDD for primary input a is shown in Figure 6.7(a). Similarly, the ROBDD for primary input b will have one vertex with index b with a low child corresponding to the 0 terminal vertex and with a high child corresponding to the 1 terminal vertex. The ROBDD for $\overline{a}$

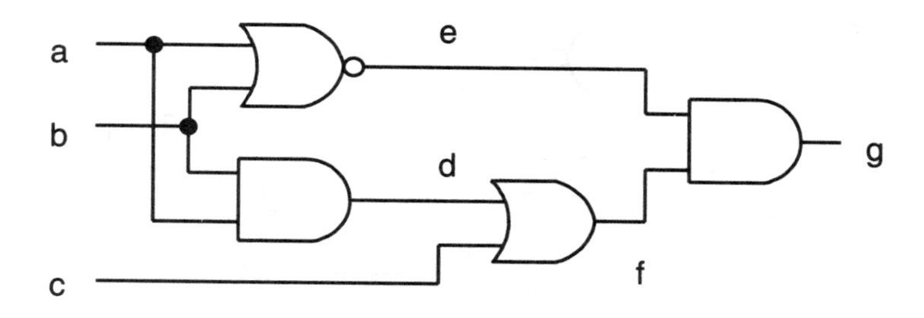

Figure 6.8: Creating ROBDDs given a multilevel circuit

is shown in Figure 6.7(b). We can create the ROBDD for the signal d in Figure 6.8 by performing an AND operation on the ROBDDs for the primary inputs a and b. This ROBDD is shown at the top left of Figure 6.1. We can create the ROBDD for signal f by performing and OR operation on the ROBDD for signal d and the ROBDD for the primary input c. In this manner we can traverse a circuit from primary inputs to the primary outputs and obtain the ROBDDs for the primary output in terms of the primary inputs.

In order to check two ROBDDs for equivalence, we can use the canonicity property of ROBDDs and perform a linear-time graph isomorphism check as per Definition 6.4.1.

6.4.8 Ordering Heuristics

The ordering of variables is very important in obtaining reasonable-sized ROBDD representations of functions. For example, if the variables in the function $f = ab + cd$ are ordered as $a = 0, b = 1, c = 2$, and $d = 3$, the resulting ROBDD has only 4 nonterminal vertices. However, if the order $a = 0, c = 1, b = 2$, and $d = 3$ is chosen, there are 7 nonterminal vertices. It has been shown that certain functions like integer multipliers, irrespective of the ordering of variables, have exponential-sized representations in terms of ROBDDs [7]. In certain cases sum-of-products representations can be more compact than ROBDDs, because of the ordering constraints [8].

Due to the sensitivity of ROBDD size to the chosen ordering finding a suitable ordering for representing a given logic function has become an important problem. Automatic heuristic ordering methods, e.g., [12], can be used to obtain reasonable-sized ROBDD representations for most commonly occurring functions.

We will not describe ordering heuristics in detail here but

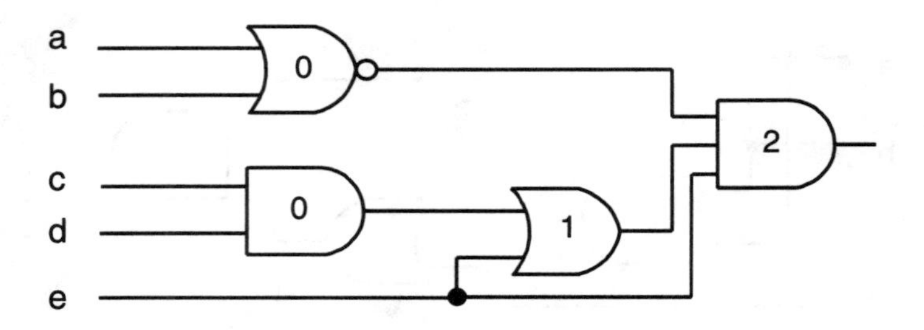

Figure 6.9: Choosing an appropriate input ordering

will touch upon a basic strategy used in choosing a good ordering. Given a circuit for which we wish to choose a primary input ordering we first levelize the circuit from primary inputs to primary outputs. Primary input signals are assigned level 0. A gate is assigned level l where l is the maximum of the levels of all the input signals to the gate. The signal connected to the output of the gate is assigned the level $l + 1$. The levelization of a small circuit is illustrated in Figure 6.9. The numbers inside the gates correspond to the assigned levels.

The gates that each primary input directly connects to are determined. A number is assigned to each primary input corresponding to the average level of all the gates that the primary input directly connects to. For example, primary input a would be assigned the number 0 since it only connects to one gate with level 0, and primary input e would be assigned the number $\frac{1+2}{2} = 1.5$ in Figure 6.9. To obtain an ordering, the primary inputs are sorted in terms of decreasing order of these assigned numbers. Ties are broken by keeping primary inputs that directly connect to the same gates together.

The above method attempts to generate an ordering such that the structure of the ROBDD under the chosen ordering mimics the circuit structure. The ordering chosen for the example of Figure 6.9 would be $e = 0$, $c = 1$, $d = 2$, $b = 3$, and $a = 4$. For this circuit, this ordering results in a minimal-size ROBDD.

6.4.9 Improvements to ROBDDs

To build the complement of a function F the ROBDD for F can be built and all the leaves (terminal vertices) can be replaced by their opposite values. If a function F and its complement are used to form another function, a copy of the ROBDD of F has to be made before modifying the leaves. This leads to unnecessary duplication of vertices for F and its complement. In addition to the memory waste,

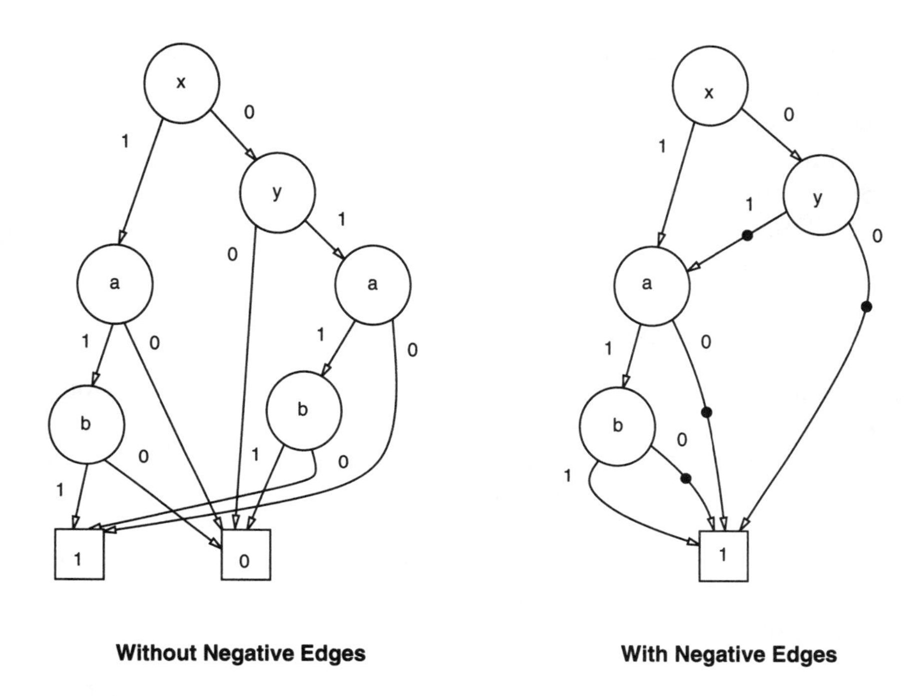

$f = a\,b\,x + \overline{a}\,y + \overline{b}\,y$

Ordering : x, y, a, b

Figure 6.10: An example ROBDD with negative edges

this leads to a waste of time in graph manipulation algorithms, most of which are polynomial in the size of the graph. To avoid this problem negative edges (or complemented edges) were first proposed in [11]. A negative edge indicates that the function rooted at the vertex that the edge points to has be complemented. The ROBDDs for a function with and without negative edges are shown in Figure 6.10. Negative edges are indicated by dots on them. Note that introducing negative edges does not destroy the canonicity of the ROBDD.

Using the *strongly canonical* form for ROBDD vertex manipulations equivalence checking can be performed in constant time. Under the strongly canonical form unreduced OBDDs are never created. A global unique table is maintained wherein every vertex rooting a unique function is given a unique label. Before creating a new vertex the table is checked to see if the function corresponding to this new vertex exists in the table. If not, the vertex is created, given a new label, and added to the unique table. If the function already exists, the vertex in the table corresponding to this function is returned. The labeling procedure in the unique table is similar to the procedure used in the reduction of an OBDD, which was described in Section 6.4.3. The description of an efficient implementation of an ROBDD package with negative edges and utilizing the strongly canonical form can be found in [2].

6.4.10 Multiplexor-Based Networks

In addition to being useful as a representation of Boolean functions, BDDs have recently aroused interest as a form of circuit representation. A multiplexor-based network is directly associated with any given BDD, and is equivalent to the logic function represented by the BDD.

We define the operation of deriving a multilevel network from a BDD and of propagating constant 0 or 1 values at the inputs of a multiplexor-based network inward.

Definition 6.4.3 *A multilevel network η is said to be derived from a BDD G if carrying out the following transformations on G results in network η.*

1. *Replace each vertex v in G that has nonterminal vertices $low(v)$ and $high(v)$ by a 2-input multiplexor corresponding to the function $f_v = \overline{x_{index(v)}} \cdot f_{low(v)} + x_{index(v)} \cdot f_{high(v)}$, where $x_{index(v)}$ is the decision variable of the vertex v.*

2. *If $low(v)$ or $high(v)$ are terminal vertices, replace v by a simplified function corresponding to f_v, where $f_{low(v)} = value(low(v))$ and $f_{high(v)} = value(high(v))$. For example, if $low(v) = 1$, then we obtain: $f_v = \overline{x_{index(v)}} + f_{high(v)}$.*

The transformation is thus a straightforward process, and can be accomplished in time linear in the size of the BDD.

Consider the two-level logic function shown in Figure 6.11(a). A BDD corresponding to the logic function is shown in Figure 6.11(b) (we have reversed the directions of the arrows). In Figure 6.11(c), we have a multilevel network derived by replacing the vertices of the BDD in Figure 6.11(b) by multiplexors. Some inputs to the multiplexors are tied to constant 0 or 1 values, and these multiplexors have been simplified. For instance, the multiplexors corresponding to

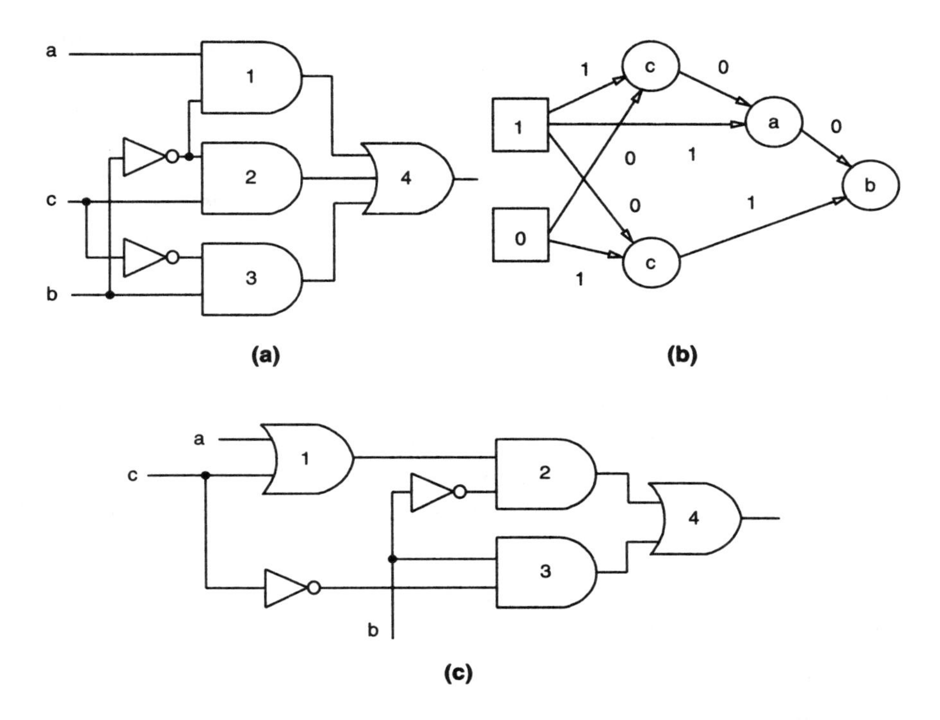

Figure 6.11: A two-level network, its BDD, and the derived multilevel network

the vertices for variable c have become direct connections to c and $\overline{c}$. Each multiplexor originally implemented a logical function $x{\cdot}p + \overline{x}{\cdot}q$, where p and q are called the *value inputs* to the multiplexor and x is the *control input.*

Problems

1. Prove that any fan-out-free multilevel circuit is unate in all its variables.

2. Generalize the procedure of obtaining a functionally equivalent multiplexor-based network from an ROBDD to the case where there are negative edges in the ROBDD, and apply it to the ROBDD with negative edges shown in Figure 6.10.

3. Convert the ROBDD of Figure 6.11(b) into an equivalent one

with the same order but with negative edges, and derive a functionally equivalent multiplexor-based network from the resultant ROBDD.

4. Cofactoring a BDD with respect to a variable corresponds to setting the variable to the appropriate constant value. Obtain an ROBDD for the cofactor with respect to $a = 1$ of the ROBDD without negative edges in Figure 6.10.

5. Give an example of a function, parameterizable over the number of inputs, whose leaf-DAG grows polynomially with the number of inputs but whose minimum sum-of-products and product-of-sums realizations grow exponentially with the number of inputs.

REFERENCES

[1] S. B. Akers. Binary Decision Diagrams. *IEEE Transactions on Computers*, C-27(6):509–516, June 1978.

[2] K. S. Brace, R. L. Rudell, and R. E. Bryant. Efficient Implementation of a BDD Package. In *Proceedings of the* 27th *Design Automation Conference*, pages 40–45, June 1990.

[3] R. Brayton, R. Rudell, A. Sangiovanni-Vincentelli, and A. Wang. MIS: A Multiple-Level Logic Optimization System. *IEEE Transactions on Computer-Aided Design of Integrated Circuits*, CAD-6(6):1062–1081, November 1987.

[4] R. K. Brayton, G. D. Hachtel, C. McMullen, and A. Sangiovanni-Vincentelli. *Logic Minimization Algorithms for VLSI Synthesis.* Kluwer Academic Publishers, Norwell, MA, 1984.

[5] M. A. Breuer and A. D. Friedman. *Diagnosis and Reliable Design of Digital Systems.* Computer Science Press, Woodland Hills, CA, 1976.

[6] R. Bryant. Graph-Based Algorithms for Boolean Function Manipulation. *IEEE Transactions on Computers*, C-35(8):677–691, August 1986.

[7] R. Bryant. On the Complexity of VLSI Implementations and Graph Representations of Boolean Functions with Application to Integer Multiplication. *IEEE Transactions on Computers*, C-40(2):205–213, February 1991.

[8] S. Devadas. Comparing Two-Level and Ordered Binary Decision Diagram Representations of Logic Functions. In *MIT Technical Report (available from the author)*, June 1991.

[9] Z. Kohavi. *Switching and Finite Automata Theory*. Computer Science Press, New York, NY, 1978.

[10] C. Y. Lee. Representation of Switching Circuits by Binary-Decision Programs. *Bell Systems Technical Journal*, 38(4):985–999, July 1959.

[11] J-C. Madre and J-P. Billon. Proving Circuit Correctness Using Formal Comparison between Expected and Extracted Behaviour. In *Proceedings of the* 25th *Design Automation Conference*, pages 205–210, June 1988.

[12] S. Malik, A. R. Wang, R. K. Brayton, and A. Sangiovanni-Vincentelli. Logic Verification Using Binary Decision Diagrams in a Logic Synthesis Environment. In *Proceedings of the International Conference on Computer-Aided Design*, pages 6–9, November 1988.

Chapter 7

Synthesis of Multilevel Circuits

Multilevel realizations are the preferred means of implementing combinational logic in *very large scale integrated* (VLSI) systems today. Because of the increased potential for reusing subcircuits, there are more degrees of freedom in implementing a Boolean function than in the two-level *programmable logic array* (PLA) case. However, this increased freedom also makes the task for automatically synthesizing the logic at a level competitive with manual design difficult.

The area of multilevel logic synthesis has blossomed since the mid-1980s. Many of the methods and algorithms developed thus far have been successfully used in commercially available computer-aided design packages. Two basic approaches have been adopted in multilevel logic synthesis. In the late 1970s, the LSS system was developed at IBM [21] using rule-based local transformations. The local transformation and rule-based methods use a set of rules which are fired when certain patterns are found in the network of logic gates. A rule transforms a pattern for a local set of gates and interconnections into another equivalent one. The transformations have limited optimization capability since they are local in nature and do not have a global perspective of the design. The LSS system has continued to evolve and now contains methods for global optimization (e.g., [9, 40]).

Beginning about 1981, in parallel with activity in two-level logic synthesis and influenced by it, an approach evolved that was based on algorithmic transformations. The algorithmic point of view uses two phases: a technology-independent step based on algorithms for manipulating general Boolean functions [15] and a technology mapping step where the design described in terms of generic Boolean functions is mapped into a set of gates that can be implemented in the design method of choice (gate arrays, standard cells, or macrocells). Both rule-based methods (e.g., [8, 20]) and algorithmic methods (e.g., [11, 13]) have been successful.

Most algorithmic logic synthesis systems use algebraic [15, 16] and Boolean [11, 22] operations in the technology-independent phase of the design, and use graph covering methods pioneered in [28] for the technology mapping phase. We will focus on the algorithmic methods for logic synthesis in this chapter, since the logic transformations used in these approaches have been the central focus of synthesis for testability strategies.

We will describe the various logic transformations used in algorithmic logic synthesis systems in Section 7.1. Implementation details of the algorithms presented here can be found in [13, 14]. We will describe the notion of division and common divisors in Section 7.2 and describe algebraic and Boolean division algorithms in Sections 7.3 and 7.5, respectively. Internal "don't-care" conditions in multilevel logic circuits are described in Section 7.6. Finally, in Sections 7.7 and 7.8 we describe algorithmic approaches to technology mapping targeting an area cost function.

In this chapter, we will focus primarily on describing the available suite of logic transformation and restructuring methods. These methods can typically be used directly for area minimization. In Chapter 8, we will focus on delay minimization.

7.1 Logic Transformations

The goal of multilevel logic optimization is to obtain a representation of the Boolean function that is optimal with respect to area, speed, testability, or power dissipation. In order to restructure a logic function the operations described in the sequel are used.

7.1.1 Decomposition

Decomposition of a Boolean function is the process of reexpressing a single function as a collection of new functions. For example, the process of translating the expression

$$F = a \cdot b \cdot c + a \cdot b \cdot d + \overline{a} \cdot \overline{c} \cdot \overline{d} + \overline{b} \cdot \overline{c} \cdot \overline{d}$$

to the set of expressions

$$\begin{aligned} F &= X \cdot Y + \overline{X} \cdot \overline{Y} \\ X &= a \cdot b \\ Y &= c + d \end{aligned}$$

is decomposition.

7.1.2 Extraction

Extraction, related to decomposition, is applied to many functions. It is the process of identifying and creating some intermediate functions and variables, and reexpressing the original functions in terms of the original as well as the intermediate variables. Extraction creates nodes with multiple fan-outs. For example, extraction applied to the following three functions

$$\begin{aligned} F &= (a+b) \cdot c \cdot d + e \\ G &= (a+b) \cdot \overline{e} \\ H &= c \cdot d \cdot e \end{aligned}$$

yields

$$\begin{aligned} F &= X \cdot Y + e \\ G &= X \cdot \overline{e} \\ H &= Y \cdot e \\ X &= a + b \\ Y &= c \cdot d \end{aligned}$$

The operation thus identifies common subexpressions among different logic functions forming a network. New nodes corresponding to the common subfunctions are created and each of the logic functions in the original network is simplified as a result of the introduction of these new nodes. The optimization problem associated with the extraction operation is to find a set of intermediate functions such that the resulting network has minimum area, delay, power dissipation, or maximum testability.

7.1.3 Factoring

Like the sum-of-products, factored forms are a way of representing Boolean functions and are perhaps a more natural way for multilevel circuits than the sum-of-products representation. A factored form is a parenthesized representation of a tree network where each internal node is an AND or an OR gate and each leaf is a literal. Factoring is the process of deriving a factored form from a sum-of-products representation of a function. For example,

$$F = a \cdot c + a \cdot d + b \cdot c + b \cdot d + e$$

can be factored to

$$F = (a+b) \cdot (c+d) + e$$

The optimization problem associated with factoring is to find a factored form with the minimum number of literals.

7.1.4 Substitution

Substitution, also called resubstitution, of a function G into F is the process of reexpressing F as a function of its original inputs and G. For example, substituting

$$G = a + b$$

into

$$F = a + b \cdot c$$

produces

$$F = G \cdot (a + c)$$

This operation creates an arc in the Boolean network connecting the node of the substituting function, namely G, to the node of the function being substituted into, namely F.

7.1.5 Elimination

Elimination, collapsing, or flattening is the the inverse operation of substitution. If G is a fan-in of F, collapsing G into F reexpresses F without G. It undoes the operation of substituting G into F. For example, if

$$\begin{aligned} F &= G \cdot a + \overline{G} \cdot b \\ G &= c + d \end{aligned}$$

then, collapsing G into F results in

$$\begin{aligned} F &= a \cdot c + a \cdot d + b \cdot \overline{c} \cdot \overline{d} \\ G &= c + d \end{aligned}$$

If the node G is not a primary output and does not fan-out to other nodes, then it may be removed from the Boolean network, resulting in a network with one less node.

Flattening a logic function could result in an exponentially larger representation. For example, consider the flattening of the nodes g_1 through g_k into F below.

$$\begin{aligned}
F &= g_1 \cdot g_2 \cdots \cdot g_k \\
g_1 &= a_1 + b_1 \\
g_2 &= a_2 + b_2 \\
&\cdot\ \cdot\ \cdot \\
g_k &= a_k + b_k
\end{aligned}$$

This will result in a representation for F with 2^k product terms. Functions that have manageable[1] sum-of-products representations are termed *flattenable* functions. Functions like F with large k above that do not have manageable sum-of-products representations are called *nonflattenable.*

7.2 Division and Common Divisors

In optimizing logic functions it is important to define operations which, when given functions f and p, find functions q and r such that $f = p \cdot q + r$, if such q and r exist. This operation is called the division of f by p generating quotient q and remainder r.

The function p is called a divisor of f if r is not null and a factor if r is null. For a given division operation the resulting q and r may be dependent upon the particular representation of f and p.

Some properties of divisors and factors are presented using the following propositions.

Proposition 7.2.1 *A logic function g is a Boolean factor of a logic function f if and only if $f \cdot \overline{g} = \phi$ (in other words, the ON-set of f is contained in the ON-set of g).*

Proposition 7.2.2 *If $f \cdot g \neq \phi$, then g is a Boolean divisor of f.*

---

[1]By manageable or reasonable sized, we mean a size of the representation that can be stored in the memory and efficiently manipulated using present day computing resources. As such, what is reasonable is dependent on the computer in use.

For any logic function there are many Boolean factors and divisors. This poses a problem in choosing a good factor since there are so many factors. If the domain is restricted to a particular subset of expressions, then the division operation is unique and much easier to carry out. A restricted version of division is called algebraic division which we describe in the next section.

7.3 Algebraic Division

We begin the description of algebraic division with some definitions. The *support* of f denoted as $sup(f)$ is the set of all variables v that occur in f as v or $\overline{v}$. For example, if $f = a \cdot \overline{b} + c$, then $sup(f) = \{a,\ b,\ c\}$. We say that f is *orthogonal* to g, written as $f \perp g$, if $sup(f) \cap sup(g) = \phi$. For example, $f = a + b$ and $g = c + d$ are orthogonal.

The function g is an *algebraic divisor* of f if there exist h and r such that $f = g \cdot h + r$, where $h \neq 0$, and $g \perp h$, and r has as few cubes as possible. The function g divides f *evenly* if $f = g \cdot h$, where $h \neq 0$, $g \perp h$ and $r = \phi$. The *quotient* f/g is the unique h such that $f = g \cdot h + r$, i.e., $f = (f/g) \cdot g + r$.

In the next section, we consider the two main problems of algebraic optimization, namely computing f/g given f and g, and determining divisors g of a given function f.

7.3.1 Computing the Quotient

Given that f and g are lists of cubes, with $f = \{b_1,\ b_2,\ \ldots,\ b_{|f|}\}$ and $g = \{a_1,\ a_2,\ \ldots,\ a_{|g|}\}$. Define $h_i = \{c_j | a_i \cdot c_j \in f\}$ for all $i = 1, 2, \ldots, |g|$, i.e., h_i corresponds to all the multipliers of the cube a_i in g that produce elements of f. It is easy to see that

$$f/g = \bigcap_{i=1}^{|g|} h_i = h_1 \cap h_2 \ \ldots \ \cap h_{|g|}$$

As an example, consider:

$$\begin{aligned} f &= a \cdot b \cdot c \ + \ a \cdot b \cdot d \ + \ d \cdot e \\ g &= a \cdot b \ + \ e \end{aligned}$$

We have $|g| = 2$ and $|f| = 3$. The number of comparison is $3 \times 2 = 6$, and they give:

$$\begin{aligned} h_1 &= \{c,\ d\} \\ h_2 &= \{d\} \end{aligned}$$

Therefore, $h_1 \cap h_2 = d$, and

$$f = (a \cdot b + e) \cdot d + a \cdot b \cdot c$$

The above strategy requires $O(|f|\ |g|)$ operations. Encoding and sorting the cubes of f and g can reduce the complexity to $O((|f| + |g|) log(|f| + |g|))$ [30].

7.3.2 Kernels and Algebraic Divisors

Given an efficient method for algebraic division, optimization can be carried out if good algebraic divisors can be found. The set of algebraic divisors is defined as $D(f) = \{g | f/g \neq \phi\}$. The *primary divisors* of f are defined as $P(f) = \{f/c | c \text{ is a cube}\}$. For example, if

$$f = a \cdot b \cdot c + a \cdot b \cdot d \cdot e$$

then

$$f/a = b \cdot c + b \cdot d \cdot e$$

is a primary divisor.

Proposition 7.3.1 *Every divisor of f is contained in a primary divisor, i.e., if g divides f, then $g \subseteq p \in P(f)$.*

Proof. Let $c \in f/g$ be a cube. Then $g \subseteq f/(f/g)$ and $f/(f/g) \subseteq f/c \in P(f)$. □

A function g is termed *cube-free* if the only cube that divides g evenly is 1. The *kernels* of f are defined as $K(f) = \{k | k \in P(f),\ k \text{ is cube-free}\}$. For example, if

$$f = a \cdot b \cdot c + a \cdot b \cdot d \cdot e$$

then

$$f/a = b \cdot c + b \cdot d \cdot e$$

is a primary divisor but not cube-free since b is a factor of f/a.

$$f/a = b \cdot (c + d \cdot e)$$

However, $f/(a \cdot b) = c + d \cdot e$ is a kernel. $a \cdot b$ is called a *cokernel.* Note that by definition, cokernels are always cubes.

The following theorem (originally proven in [15]) is the basis of algebraic optimization methods.

Theorem 7.3.1 *f and g have a nontrivial common divisor d (i.e., d is not a cube) if and only if there exist kernels $k_f \in K(f)$ and $k_g \in K(g)$ such that $k_f \cap k_g$ is nontrivial with 2 or more terms (not a cube).*

Proof. It is clear that $k_f \cap k_g$ is a common divisor. It remains to prove the "only if" part.

Let d divide f, and let d divide g with d being nontrivial with 2 or more terms. Then there is a cube-free e such that e divides d. Now e divides f so by Proposition 7.3.1, $e \subseteq k_f \in P(f)$ and $e \subseteq k_g \in P(g)$ for some k_f and k_g. Since e is cube-free, k_f and k_g are as well. Hence, $k_f \in K(f)$ and $k_g \in K(g)$. Finally, since $e \subseteq k_f \cap k_g$, we have $k_f \cap k_g$ being nontrivial. □

We can therefore use the kernels of f and g to locate common divisors. Note that these are not the only common divisors of f and g, but they are good common divisors to consider during logic optimization. We compute the set of kernels for each logic expression, then form nontrivial intersections among kernels from the different logic expressions. If this intersection set is empty, then by Theorem 7.3.1, we need only look for divisors consisting of single cubes. If we find a nontrivial intersection, then we have found a algebraic divisor common to two or more expressions.

7.3.3 Computing the Kernels

The kernels of a function f can be computed using the algorithm of Figure 7.1.

The kernel generation algorithm first makes f cube-free by finding its largest cube factor. It then selects the literals of f in lexicographical order and divides them into f; the resulting quotient is a kernel if it is cube-free. (Note that this kernel might contain other kernels too.) If it is not cube-free, then it is made cube-free by selecting its largest cube factor. Note that in this context the largest cube is the cube with the most number of literals. The procedure is repeated on the resulting functions until functions with no kernels (called the level-0 kernels of f) are found. A major efficiency is obtained by noting that if the largest cube factor extracted contains an already selected literal, then the current branch can be terminated, since all the kernels that can be found by continuing have already

KERNELS(f):
{
  c_f = largest cube (with maximum number of literals) factor of f ;
  K = **KERNEL1**(0, f/c_f) ;
  if (f is cube-free)
    return($f \cup K$) ;
  return(K) ;
}

KERNEL1(j, g):
{
  $R = g$;
  N = Maximum index of variables in g ;
  for($i = j+1$; $i \leq N$; $i = i+1$) {
    if (l_i in 1 or no cubes of g) continue ;
    c = largest cube dividing g/l_i evenly ;
    if (for all $k \leq i$, $l_k \notin c$) /* Pruning Condition */
      $R = R \cup$ **KERNEL1**(i, $g/(l_i \cap c)$) ;
  }
  return(R) ;
}

Figure 7.1: Procedure to determine all the kernels of a single-output logic function

been generated. This leads to an algorithm in which no cokernel is duplicated.

As an example consider:

$$f = a \cdot b \cdot c \cdot d + a \cdot b \cdot c \cdot e + a \cdot b \cdot e \cdot f$$

In the routine **KERNELS** $c_f = a \cdot b$. Therefore,

$$f/c_f = c \cdot d + c \cdot e + e \cdot f$$

In the next step we call **KERNEL1**(0, $c \cdot d + c \cdot e + e \cdot f$).

In **KERNEL1** we set $R = \{c \cdot d + c \cdot e + e \cdot f\}$. Since the ordering is lexicographic, we have $l_1 = a$, $l_2 = b$, etc. Note that $N = 6$. The literals l_1 and l_2 are in none of the terms of R, and we move to $l_3 = c$. The largest cube dividing $(c \cdot d + c \cdot e + e \cdot f)/c$ which is $d + e$ is 1.

We therefore make a recursive call to **KERNEL1**(3, $(c \cdot d + c \cdot e + e \cdot f)/(c \cap 1)$). This call returns with $\{d + e\}$. In the parent **KERNEL1** R is set to $\{c \cdot d + c \cdot e + e \cdot f,\ d + e\}$. We skip $l_4 = d$ and move to $l_5 = e$. The largest cube evenly dividing $(c \cdot d + c \cdot e + e \cdot f)/e$ which is $c + f$ is 1. We next call **KERNEL1**(5, $(c \cdot d + c \cdot e + e \cdot f)/(e \cap 1)$). This returns with $c + f$.

We end with $K = R = \{c \cdot d + c \cdot e + e \cdot f,\ d + e,\ c + f\}$.

A second example illustrates the pruning condition: If the largest cube factor extracted contains an already selected literal, then the current branch can be terminated, since all kernels that can be found by continuing have already been generated. Consider:

$$f = a \cdot b \cdot c \cdot (d + e) \cdot (k + l) + a \cdot f \cdot g + h$$

In the first call to **KERNEL1**, we will generate the kernels corresponding to:

$$f/a = b \cdot c \cdot (d + e) \cdot (k + l) + f \cdot g$$

KERNEL1 calls itself recursively to compute

$$f/(a \cdot b) = c \cdot (d + e) \cdot (k + l)$$

Since $f/(a \cdot b)$ is not cube-free, the next recursive call to **KERNEL1** will use $(d + e) \cdot (k + l)$. All the kernels of this expression will be generated.

We move up one level in the recursion and compute

$$f/(a \cdot c) = b \cdot (d + e) \cdot (k + l)$$

At this stage, we note that $f/(a \cdot c)$ is not cube-free, and the largest cube dividing this expression evenly is b. However, b is an already selected literal implying that we have already generated the kernels for the cube-free expression $(d + e) \cdot (k + l)$. We do not have to recursively call **KERNEL1** for this branch and can go ahead to $f/(a \cdot d)$.

It is possible to modify the **KERNEL1** procedure to generate only the level-0 kernels which do not contain other kernels. This modification is based on the observation that if no kernels of g are found in the **for** loop, then g is a level-0 kernel.

```
GFACTOR(f):
{
    if (number of terms in f is 1)
        return(f) ;
    g = CHOOSE_DIVISOR(f) ;
    (h, r) = DIVIDE(f, g) ;
    f = GFACTOR(g) · GFACTOR(h) + GFACTOR(r) ;
    return(f) ;
}
```

Figure 7.2: Procedure to algebraically factor a function

7.3.4 Factoring Algorithm

A function can be algebraically factored using the notions presented in previous sections. A generic factoring algorithm is shown in Figure 7.2.

The procedure **DIVIDE** performs algebraic division as described in Section 7.3.1, and reexpresses f as $g \cdot h + r$. The procedure **CHOOSE_DIVISOR** is critical to obtaining a good factorization. One alternative is to select an arbitrary level-0 kernel as a divisor. This may not produce the best final result. Another alternative is to select a kernel which when substituted into the original function maximally reduces the total number of literals. For example, given:

$$X = a \cdot c + a \cdot d + a \cdot e + a \cdot g + b \cdot c + b \cdot d + b \cdot e + b \cdot f + c \cdot e + c \cdot f + d \cdot f + d \cdot g$$

if, in the procedure **CHOOSE_DIVISOR**, we choose literals in lexicographical order we obtain:

$$X = a \cdot (c + d + e + g) + b \cdot (c + d + e + f) + c \cdot (e + f) + d \cdot (f + g)$$

However, if we choose kernels, we obtain a better factorization with fewer literals:

$$X = (c + d + e) \cdot (a + b) + f \cdot (b + c + d) + g \cdot (a + d) + c \cdot e$$

7.3.5 Extraction and Resubstitution Algorithm

To identify cube-free expressions that occur in several functions $\{f_i\}$ we do the following:

1. Generate kernels for each f_i.

2. Select a pair of kernels $k_1 \in K(f_i)$ and $k_2 \in K(f_j)$ where $i \neq j$ such that $k_1 \cap k_2$ is not a cube. If no such pair exists, stop.

3. Set the new variable v to $k_1 \cap k_2$.

4. Update the associated functions to:

$$f_i = v \cdot (f_i/(k_1 \cap k_2)) + r_i$$

Common cubes are extracted as follows.

1. Select a pair of cubes $c_i \in f_i$, $c_2 \in f_j$ with $i \neq j$ such that $c_1 \cap c_2$ consists of 2 or more literals. If no such pair exists, stop.

2. Set the new variable x to $c_1 \cap c_2$.

3. Update each function f_i with the new variable wherever possible in the network.

Consider as an example the factored functions:

$$\begin{aligned} X &= a \cdot b \cdot (c \cdot (d + e) + f + g) + h \\ Y &= a \cdot i \cdot (c \cdot (d + e) + f + j) + k \end{aligned}$$

We have $d+e$ being a level-0 kernel of both functions, and extraction results in:

$$\begin{aligned} L &= d + e \\ X &= a \cdot b \cdot (c \cdot L + f + g) + h \\ Y &= a \cdot i \cdot (c \cdot L + f + j) + k \end{aligned}$$

Now, we select $c \cdot L + f + g$ as a level-0 kernel of the reexpressed X and $c \cdot L + f + j$ as a level-0 kernel of reexpressed Y. We obtain:

$$\begin{aligned} M &= c \cdot L + f \\ L &= d + e \\ X &= a \cdot b \cdot (M + g) + h \\ Y &= a \cdot i \cdot (M + j) + k \end{aligned}$$

Now X and Y have no kernel intersections that are nontrivial. We now extract common cubes. The cubes $a \cdot b \cdot M$ in X and $a \cdot i \cdot M$ in Y have 2 literals in common. Extraction produces:

$$\begin{aligned} N &= a \cdot M \\ M &= c \cdot L + f \\ L &= d + e \\ X &= b \cdot (N + a \cdot g) + h \\ Y &= i \cdot (N + a \cdot j) + k \end{aligned}$$

Because we are continually recomputing level-0 kernels on the reexpressed functions, it is possible to obtain decompositions corresponding to level-k > 0 kernels. If we collapse L into M into N above, we obtain:

$$\begin{aligned} N &= a \cdot (c \cdot (d + e) + f) \\ X &= b \cdot (N + a \cdot g) + h \\ Y &= i \cdot (N + a \cdot j) + k \end{aligned}$$

where N contains a level-1 kernel of the original X and Y, since it contains the level-1 kernel M which contains the level-0 kernel $d + e$.

7.3.6 Algebraic Resubstitution with Complement

Algebraic factorization and resubstitution can be performed with the complement of a given divisor. For example, consider:

$$f = a \cdot b + a \cdot c + \overline{b} \cdot \overline{c} \cdot d$$

where we choose $b + c$ as a level-0 kernel of f and decompose f as shown below.

$$\begin{aligned} f &= a \cdot X + \overline{b} \cdot \overline{c} \cdot d \\ X &= b + c \end{aligned}$$

In many cases it is useful to check if the complement of the new variable is an algebraic divisor for the function. In this case we can obtain:

$$\begin{aligned} f &= a \cdot X + \overline{X} \cdot d \\ X &= b + c \end{aligned}$$

7.4 Rectangles and Rectangle Covering

One of the key problems in algebraic optimization is the identification of good (common) divisors. We have described the use of kernels for determining a good set of divisors for algebraic factoring, decomposition, and extraction. The problem of finding a kernel and finding a single-cube or multiple-cube divisor can be reduced to the combinatorial optimization problem of rectangle covering. This formulation of the problem is not only elegant, but it is also favors the development of fast and effective algorithms. In this section, we closely follow the methods and algorithms of [36].

7.4.1 Definitions

A *rectangle* (R, C) of a matrix B, $B_{ij} \in \{0, 1, *\}$ is a subset of rows R and subset of columns C such that $B_{ij} \in \{1, *\}$ for all $i \in R$ and $j \in C$. Note that the rows and columns forming the rectangle do not have to be contiguous.

A rectangle (R_1, C_1) is said to *strictly contain* rectangle (R_2, C_2) if $R_2 \subseteq R_1$ and $C_2 \subset C_1$ or $R_2 \subset R_1$ and $C_2 \subseteq C_1$.

A rectangle (R, C) of B is said to be a *prime rectangle* if it is not strictly contained in any other rectangle of B.

The *corectangle* of a rectangle (R, C) is the pair (R, C') where C' is the set of columns not in C.

A set of rectangles $(\{R^K, C^K\})$ form a rectangle cover of matrix B if $B_{ij} = 1$ implies that $i \in R^k$ and $j \in C^k$ for some k. Thus, each 1 in B must be covered by at least one rectangle from the cover. A covering need not be disjoint, and therefore a 1 in B can be covered by more than one rectangle. The points of B which are labeled $*$ are not required to be covered by any rectangle in the cover and therefore represent don't-care points in the matrix.

In the following matrix

	1	2	3	4	5
1	1	1	1	0	0
2	1	*	1	0	*
3	0	1	1	0	1
4	1	0	1	1	1

$(\{1, 2\}, \{2, 3\})$ is a rectangle, but it is not prime as it is contained in the prime rectangle $(\{1, 2\}, \{1, 2, 3\})$. $(\{2, 4\}, \{1, 3, 5\})$ is another prime rectangle while $(\{2, 3\}, \{1, 2\})$ is not a rectangle.

Each rectangle (R^k, C^k) has an associated weight or cost defined by a weight function $w(R^k, C^k)$. The weight of a rectangle cover is then defined as the sum:

$$\sum_k w(R^k, C^k)$$

The minimum weighted rectangle covering problem is that of finding a rectangle cover of a matrix with minimum total weight.

7.4.2 Rectangles and Kernels

Rectangles provide an alternate way of looking at kernels of a function. Consider the expression $g = a \cdot b \cdot e + a \cdot c \cdot d + b \cdot c \cdot d$. This Boolean expression can be represented using a *cube-literal matrix* shown below where each row corresponds to a cube in the expression and the columns correspond to all the distinct literals.

	a	b	c	d	e
$a \cdot b \cdot e$	1	1	0	0	1
$a \cdot c \cdot d$	1	0	1	1	0
$b \cdot c \cdot d$	0	1	1	1	0

Consider the prime rectangle $(R, C) = (\{2, 3\}, \{3, 4\})$ and its corectangle $(R, C') = (\{2, 3\}, \{1, 2, 5\})$. The rectangle obviously corresponds to a cube $c \cdot d$ that is common to all the product terms corresponding to rows in R. Since the rectangle is prime, it is the largest cube common to all the product terms in R. If this cube is extracted from these product terms, the resulting expression is cube-free and is also a divisor of the original function g. In other words, the resulting expression is a kernel of g. The expression resulting from the extraction of the cube corresponds to the corectangle $(R, C') = (\{2, 3\}, \{1, 2, 5\})$, which is $a + b$. Therefore, each prime rectangle is a cokernel, while each corectangle of a prime rectangle is a kernel of the expression represented by the matrix.

From the rectangle interpretation of kernels, it is also possible to understand more clearly the notion of the level of a kernel. A level-0 kernel is the corectangle of a prime rectangle which has no other rectangle containing its column set, i.e., a rectangle of maximal width. The corectangle of a prime rectangle of maximal height, i.e., one whose row set is not contained in any other rectangle, corresponds to a kernel of maximal level.

7.4.3 Common-Cube Extraction

Common-cube extraction is the process of finding cubes common to two or more expressions and extracting the common cube to simplify each of the expressions. To optimize the network it is necessary to find the particular cubes to introduce that provide an optimal decomposition. The optimal decomposition can be defined as minimizing the total number of literals summed over all expressions or minimizing the total number of literals given a bound on the number of levels of logic in the final circuit.

Common cubes can be easily identified using the cube-literal matrix described above. Consider the following equations:

$$\begin{aligned} F &= a \cdot b \cdot c + a \cdot b \cdot d + e \cdot g \\ G &= a \cdot b \cdot f \cdot g \\ H &= b \cdot d + e \cdot f \end{aligned}$$

The cube-literal matrix for this expression is:

		a	b	c	d	e	f	g
F_1	$a \cdot b \cdot c$	1	1	1	0	0	0	0
F_2	$a \cdot b \cdot d$	1	1	0	1	0	0	0
F_3	$e \cdot g$	0	0	0	0	1	0	1
G_1	$a \cdot b \cdot f \cdot g$	1	1	0	0	0	1	1
H_1	$b \cdot d$	0	1	0	1	0	0	0
H_2	$e \cdot f$	0	0	0	0	1	1	0

The rectangle $(\{1, 2, 4\}, \{1, 2\})$ corresponds to common cube ab which is present in functions F and G. If this common cube is extracted as a new function X, the equations can be rewritten as:

$$\begin{aligned} F &= X \cdot c + X \cdot d + e \cdot g \\ G &= X \cdot f \cdot g \\ H &= b \cdot d + e \cdot f \\ X &= a \cdot b \end{aligned}$$

The process of extracting a cube modifies a Boolean network. A new node is added to the Boolean network with a logic function which is the common-cube divisor. All functions which the cube divides are replaced with the algebraic division of the function by the single cube. In order to extract cubes efficiently in an iterative algorithm it is necessary to modify the cube-literal matrix incrementally to reflect the extraction of the cube. The advantage is that the cube-literal matrix does not have to be recreated as each cube is extracted.

The modifications required to form the new cube-literal matrix are the following. A new row is added to reflect the new single cube expression added to the network. The entries covered by the rectangle are marked with a $*$ to reflect that the position has been covered. However, the $*$ allows other rectangles to cover the same position.

The choice of the weight function for a rectangle measures the optimization goal for cube extraction. To minimize the total number of literals in the network the weight of a rectangle is chosen so that the weight of a rectangle cover of the cube-literal matrix equals the total number of literals in the network after the new single-cube functions are added to the network. Hence, a minimum weighted rectangle cover corresponds to the optimal simultaneous extraction of a collection of cubes. The weight of a rectangle is defined as:

$$w(R, C) = \begin{cases} |C| & if \quad |R| = 1 \\ |C| + |R| & if \quad |R| > 1 \end{cases}$$

If there is a single row in the rectangle, then it corresponds to leaving the cube unchanged in the network. Hence, the weight of the rectangle counts the number of literals in the cube, which equals the number of columns. When the number of rows is greater than one, this corresponds to creating a new single cube function with $|C|$ literals and substituting this new function into $|R|$ other cubes at a cost of $|R|$ literals.

Note that the above weight does not reflect the savings obtained in terms of the number of literals by extracting a common cube. Therefore, when searching for a cube to extract it is useful to define a second function called the *value* of the rectangle. For cube extraction the value of the rectangle should indicate the savings obtained from extracting the corresponding cube. Since the number of literals before cube extraction is the number of 1s in the rectangle and the number of literals after cube extraction is the weight of the rectangle, the value $v(R, C)$ of a rectangle is defined as:

$$v(R, C) = |\{(i, j) | B_{ij} = 1, i \in R, j \in C\}| - w(R, C)$$

For example, for the rectangle $(\{1, 2, 4\}, \{1, 2\})$ in the cube-literal matrix shown above, its weight is the number of rows plus the number of columns, which equals 5. There are 6 positions in this rectangle and each of them has a 1. Therefore, the value of the rectangle is $6 - 5 = 1$. Therefore only one literal can be saved by extracting this rectangle, as illustrated in the example above.

7.4.4 Kernel Intersection

As described previously, intersections among the kernels of a collection of expressions are useful for finding common multiple-cube divisors between two or more expressions. If two functions share a common multiple-cube divisor, then the common divisor can be found as the intersection of a kernel from each of the functions.

The Boolean matrix associated with the optimal kernel intersection problem is called the *cokernel-cube matrix.* A row in this matrix corresponds to a cokernel (and its associated kernel) and each column corresponds to a cube present in some kernel, called a *kernel-cube.* The entry B_{ij} is set to 1 if the kernel associated with row i contains the cube associated with column j.

Consider the functions:

$$\begin{aligned} F &= a \cdot f + b \cdot f + a \cdot g + c \cdot g + a \cdot d \cdot e + b \cdot d \cdot e + c \cdot d \cdot e \\ G &= a \cdot f + b \cdot f + a \cdot c \cdot e + b \cdot c \cdot e \\ H &= a \cdot d \cdot e + c \cdot d \cdot e \end{aligned}$$

The kernels and cokernels of each of the functions are shown below:

Function	Cokernel	Kernel
F	a	$d \cdot e + f + g$
F	b	$d \cdot e + f$
F	$d \cdot e$	$a + b + c$
F	f	$a + b$
F	c	$d \cdot e + g$
F	g	$a + c$
G	a	$c \cdot e + f$
G	b	$c \cdot e + f$
G	f	$a + b$
G	$c \cdot e$	$a + b$
H	$d \cdot e$	$a + c$

Note that functions F and G are themselves kernels but have not been shown above for the ease of presentation. Let us number the cubes in the original function from 1 to 13, with $a \cdot f$ being 1, $b \cdot f$ being 2, and so on. The cokernel-cube matrix for this set of kernels is shown below. Note that instead of 1s in the matrix, we have numbers. These numbers indicate a cube of the original functions formed by multiplying the cokernel corresponding to a row and the cube corresponding to a column. For example, in the third row under column a we have the number 5 corresponding to the the fifth cube $a \cdot d \cdot e$.

		a	b	c	ce	de	f	g
F	a	0	0	0	0	5	1	3
F	b	0	0	0	0	6	2	0
F	$d \cdot e$	5	6	7	0	0	0	0
F	f	1	2	0	0	0	0	0
F	c	0	0	0	0	7	0	4
F	g	3	0	4	0	0	0	0
G	a	0	0	0	10	0	8	0
G	b	0	0	0	11	0	9	0
G	f	8	9	0	0	0	0	0
G	$c \cdot e$	10	11	0	0	0	0	0
H	$d \cdot e$	12	0	13	0	0	0	0

A rectangle of the cokernel-cube matrix identifies an intersection of kernels. The columns of the rectangle identify the cubes in the subexpression, and the rows in the rectangle identify the particular functions the subexpression divides. For example, rectangle $(\{3,4,9,10\},\{1,2\})$ identifies the subexpression $a+b$. This corresponds to the factorization of the equations into the form:

$$
\begin{aligned}
F &= d \cdot e \cdot X + f \cdot X + a \cdot g + c \cdot g + c \cdot d \cdot e \\
G &= c \cdot e \cdot X + f \cdot X \\
H &= a \cdot d \cdot e + c \cdot d \cdot e \\
X &= a + b
\end{aligned}
$$

Whenever a new subexpression is identified, it is inserted into the Boolean network. This consists of adding a new node to the network and dividing the node into each of the expressions which this node divides. A new cokernel-cube matrix is then created for the modified Boolean network.

To reduce the complexity of extracting each factor from the network it is desirable to modify the cokernel-cube matrix incrementally as each subexpression is identified. To do this new rows are added to the cokernel-cube matrix for each kernel of the new subexpression. The cubes which are formed by the insertion of this new factor into the network are then marked as covered. This includes the points directly contained in the rectangle and other points which are labeled with the same number. These points are marked $*$ so that other rectangles can cover them.

The weight of a rectangle of the cokernel-cube matrix is chosen to reflect the number of literals in the network if the corresponding common subexpression is inserted into the network. A minimum weighted rectangle cover of the cokernel-cube matrix then corresponds to a simultaneous selection of a set of subexpressions to add to the network in order to minimize the total number of literals.

Let w_j^c be the number of literals in the kernel-cube for column j. w_j^c is also called the column weight of column j. If a rectangle (R, C) is used to identify a subexpression, then a new function is formed from the columns of C. This new function has $\sum_{j \in C} w_j^c$ literals. Let w_i^r be 1 plus the number of literals in the cokernel corresponding to row i. w_i^r is also called the row weight of row r. The chosen subexpression divides the expressions indicated by the rows R of the rectangle. After algebraic division by the subexpression each of these expressions consists of a sum of the corresponding cokernel cubes multiplying the literal for the new expression. The number of literals in the affected functions after the extraction of the subexpression corresponding to the rectangle is $\sum_{i \in R} w_i^r$. Therefore, the weight of a rectangle (R, C) in the cokernel-cube matrix is defined as:

$$w(R, C) = \sum_{i \in R} w_i^r + \sum_{j \in C} w_j^c$$

The value of a rectangle measures the difference in the number of literals in the network if the particular rectangle is selected. The number of literals after the rectangle is selected is the weight of the rectangle as defined above. Let V_{ij} be the number of literals in the cube which is covered by position (i, j) of the cokernel-cube matrix. Then the number of literals before extraction of the rectangle is simply $\sum_{i \in R, j \in C} V_{ij}$. As elements of the cokernel-cube matrix are covered, their values V_{ij} are set to 0. This includes the elements V_{ij} covered by the matrix and all other elements which represent the same cube in the network. The value of a rectangle (R, C) of the cokernel-cube matrix is thus defined as:

$$v(R, C) = \sum_{i \in R, j \in C} V_{ij} - w(R, C)$$

For example, for the rectangle $(\{3, 4, 9, 10\}, \{1, 2\})$ in the cokernel-cube matrix above, $\sum_{i \in R, j \in C} V_{ij} = 3 + 3 + 2 + 2 + 2 + 2 + 3 + 3 = 20$,

$\sum_{i \in R} w_i^r = 3 + 2 + 2 + 3 = 10$, $\sum_{j \in C} w_j^c = 1 + 1 = 2$. Therefore, the value of the rectangle is $value = 20 - 10 - 2 = 8$. Eight literals can be saved by extracting the expression corresponding to the rectangle, as can be verified in the example above.

7.4.5 Rectangle Algorithms

There are two types of algorithms for rectangle covering. The first type of algorithm is greedy and selects one rectangle at a time and modifies the matrix to reflect the extraction of the rectangle. The advantage of this technique is that it immediately takes into account common factors between the newly extracted function and the rest of the logic network. The disadvantage of this approach is that it selects only one rectangle at a time and does not easily account for the simultaneous extraction of multiple rectangles. The second type of algorithm finds the best collection of factors to extract at each step by solving the minimum-weighted rectangle covering problem heuristically. First, all the prime rectangles are generated, and a collection of rectangles are then extracted. Second, the matrix is updated, and the entire process is repeated to find factors between the new expressions and the remainder of the logic network.

In this section we will first discuss a heuristic algorithm called **PING-PONG** that can be used to generate a good valued rectangle without generating all prime rectangles. Such an algorithm is useful for the extraction of a single good factor.

The inputs to the algorithm are the matrix and the value function which computes the value of a rectangle. The value function is defined in terms of the row and column weights. The algorithm chooses two seed rectangles — one corresponding to the best row and the other corresponding to the best column. These two seed rectangles are expanded until they are prime. From these expanded rectangles the one with the larger value is returned.

We will illustrate this algorithm with an example. Consider the function

$$F = a \cdot c + a \cdot d + a \cdot e + a \cdot g + b \cdot c + b \cdot d$$

$$+ b \cdot e + b \cdot f + c \cdot e + c \cdot f + d \cdot f + d \cdot g$$

Its kernels (and cokernels) are $c + d + e + g(a)$, $c + d + e + f(b)$, $a + b + e + f(c)$, $a + b + f + g(d)$, $a + b + c(e)$, $b + c + d(f)$, and $d(g)$. The cokernel-cube matrix with the value at each position is shown below. Note each row has a weight of 2 (because each cokernel has one literal) and each column has a weight of 1 (as there is one literal in each kernel-cube).

	a	b	c	d	e	f	g
a	0	0	2	2	2	0	2
b	0	0	2	2	2	2	0
c	2	2	0	0	2	2	0
d	2	2	0	0	0	2	2
e	2	2	2	0	0	0	0
f	0	2	2	2	0	0	0
g	0	0	0	2	0	0	0

We will illustrate the **PING-PONG** procedure using one of the seed rectangles, the best row. The dual of the procedure with the role of rows and columns interchanged has to be followed for the other seed rectangle, the best column. In this example there are four rows with the same value, and we arbitrarily choose row 1 as the seed. Therefore, the seed rectangle is $(\{1\}, \{3,4,5,7\})$ with a value 2. This value is derived by adding the numbers at all points in the rectangle $(2 \times 4 = 8)$ and subtracting the row weight (2) and the column weights $(1 \times 4 = 4)$ from it $(8 - 4 - 2 = 2)$. We now try to add another row to this rectangle by examining all the other rows and selecting a row which if added maximizes the value of the seed rectangle. This row is added, and only columns that are both in the seed rectangle column set and the row selected are retained. In this example addition of row 2 maximizes the value of the rectangle to 5, and therefore the row is added to the seed rectangle. The new rectangle is $(\{1,2\}, \{3,4,5\})$. Note that column 7 was deleted from the initial rectangle and row 2 was added to form this new rectangle. A new row is selected for addition to this rectangle. This time it is row number 6, which produces a rectangle $(\{1,2,6\}, \{3,4\})$ with value 4. This process is repeated until the column set of the rectangle consists of only one column. This happens when row 5 is added to the set, producing the rectangle $(\{1,2,5,6\}, \{3\})$ with a value -1. From all the rectangles produced in the process the best one, which is $(\{1,2\}, \{3,4,5\})$, is chosen.

This rectangle can now be improved. One column of this rectangle with the largest value is chosen as a seed column. In this case it is column 3. Now the dual of the procedure described above is followed with the role of rows and columns interchanged, i.e., columns are added to this seed column to form new rectangles. Each time, a column that maximizes the value of the rectangle is chosen, and only rows that are both in the row set of the rectangle and the column chosen are retained. The process continues until there is only one row left in the rectangle. The rectangle with the largest value (including the initial one) is returned. In this example the successive rectangles that are produced are $(\{1,2,5,6\},\{3\})$ with value -1, $(\{1,2,6\},\{3,4\})$ with value 4, $(\{1,2\},\{3,4,5\})$ with value 5, and $(\{1\},\{3,4,5,7\})$ with value 2. The best rectangle is still the initial one, which is $(\{1,2\},\{3,4,5\})$.

If the best rectangle chosen changed, then another pass is initiated where a row of the best rectangle is chosen as the seed row and is grown into a rectangle using the method described previously. This method of growing seed rows and columns is iterated until no improvement can be obtained. In this example the best rectangle chosen would be $(\{1,2\},\{3,4,5\})$, corresponding to the kernel intersection $c+d+e$.

We will now illustrate a method to generate all prime rectangles given a matrix. The worst-case complexity of this method is exponential in the size of the matrix. However, when the matrix is sparse, it is often feasible to enumerate all prime rectangles. We use as an example the expression

$$F = a\cdot b\cdot c\cdot d\cdot g + a\cdot b\cdot c\cdot d\cdot h + a\cdot b\cdot c\cdot e + a\cdot b\cdot c\cdot f + a\cdot b\cdot i$$

whose cube-literal matrix is shown below.

	a	b	c	d	e	f	g	h	i
1	1	1	1	1	0	0	1	0	0
2	1	1	1	1	0	0	0	1	0
3	1	1	1	0	1	0	0	0	0
4	1	1	1	0	0	1	0	0	0
5	1	1	0	0	0	0	0	0	1

The first step in generating prime rectangles is to process and record the trivial prime rectangles. A single row (column) is a prime rectangle if it is not contained in any other row (column). Also, if the matrix does not contain any columns (rows) with all 1s, then the rectangle consisting of all the rows (columns) and no columns (rows) is also prime. In our example above we have no trivial prime rectangles.

The recursive procedure described in Figure 7.3 is qualitatively similar to the kernel generation procedure of Figure 7.1. It

```
GENERATE_RECTANGLES(M):
{
    Find all trivial rectangles and store ;
    rect = new rectangle ;
    RECTANGLE1(M, 0, rect) ;
}

RECTANGLE1(M, index, rect):
{
    for( each column c of M ) {
        if ( c has 2 or more elements and c.index ≥ index ) {
            M1 = new matrix ;
            for( each element p in column c )
                copy row corresponding to p in M to M1 ;

            rect1 = new rectangle ;
            Duplicate columns of rect in rect1 ;
            Rows of rect1 correspond to elements of c in M ;

            if ( indices of all columns of all 1s in M1 ≥ c.index ) {
                Delete columns of all 1s in M1 and add to rect1 ;
                Store rect1 in prime rectangle set ;
                RECTANGLE1( M1, c.index, rect1) ;
            }
        }
    }
}
```

Figure 7.3: Procedure to determine all the prime rectangles of a matrix

keeps track of the current submatrix which is being searched for prime rectangles, namely M, the current column index, namely *index*, and the rectangle found up to this point, namely *rect*. The recursive assumption made is that all the rows of *rect* contain a 1 or $*$ for all columns of *rect*.

The procedure will search the submatrix M to find all of the prime rectangles with fewer rows but more columns. Each column c

with an index greater than the starting index is examined as a column to include in the rectangle. If the column has only a single element, then it cannot create a nontrivial rectangle, so only columns with two or more elements are of interest. A submatrix M_1 of the original matrix M is created by selecting only the rows where the column c has a nonzero value, and a new rectangle $rect_1$ is formed from the columns of the old rectangle and the rows for which c has a nonzero value.

Any column of the submatrix M_1 including c which is now all 1s can also be added to the rectangle $rect_1$. At this point, $rect_1$ represents a new prime rectangle of the matrix, and M_1 is a new submatrix to be searched for more prime rectangles.

A pruning operation can be performed when $rect_1$ is constructed. If a column of 1s occurs for a column index less than the starting index, then all rectangles in the current submatrix M_1 have already been examined when that column index was processed. Hence, if this condition is detected, it is not necessary to do another recursion. This condition is equivalent to the pruning condition in Figure 7.1.

In our example we begin in **RECTANGLE1** with the original matrix M, the starting column index 0, and with *rect* being a new rectangle. We pick column 1 in the loop. There are more than two elements in column 1 so we create a new matrix M_1. M_1 will be set to M since column 1 has all five elements. $rect_1$ is created as a new rectangle and is initially set to $(\{1,2,3,4,5\},\{1\})$. Next, column 2 which has all 1s in M_1 is added to $rect_1$ since its index is greater than 1. We now have a prime rectangle $(\{1,2,3,4,5\},\{1,2\})$ that is added to the set of prime rectangles. This rectangle corresponds to the kernel (cokernel) $c \cdot d \cdot g + c \cdot d \cdot h + c \cdot e + c \cdot f + i$ $(a \cdot b)$.

We enter the recursive call of **RECTANGLE1** with M shown below:

	c	d	e	f	g	h	i
1	1	1	0	0	1	0	0
2	1	1	0	0	0	1	0
3	1	0	1	0	0	0	0
4	1	0	0	1	0	0	0
5	0	0	0	0	0	0	1

The variable *index* in this call is 1 and *rect* is $(\{1,2,3,4,5\},$ $\{1,2\})$. We first pick column 3 in M and create M_1 which is M

but with row 5 deleted. $rect_1$ is set to $(\{1,2,3,4\},\{1,2\})$. We add the column of all 1s, namely column 3, to $rect_1$ to make it a prime rectangle. The prime rectangle $(\{1,2,3,4\},\{1,2,3\})$ is added to the set. This rectangle corresponds to the kernel (cokernel) $d{\cdot}g{+}d{\cdot}h{+}e{+}f$ $(a \cdot b \cdot c)$.

In the next recursive call we will discover the prime rectangle $(\{1,2\},\{1,2,3,4\})$ corresponding to the kernel (cokernel) $g+h$ $(a \cdot b \cdot c \cdot d)$. We have now discovered all the kernels of the expression.

Going back to the first call of **RECTANGLE1** we will pick column 2 with the original matrix M, the starting column index 0, and *rect* being a new rectangle. We create M_1 as a new matrix with all the rows and columns of M. $rect_1$ is set to $(\{1,2,3,4,5\},\{2\})$. When we check for the pruning condition we see that column 1 is a column of all 1s with an index less than 2, and we will not perform any recursive calls.

7.5 Boolean Division

So far we have primarily described algebraic optimization methods. In this section we will focus on Boolean division. As noted in Section 7.2 there are many Boolean divisors of a function. If a good Boolean divisor can be found, then two-level minimization under an appropriate don't-care set can be used to compute $f//g$, where we use $//$ to denote Boolean division.

A procedure to compute $f//g$, given sum-of-products representations for f and g is given below. The procedure is restricted to the case where $sup(g) \subseteq sup(f)$.

1. Add an extra input to f corresponding to g. Call it G.

2. Construct a function F whose *DC*-set D is $G \cdot \overline{g} + \overline{G} \cdot g$, *ON*-set is $f \cap \overline{D}$, and *OFF*-set is $\overline{f \cup D}$.

3. Minimize F for minimum literals in sum-of-products form.

The specification of the don't-care set in the above procedure is key. The don't-care set corresponds to the set of inputs to f that cannot occur. Since G is itself a function of the primary inputs, G can never be a 0 when a minterm in the *ON*-set of g is applied. Similarly for the *OFF*-set.

Consider an example of the application of Boolean division.

$$\begin{aligned} f &= a \cdot b \cdot c + a \cdot \overline{b} \cdot \overline{c} + \overline{a} \cdot b \cdot \overline{c} + \overline{a} \cdot \overline{b} \cdot c \\ g &= a \cdot b + \overline{a} \cdot \overline{b} \end{aligned}$$

The specifications for the *DC*-set and *ON*-set for F are:

$$\begin{aligned} F^{DC} &= a \cdot b \cdot \overline{G} + \overline{a} \cdot \overline{b} \cdot \overline{G} + \overline{a} \cdot b \cdot G + a \cdot \overline{b} \cdot G \\ F^{ON} &= a \cdot b \cdot c \cdot G + \overline{a} \cdot \overline{b} \cdot c \cdot G + \overline{a} \cdot b \cdot \overline{c} \cdot \overline{G} + a \cdot \overline{b} \cdot \overline{c} \cdot \overline{G} \end{aligned}$$

Minimization produces:

$$F = c \cdot G + \overline{c} \cdot \overline{G}$$

Note that G has been used in true and complemented forms.

In general the above procedure works well, but the minimization at Step 3 is not guaranteed to result in a representation for F that uses the extra input G. In order to ensure that G appears in the minimized sum-of-products representation for F during heuristic two-level minimization using ESPRESSO, a modification is made. In the **EXPAND** step the literals in cubes corresponding to the inputs to F other than G are expanded first. In the **REDUCE** step the literals corresponding to the input G are reduced first.

There has been little work done in identifying Boolean divisors. A method based on multiple-valued minimization has been proposed [22]. Another technique that is used presently is to identify a good algebraic divisor and then use the Boolean division algorithm. This will produce at least as good results as algebraic division using the same divisor.

7.6 Don't-Care-Based Optimization

The don't-care conditions arising in multilevel logic can either be specified by the user or can be an artifact of the network structure. User specified don't-cares or don't-cares derived from considerations other than the network structure are called *external* don't-cares.

Internal don't-cares arise in multilevel logic because of the structure of a Boolean network. They are divided into satisfiability and observability don't-cares.

7.6.1 Satisfiability Don't-Cares

Satisfiability don't-cares are a result of the existence of the additional intermediate variables y_j introduced at the intermediate nodes of a Boolean network. As an example consider the network:

$$e = \overline{a} \cdot \overline{b}$$

$$g = \overline{c} \cdot \overline{d}$$

$$f = \overline{e} \cdot \overline{g}$$

which implements $f = (a+b) \cdot (c+d)$. For any node that uses the intermediate variables e and g we have the option of eliminating e and g or expanding the Boolean space to include these variables. If we do the latter there are combinations of variables which will never occur. For example, the combination $e = 0$, $a = 0$, and $b = 0$ will never occur. In this case the combinations that will never occur are expressed by the logic function:

$$e \cdot (a+b) + \overline{e} \cdot \overline{a} \cdot \overline{b}$$

The intermediate nodes of a Boolean network impose the relation

$$y_j = F_j(x, y)$$

where x is the set of primary inputs and y is the set of intermediate variables. Of course, since the Boolean network is acyclic, F_j depends only on a subset of the y variables. In the space over the primary inputs and intermediate variables $B^{|x|+|y|}$, the *satisfiability don't-care set* is given by:

$$SDC = \sum_j (y_j \cdot \overline{F_j} + \overline{y_j} \cdot F_j)$$

The SDC gives all the internal patterns of signals that will never occur due to the network structure and is so called because each of the relations

$$y_j = F_j(x, y)$$

must be satisfied during the correct operation of the network. The part in the SDC contributed by the function F_j, namely $y_j \cdot \overline{F_j} + \overline{y_j} \cdot F_j$ is called the satisfiability don't-care set SDC_j given by F_j.

In order to optimize a given node j we are typically interested in the satisfiability don't-care sets given by the nodes that fan into F_j.

7.6.2 Observability Don't-Cares

Observability don't-cares occur in a network because at each node there is a network structure that limits the observability of the value of the node as seen at a primary output. To discuss this we need to extend the notion of a cofactor of a function to that of a Boolean network η with respect to a literal x. This results in the cofactored network denoted by η_x, which is obtained by cofactoring each node function F_j of η with respect to the literal x. However, there is a subtle distinction to be noted here. Each fan-out of x, F_j becomes F_{j_x}, a function independent of x. If a node did not depend explicitly on x, then it is unaltered in the new network, i.e., $F_{j_x} = F_j$. However, if it implicitly depended on x, then F_{j_x} as a function of the primary inputs is different from F_j.

The observability of a node j at a primary output k of a network is the notion that F_k is implicitly dependent on y_j. This is given by the function:

$$\frac{\partial F_k}{\partial y_j} \equiv F_{k y_j} \oplus F_{k \overline{y_j}}$$

where $\oplus$ is the XOR operator. The meaning of this is that $\frac{\partial F_k}{\partial y_j}$ gives the input conditions under which the output F_k differs in the two networks η_{y_j} and $\eta_{\overline{y_j}}$, i.e., the conditions under which the value of y_j can be observed at output k. The preceding expression is called the Boolean difference of F_k with respect to y_j.

There may be conditions under which the value of y_j cannot be observed at any of the outputs. If the external don't-care set is empty, then we have:

$$ODC_j = \prod_{all\ outputs} (F_{k y_j} = F_{k \overline{y_j}}) = \prod_{all\ outputs} \overline{\frac{\partial F_k}{\partial y_j}}$$

This is called the *observability don't-care set* for the signal y_j.

The observability and satisfiability don't-care sets can be used to minimize nodes in the Boolean network, so as to reduce the area of a network. Two-level representations of the don't-care sets are obtained, and two-level Boolean minimization is applied to the different nodes in the Boolean network under the appropriate don't-care sets [7, 11, 37]. These don't-care sets are obtained for each node in terms of the immediate inputs of the node. We will illustrate the generation of don't-care sets with an example in the next section.

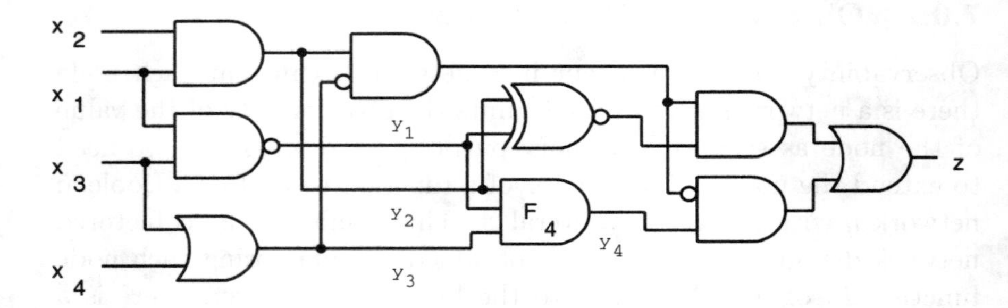

Figure 7.4: Example illustrating don't-care generation

7.6.3 Don't-Care Generation

Consider the circuit of Figure 7.4. The primary inputs of the circuit are x_1, x_2, x_3, and x_4. The circuit has a single output marked z. Focus on the AND gate 4 whose output is marked y_4, and whose inputs are marked y_1, y_2, and y_3.

We will first use satisfiability don't-cares to optimize the 4. The functions driving the immediate inputs to node 4 are $y_1 = x_1 \cdot x_2$, $y_2 = \overline{x_1 \cdot x_3}$, and $y_3 = x_3 + x_4$. There are 8 possible input combinations over y_1, y_2, y_3. Since $y_3 = 0 \Rightarrow y_2 = 1$, the combinations 000 and 100 never appear on the y_j's. The *local satisfiability don't-care set* for node 4 is a part of the satisfiability don't-care set SDC and is represented in terms of the immediate inputs to node 4.

$$SDC_4 \;=\; \overline{y_2} \cdot \overline{y_3}$$

We can add this don't-care set to the sum-of-products representation of $F_4 = y_1 \cdot y_2 \cdot y_3$. However, minimization under the don't-care set does not change the implementation of F_4.

Now consider the observability don't-care set for node 4. The function at the primary output z can be written as:

$$z \;=\; \overline{x_1 \cdot x_2 \cdot \overline{x_3} \cdot \overline{x_4}} \cdot y_4 \;+\; \overline{x_1 \oplus x_2} \cdot x_1 \cdot x_2 \cdot \overline{x_3} \cdot \overline{x_4}$$

We have expressed z in terms of y_4 and the primary inputs. This will enable us to compute the Boolean difference of z with respect to y_4. First, simplifying the expression for z gives:

$$z \;=\; (\overline{x_1} + \overline{x_2} + x_3 + x_4) \cdot y_4 \;+\; x_1 \cdot x_2 \cdot \overline{x_3} \cdot \overline{x_4}$$

The cofactors of z with respect to y_4 are:

$$z_{y_4} \;=\; \overline{x_1} + \overline{x_2} + x_3 + x_4 + x_1 \cdot x_2 \cdot \overline{x_3} \cdot \overline{x_4} \;=\; 1$$

$$z_{\overline{y_4}} = x_1 \cdot x_2 \cdot \overline{x_3} \cdot \overline{x_4}$$

Finally, ODC_4 can be expressed as:

$$ODC_4 = \overline{z_{y_4} \oplus z_{\overline{y_4}}} = x_1 \cdot x_2 \cdot \overline{x_3} \cdot \overline{x_4}$$

We now have the observability don't-care set for node 4 in terms of the primary inputs. In order to minimize F_4 we have to find the *local observability don't-care set* in terms of the immediate y_1, y_2, and y_3 inputs to node 4. The *local care set* of y_1, y_2, and y_3 combinations corresponds to all the combinations generated by primary input combinations in $\overline{ODC_4}$. The generated combinations over y_1, y_2, and y_3 are 001, 010, 011, 101, and 111. (The primary input combination $x_1 \cdot x_2 \cdot \overline{x_3} \cdot \overline{x_4}$ is in ODC_4, therefore the y_j combination that it produces, namely 110, is not in the local care set.) The *local satisfiability and observability don't-care set* or local don't-care set for node i is the complement of the local care set. In terms of the immediate inputs the local don't-care set for node 4 can be expressed as:

$$SDC_4 \cup ODC_4 = \overline{y_2} \cdot \overline{y_3} + y_1 \cdot y_2 \cdot \overline{y_3}$$

Minimizing $F_4 = y_1 \cdot y_2 \cdot y_3$ with respect to the above don't-care set results in an implementation $F_4 = y_1 \cdot y_2$. The connection corresponding to the y_3 input to F_4 can be removed in Figure 7.4 without changing the functionality of the circuit.

The don't-care generation and optimization procedure can be summarized as follows:

1. Select a node i in the Boolean network.

2. For each primary output of the network z_k, compute the cofactors of z_k with respect to y_i and $\overline{y_i}$.

3. Compute

$$C_i = \sum_{all\ outputs\ k} z_{k y_i} \oplus z_{k \overline{y_i}}$$

   This is a function over the primary inputs to the network.

4. Given the care set C_i for node i we find the local care set for node i in terms of its immediate inputs. Call this LC_i. LC_i corresponds to all combinations at the inputs to node i generated by primary inputs in C_i.

5. F_i, the cover for node i, is minimized with respect to its local don't-care set $SDC_i \cup ODC_i = \overline{LC_i}$.

7.6.4 ROBDD implementation

While the local don't-care set of a node in terms of its immediate inputs typically has a manageable sum-of-products representation, the procedure described in the previous section may require the collapsing of the entire network into sum-of-products form while generating the don't-cares. In particular computing C_i in terms of the primary inputs is not possible for circuits that cannot be collapsed to two levels.

The use of ROBDDs alleviates the above problem. The procedure of don't-care generation first creates ROBDDs for every node in the network in terms of the primary inputs. Next a node i is selected, and ROBDDs are created for each primary output, assuming that y_i is a 0 and that y_i is a 1. An ROBDD representation for C_i is obtained by performing a XOR operation on the two cofactors of the circuit with respect to y_i as described in Step 3 of the procedure.

We now have an ROBDD representation for C_i. To compute LC_i for node i in Step 4 of the procedure, we have to compute the set of all possible combinations of the immediate inputs to node i given the primary input combinations C_i. This is termed *range computation.* We will describe how this is done using ROBDDs in the next section. Once an ROBDD representation for $\overline{LC_i}$ has been obtained, it can be converted into a sum-of-products representation to minimize the node.

7.6.5 Range Computation

We first present some terminology. In the following description B is the set $\{0, 1\}$. For any function $f : B^N \rightarrow B^M$ the input variables are denoted as $x = (x_1, x_2, \ldots, x_N)$ and the output variables as $y = (y_1, y_2, \ldots, y_M)$.

Let $f : B^N \rightarrow B^M$ be a Boolean function of N inputs and M outputs and let $A \subseteq B^N$. The *image* of A by f is the set $f(A) = \{y : y \in B^M \ \ and \ \ \exists x \in A(y = f(x))\}$. The complete input space B^N is called the *domain* of the function f. The image of the domain is called the *range* of the function. If $C \subseteq B^M$, the *inverse image* of C by f is the set $f^{-1}(C) = \{x : x \in B^N \ \ and \ \ \exists y \in C(y = f(x))\}$.

Let $A \subseteq B^N$ be a set. The *characteristic function* of A is a function $\eta_A : B^N \rightarrow \{0, 1\}$ which is defined as $\eta_A(x) = 1$ if $x \in A$, else $\eta_A(x) = 0$. Characteristic functions are functional representations of subsets of sets. Characteristic functions can be defined for functions

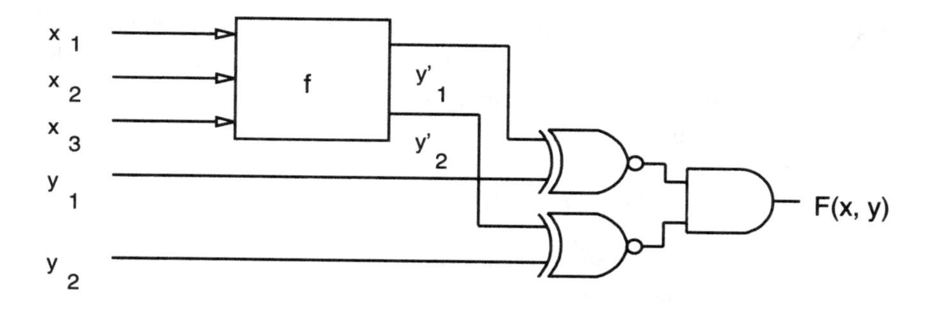

Figure 7.5: Characteristic function

and relations as well. Let $R : A \rightarrow C$ be a relation. Therefore, $R \subseteq A \times C$ and the characteristic function of R is a function $\eta_R : A \times C \rightarrow \{0, 1\}$ defined as $\eta_R(x, y) = 1$ if $\exists x \in A \;\; \exists y \in C(R(x, y))$. Any set can be represented by its characteristic function, which in turn can be represented using ROBDDs.

Let $f : B^N \rightarrow B$ be a Boolean function and $x = (x_1, x_2, \ldots, x_N)$ be a set of input variables of f. The *smoothing* of f by x is defined as:

$$\begin{aligned} S_x f &= S_{x_1} \ldots S_{x_N} f \\ S_{x_i} f &= f_{x_i} + f_{\overline{x_i}} \end{aligned}$$

where f_a designates the cofactor of f by the literal a. The smoothing operation can be performed efficiently on ROBDDs. Multiple variables can be smoothed away in one pass by performing a bottom-up traversal of the ROBDD.

A method for range computation was proposed in [17]. Given a Boolean function $f : B^N \rightarrow B^M$, let $x = (x_1, \ldots, x_N)$ be the set of input variables and $y = (y_1, \ldots, y_M)$ be the set of output variables, i.e., $y_i = f_i(x)$. The characteristic function of the *transition relation* [17, 39] associated with f, denoted as $F : B^N \times B^M \rightarrow B$, is defined as $F(x, y) = 1$ if and only if $(x, y) \in B^N \times B^M$ and $y = f(x)$, and $F(x, y) = 0$ otherwise. Equivalently, in terms of Boolean operations:

$$F(x, y) = \prod_{1 \leq i \leq M} \overline{(y_i \oplus f_i(x))}$$

The characteristic function is illustrated in Figure 7.5 for a function with $N = 3$ and $M = 2$.

The characteristic function F can be used to compute the image of a set $A \subseteq B^N$ by f. The image is defined as the set

$f(A) = \{y : \exists x(x \in A) \wedge F(x, y) = 1\}$. Replacing the existential quantification by the smoothing operator and the logical AND by the Boolean AND operator, the image can be represented in terms of Boolean operations by:

$$f(A)(y) = S_x(F(x, y) \cdot A(x))$$

In terms of ROBDD operations, this is achieved by a Boolean AND and a smooth. The smoothing and a Boolean AND can be done in one pass on the ROBDDs, thereby reducing the need for intermediate storage [17]. The inverse image of a subset A of B^M by f can be computed similarly:

$$f^{-1}(A)(x) = S_y(F(x, y) \cdot A(y))$$

In the don't-care generation example of Figure 7.4 we have y_1, y_2, and y_3 as the immediate inputs to node 4.

$$\begin{aligned} y_1 &= x_1 \cdot x_2 \\ y_2 &= \overline{x_1} + \overline{x_3} \\ y_3 &= x_3 + x_4 \end{aligned}$$

To compute the range of the function $y = f(x)$ where $N = 4$ and $M = 3$ we first compute the characteristic function $F(x, y)$ as illustrated in Figure 7.5. We have

$$F(x, y) \; = \; \overline{y_1 \oplus (x_1 \cdot x_2)} \cdot \overline{y_2 \oplus (\overline{x_1} + \overline{x_3})} \cdot \overline{y_3 \oplus (x_3 + x_4)}$$

$F(x, y)$ is intersected with $A(x)$, which corresponds to C_i in our don't-care generation procedure.

$$C_4 \; = \; A(x) \; = \; \overline{x_1} + \overline{x_2} + x_3 + x_4$$

The local care set is computed as:

$$LC_4 \; = \; f(A)(y) \; = \; S_x(F(x, y) \cdot A(x))$$

and $\overline{LC_4}$ is expressed in sum-of-products form as:

$$\overline{LC_4} \; = \; \overline{y_2} \cdot \overline{y_3} + y_1 \cdot y_2 \cdot \overline{y_3}$$

7.7 Technology Mapping

7.7.1 Introduction

Technology mapping is the problem of implementing the Boolean network using the gates of a particular technology library. Typically, the goal is to make optimal use of all of the gates in the library to produce a circuit with critical-path delay less than a target value and minimum area.

In its broadest interpretation, this statement of the technology mapping problem appears to be a restatement of the general logic optimization problem. The difference is the assumption that the Boolean network has undergone significant technology-independent optimization. It is assumed that the equations provide a good structure for the final circuit. The role of technology mapping is to finish the synthesis of the circuit by performing the final gate selection from a particular library. The algorithms chosen for technology mapping are simplified because they are constrained by the structure of the equations produced by the technology-independent optimizations. It is not the role of technology mapping to change the structure of the circuit radically, for example, by finding common subexpressions between two or more parts of the circuit. Likewise, it is not the role of technology mapping to reduce the number of levels of logic along the critical path. The role of technology mapping is the actual gate choice to implement the equations – for example, choosing the fastest gates along the critical path and using the most area-efficient combination of gates off the critical path.

An algorithm for technology mapping should ideally satisfy several goals. The technology mapping algorithm should be able to adapt to a variety of different libraries. This is difficult because many libraries have an irregular collection of logic functions available as primitives. An algorithm which depends on characteristics of a particular library, for example, availability of a complete set of *complementary metal oxide semiconductor* (CMOS) AND-OR-INVERT gates or the existence of a dual of each function, is of limited use. Also, an algorithm which is geared to a subset of the gates in a library is limited in its optimization potential. To practically achieve this goal of adaptability an approach to technology mapping must be *user-programmable*. A user must be able to provide new gates to the technology mapper without understanding its detailed operation, and these gates should be used effectively.

During technology mapping, simple cost functions such as transistor count or levels of logic will not provide high-quality circuits. Instead, it is necessary to consider more detailed models for the cost of a gate in the actual target technology. This detailed level of modeling, coupled with gates which have irregular area and delay cost functions, greatly complicates the technology mapping process. Therefore, to provide high-quality results for different libraries and circuits, a technology mapping algorithm must make few assumptions about the relative cost and performance of the gates in a library, and must be prepared to model accurately user-specified cost functions that are to be optimized.

This section is organized as follows: In Section 7.7.2 we describe the technology libraries into which circuits are mapped. In Section 7.7.3 we briefly describe various cost models that are used in technology mapping. In Section 7.7.4 we give a formulation of technology mapping as a graph covering problem. In the next section, Section 7.8, we present in detail the locally optimum solution to graph covering using tree covering.

7.7.2 Technology Libraries

The algorithms described thus far operate on a Boolean network, where each node can be an arbitrary two-level function. Gates in VLSI circuits (especially for application-specific integrated circuits) are usually restricted to be from a *technology library* of *gates.* A *gate* is the primitive element which is available in a particular implementation technology and a *technology library* is a collection of these gates. A technology library is assumed to consist of a finite collection of gates. For example, the gates in a static CMOS gate-array (or standard-cell) design typically include inverters, NAND gates, NOR gates, and a variety of complex gates, whereas the gates in an ECL gate-array are typically NOR gates and XOR gates. We consider only gates with a single output; if a gate has multiple outputs, only a single output is used.

These libraries are typically composed of a few hundred gates and sequential elements like latches and flip-flops for which highly optimized layouts have been manually designed for a particular technology. The logic designers are then restricted to using these gates in their logic circuits. For the purposes of illustration the combinational subset of a very simple library is shown in Figure 7.6. The library cell names, associated area costs, their functions and their

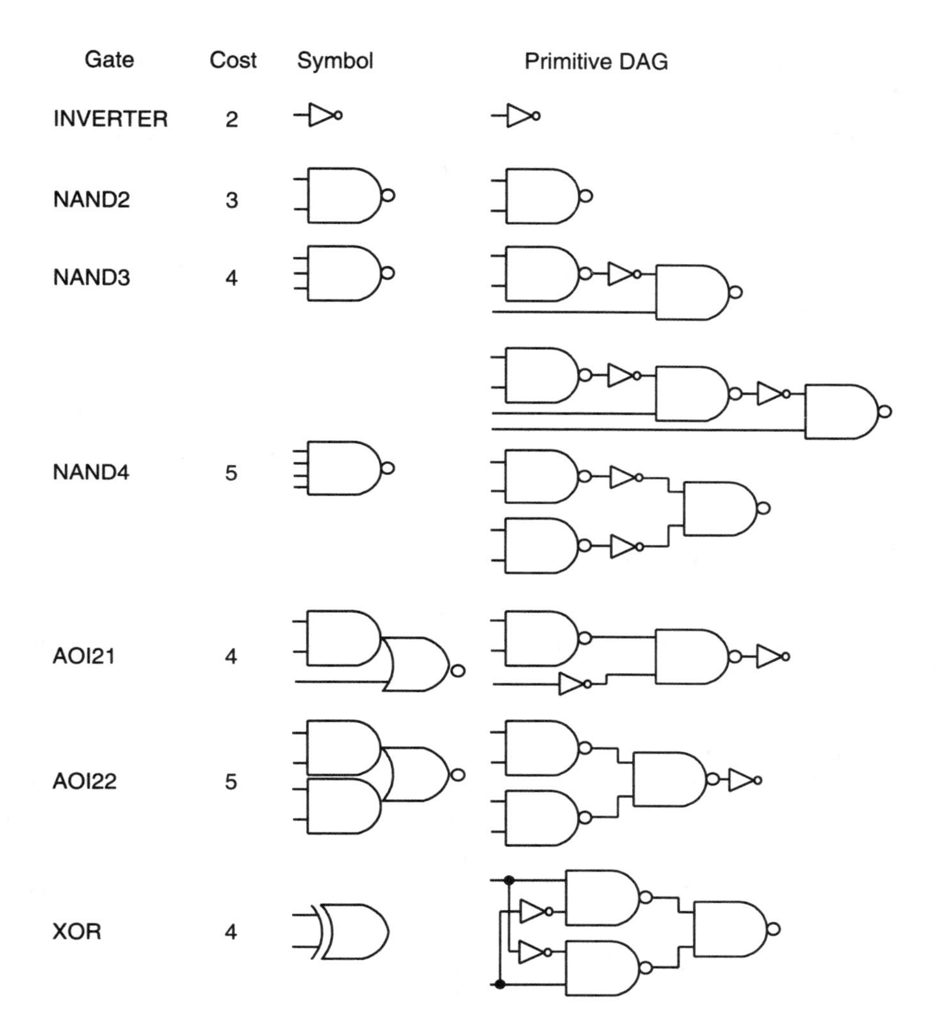

Figure 7.6: Gate library

representations in terms of two-input NAND gates and inverters are shown.

7.7.3 Cost Models

Each gate is assigned a number of values associated with the different cost functions under which it will be optimized. For example, each gate is assigned a value called the *area* of the gate. This represents the physical area occupied by the gate, and can be measured in, for example, *square microns* or *routing grids*. To capture the effect of physical design the area model can include an approximate cost for the wiring. For example, in standard-cell and compacted-array design the average routing area is correlated to the number of pins on a net and the number of cells in the design. Hence, for these implementation styles, the wiring area can be estimated by adding an area penalty for each net.

Models used in technology mapping for delay will be treated in Section 8.6.

Given a technology library, the problem of technology mapping is as follows: find an equivalent multilevel circuit to the given Boolean network that is comprised of gates in the library and which has minimum cost. The cost could be the area, delay, testability, or power consumption of the resulting circuit.

7.7.4 Graph Covering

A systematic approach to technology mapping is based on the notion of graph covering. With this formulation, the technology mapping problem can be viewed as the optimization problem of finding a minimum cost covering of the subject graph by choosing from the collection of pattern graphs for all gates in the library. A *cover* is a collection of pattern graphs such that every node of the subject graph is contained in one (or more) of the pattern graphs. One restriction of any cover is that only the leaves of one pattern may overlap the root of another pattern. Interpreted from a circuit standpoint this means that the inputs of one cell in the covering must be the outputs of some other cell in the covering. Overlap would imply that the inputs of one cell come from internal nodes in another cell. As these internal signal values are not visible outside the cell, this would not be meaningful.

In the graph covering method the Boolean network is represented in a normal form.[2] For instance, each node in the Boolean network can be replaced by two levels of NAND gates possibly with input inverters. The canonical form for the Boolean network is termed the *subject directed acyclic graph* (DAG). Each library gate is also represented in this canonical form. The canonical realizations, also called patterns, for each library gate in terms of two-input NAND gates and inverters are shown in Figure 7.6. Each realization is termed a *primitive* DAG. Note that a gate may have more than one associated

[2]This form has been called a canonical form in the literature. Strictly speaking there is nothing canonical about this representation.

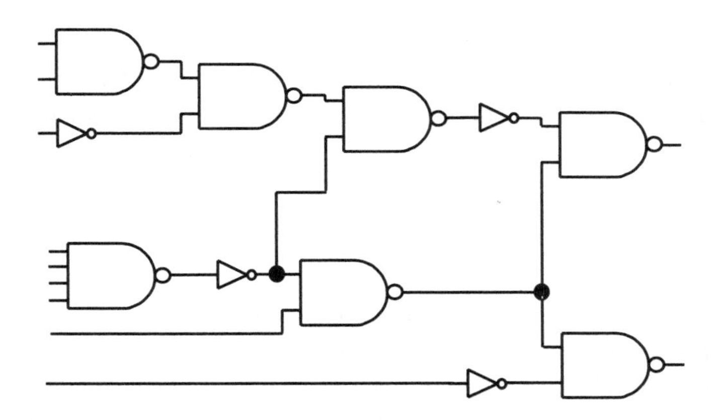

Figure 7.7: Example subject DAG

primitive DAG (e.g., the four-input NAND gate in Figure 7.6). The optimization problem can now be stated as: find a minimum cost covering of the subject DAG by the primitive DAGs.

An example subject DAG is shown in Figure 7.7.

7.7.5 Choice of Atomic Pattern Set

The choice of which atomic patterns to use for the subject and pattern graphs is an important consideration for graph covering algorithms. This decision influences the range of solutions for the covering problem and the number of patterns needed.

Recall that a Boolean network is a DAG where each node has an associated logic function, which can be represented, for example, as a two-level logic function. We define a Boolean network where each node is restricted to be a NAND function of any number of inputs (including one) a NAND-*network*. A NAND-network where each node has fan-in of two or less is called a NAND2-network.

Here we use a two-input NAND gate and an inverter as the atomic operations. Hence, the subject graph and pattern graphs will be represented as NAND2-networks. This choice is motivated by the following observations. Adding additional functions such as a two-input NOR gate, a two-input AND gate, or a two-input OR gate cannot provide higher-quality solutions. Likewise, adding NAND, NOR, AND, or OR gates with more than two inputs cannot provide higher-quality solutions. This observation is based on the fact that given a cover for a subject graph using a larger set of functions, it is possible to show an equivalent cover where each function is replaced by an equivalent set of two-input NAND gates and inverters.

Restricting ourselves to only a two-input NAND gate and inverter does come at the price of increasing the number of patterns needed to represent some logic functions. With this approach, the logic function

$$f = \overline{a \cdot b \cdot c \cdot d + e \cdot f \cdot g \cdot h + i \cdot j \cdot k \cdot l + m \cdot n \cdot o \cdot p}$$

requires only one pattern corresponding to a tree of five four-input NAND gates. Representing all patterns for this same function using two-input NAND gates and inverters requires 18 patterns. We saw an example of the four-input NAND having two different patterns in Figure 7.6.

Experience has shown that the increase in the number of patterns (and hence the increase in the memory and time required for technology mapping) is not significant.

There is the problem of converting the optimized Boolean network into a NAND2-network before covering. We will address this problem in more detail after we discuss the tree covering approach we use to solve graph covering.

7.8 Technology Mapping by Tree Covering

7.8.1 Tree Covering Approximation

One technique for solving the graph covering problem is to formulate it as a binate-covering problem [36] (a special case of an integer-linear program). Following the paradigm established in the domain of code generation [3] an alternative approach is to partition the subject graph into a forest of trees and solve the covering problem on each of the trees. A tree is a DAG where every node (including primary inputs) has a fan-out of 1. The tree necessarily has a single sink (primary output) called the *root* and the sources (primary inputs) of the tree are called the *leaves* of the tree. The motivation for looking at the problem of tree covering is the existence of an efficient algorithm for the optimal tree covering problem.

The application of the tree covering to technology mapping proceeds as follows. The first step is to convert the Boolean network into a NAND-network. This NAND gate network is partitioned into a forest of trees by cutting the graph at each multiple-fan-out stem. The resulting trees are optimally covered one tree at a time. This

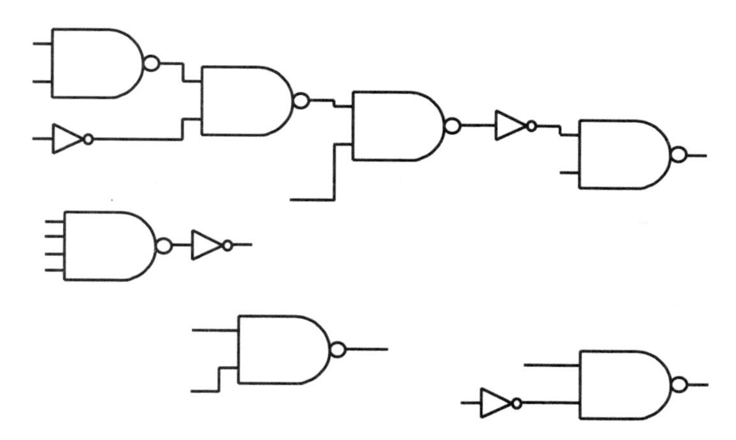

Figure 7.8: Decomposing a subject DAG into a forest of trees

starts by converting each tree into a NAND2-network. The conversion from the NAND-network to the NAND2-network is called the *technology decomposition* problem. Finding the optimum covering of a tree is done by generating the complete set of matches for each node in the tree (that is, the set of tree patterns which are candidates for covering a particular node) and then selecting the optimum match from among the candidates using a dynamic programming algorithm. We describe these steps in the following sections.

7.8.2 Partitioning the Subject Graph

To apply the tree covering approximation the subject graph must first be converted into a forest of trees. One approach is to break the graph at each multiple-fan-out point. Each node with fan-out greater than one becomes a root of a tree, and each fan-out of this node becomes a leaf of a tree. With this technique no nodes in the subject graph are duplicated. Other heuristics could be used to partition the subject graph; however, all other partitions increase the number of nodes in the forest of trees. This can be useful, however, to improve the quality of the final cover. We note that, in the limit, the optimum graph cover can be found by applying tree covering after choosing the correct nodes to duplicate.

The partitioning of the subject DAG of Figure 7.7 into a forest of trees is illustrated in Figure 7.8.

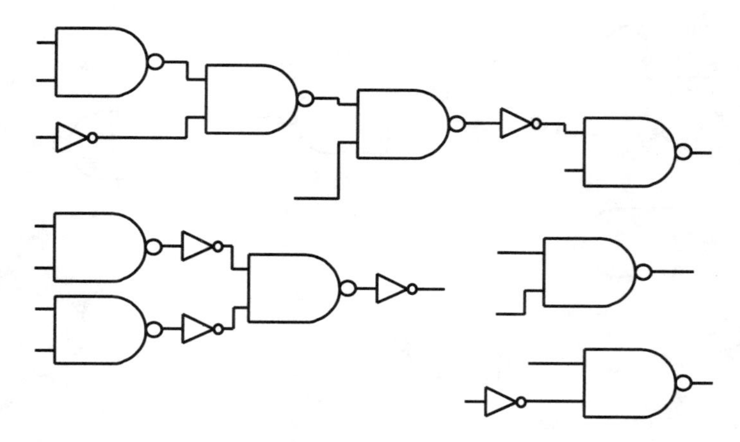

Figure 7.9: Converting the forest of trees into a normal form

7.8.3 Technology Decomposition

The NAND-tree is next decomposed into a NAND2-tree. This is called the technology decomposition step. Note that in general, there are many decompositions of the NAND-tree into different NAND2-trees, and each of these leads to a graph cover of a different cost. Even if the resulting tree covering problem is solved exactly, every one of these starting points would have to be considered for an optimum solution.

Most commonly, each node in the NAND-tree is replaced with a balanced decomposition of the n-input NAND gate into a tree of two-input NAND gates. This is done recursively as follows. A two-input NAND gate is placed at the root, the number of leaves is divided in half, and a tree is built recursively for that many leaves. The recursion terminates with a tree of one leaf. Note that this balanced-tree decomposition is not necessarily the optimal form of the subject graph, and depending on the cost function other decompositions may be considered.

The forest of trees of Figure 7.8 is converted into the normal form, i.e., a NAND2-network, in Figure 7.9.

7.8.4 Tree Matching Techniques

A straightforward solution to establishing the initial set of candidate matches for a tree is to attempt to match each pattern at each node in the tree. If there are p patterns in the pattern set and n nodes in the subject graph, then this approach has complexity $O(n \cdot p)$. There are a number of more sophisticated approaches to solving this problem. Because a tree can also be represented by a string (a sequence of symbols), one natural approach to take is to reduce the tree matching problem to the problem of string matching.

A number of fast string matching algorithms have been proposed, such as the Aho-Corasick algorithm [2], the Knuth-Morris-Pratt algorithm [29], and the Boyer-Moore algorithm [12]. Using the Aho-Corasick string matching algorithm, it is possible to find all of the strings which match a given string in time proportional to the length of the longest string in the pattern set. This algorithm provides a significant improvement over the naive algorithm for finding all matching strings.

7.8.5 Optimal Tree Covering

Having generated a set of candidate matches for each node in the subject graph, an optimal tree cover must then be selected from among the candidates. Dynamic programming can be used for this purpose. Dynamic programming is a general technique for algorithm design which can be applied when the solution to a problem can be built from the solutions of a number of subproblems.

Consider the problem of finding a minimum area cover for a subject tree T. A scalar cost is assigned to each tree pattern, and the cost for a cover is the sum of the costs for each pattern in the cover. The key observation is this: The minimum-area cover for a tree T can be derived from the minimum-area covers for every node below the root of T. This is the *principle of optimality* for tree covering and is used as follows to find an optimal cover for T. For every match at the root of the tree the cost of an optimal cover containing that match equals the sum of the cost of the corresponding gate and the sum of the costs of the optimal covers for the nodes which are inputs to the match.[3] Note that the optimal covers for each input to the match at the root can be computed once and stored; it is not necessary to recompute the optimal cover for each input of each match.

Note that each node in the tree is visited only once. Therefore, the complexity of this algorithm is proportional to the number of nodes in the subject tree times the maximum number of matches at any node in the subject tree. The maximum number of matches is a function of the library and is therefore a constant independent of the

[3]The reader may wish to review the rules for legal coverings given in Section 7.7.4.

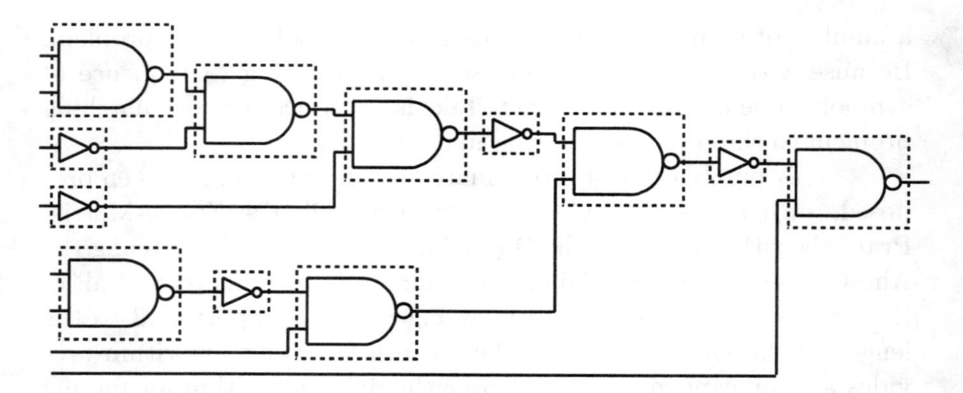

Figure 7.10: Trivial covering

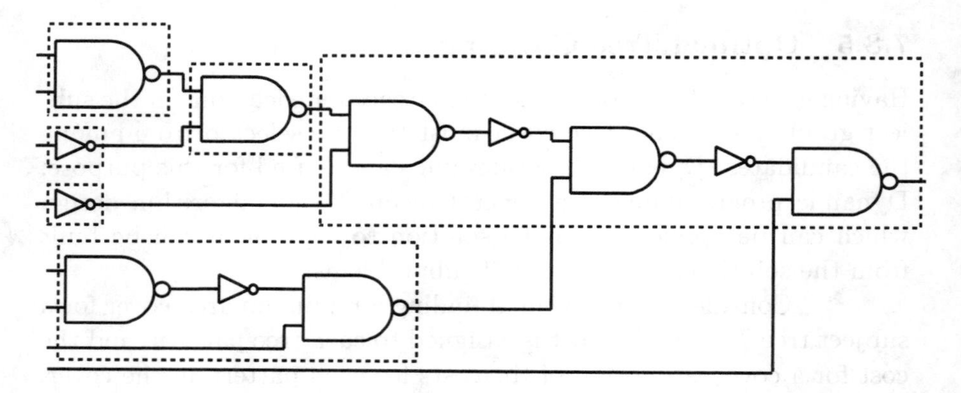

Figure 7.11: Better covering

subject tree size. As a result the covering algorithm has linear complexity in the size of the subject tree, and the memory requirements are also linear in the size of the subject tree.

We now demonstrate this approach to technology mapping with an example. Consider a Boolean network given by:

$$\begin{aligned} Z &= X + \overline{Y} + h \\ Y &= W \cdot \overline{d} \\ X &= e \cdot f \cdot g \\ W &= a \cdot b + c \end{aligned}$$

A normal form representation of the Boolean network is given in Figure 7.10. The trivial covering of the subject DAG by primitive DAGs from the library of Figure 7.6 is also illustrated in Figure 7.10. The cost of this trivial covering corresponds to the cost for seven two-input

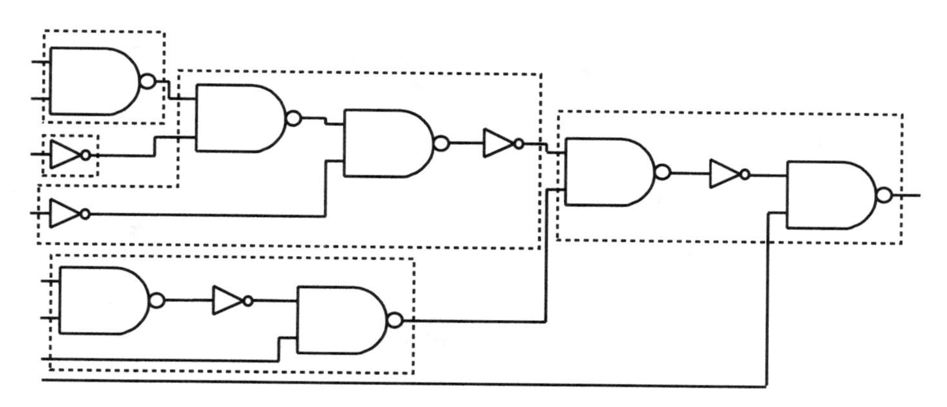

Figure 7.12: Optimum covering

NAND gates and five inverters, giving a cost of 31. A substantially better covering that exploits the larger gates in the library is shown in Figure 7.11. The cost of this covering is the cost of two inverters, two two-input NAND gates, one three-input NAND and one four-input NAND for a total cost of 19. A covering which utilizes an AOI gate with a lower cost of 17 is shown in Figure 7.12.

7.8.6 Inverter-Pair Heuristic

A simple way to improve the quality of circuits produced by the tree covering algorithm is the *inverter-pair heuristic.* Redundant inverters are added to each tree to improve the number of patterns which can match at each node. This leads to an examination of more possible covers for each tree, leading directly to an improvement in the optimization quality.

The technique works as follows. Each edge in the subject tree and each edge in a pattern which connects two NAND gates is replaced with a pair of inverters. An extra pattern consisting of a pair of inverters is added to the matching patterns. This extra pattern is given zero area cost and zero delay cost. The tree covering algorithm is then applied unmodified.

Because of the optimality of the tree covering algorithm adding these extra inverters cannot lead to a cover with a greater cost. Each pair of inverters can be covered by the inverter-pair pattern, which leads to the solution which existed before the inverters were added. However, the advantage is that the tree covering algorithm is able to make the optimal choice between covering the extra

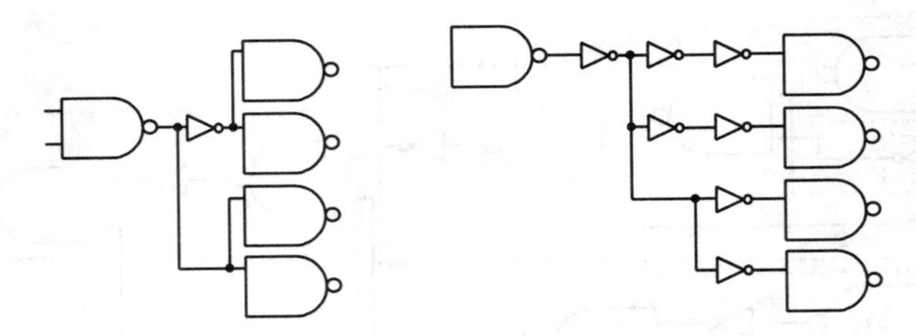

Figure 7.13: Adding inverter-pairs at a branch point

inverters with the inverter-pair pattern at no cost or splitting the inverters between two patterns if this leads to a cover with less cost. The only disadvantage is that the number of nodes in the subject tree and the pattern trees has increased. The increase in the number of nodes is bounded by a factor of three (two extra inverter nodes for each node in the subject tree); however, the actual increase is typically less because redundant inverters are added only at the output of a NAND gate and not at the output of each inverter in the subject tree.

Adding inverters within a tree is straightforward. If a NAND gate has a fan-out of one and feeds another NAND gate, then a pair of inverters is added after the NAND gate. However, a slightly more complicated approach is taken at the tree boundaries. At a tree boundary, extra inverters are added so that each tree is connected to the branch node through either one or two inverters. Figure 7.13 shows in general how inverters are added at a multiple-branch node.

The motivation for adding the extra inverters in this fashion is again to increase the number of possible matches. Every tree should have one or two inverters at each leaf to allow more patterns to match near the leaves of the tree. The root of each tree could be constructed with no inverter at the root, but leaving one inverter in the previous tree increases the number of matches at the root of the tree.

7.8.7 Extension to Nontree Patterns

Some gates in a technology library cannot be represented in tree form. Common examples are the two-input XOR gate shown at the bottom of Figure 7.6, a two-to-one multiplexor, and a three-input majority gate (logic function $f = a \cdot b + a \cdot c + b \cdot c$). However, a simple extension allows these patterns to be included.

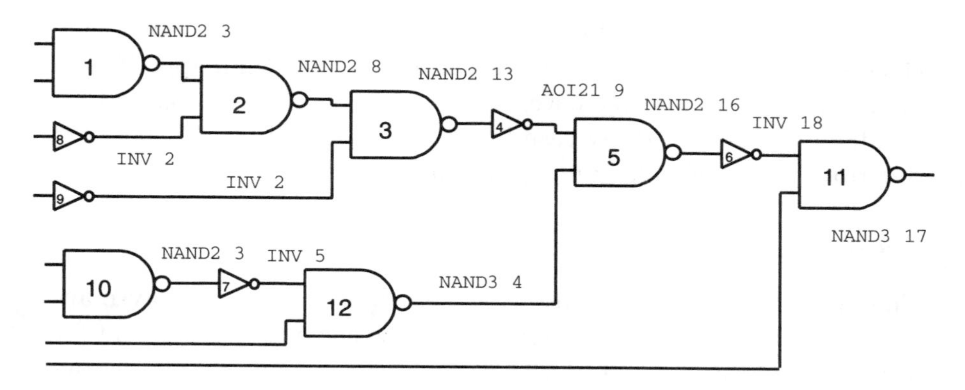

Figure 7.14: Determining the optimum covering

A leaf-DAG is a DAG where the only nodes with fan-out greater than one are the primary inputs. Patterns which are trees, and patterns which are leaf-DAGs can be used directly by the tree covering algorithm. This includes the XOR pattern shown in Figure 7.6. Note, however, that because of the multiple-fan-out of one of these matches, the XOR gate must match at the leaves of the tree.

We will now illustrate the optimum covering algorithm on the tree of Figure 7.14. We walk from the primary inputs to the primary output of the tree and determine the best match at each gate output. At each gate output the match selected for the subtree whose root is the gate output has been shown along with the total cost of the optimal cover for this subtree. For the first level gates, only two-input NAND gate and inverter matches are possible. At the output of gate **2** the only match is with a two-input NAND, and therefore the total cost is 8. At the output of gate **12** two matches are possible, with a two-input NAND or with a three-input NAND. The former will result in a cost of 8, so we pick the latter which has a cost of 4. At the output of gate **4** the best match corresponds to an AOI gate with a cost of 9. The final cost at the primary output is 17. The optimum covering corresponds to that of Figure 7.12.

7.8.8 Delay Optimization

Under the assumption that the delay of a gate is independent of the fan-out of the gate, the tree covering algorithm provides the minimum arrival time cover, if we compute and store the arrival time at each node and choose the minimum arrival time match at each node. For a more general delay model the principle of optimality does not apply, since the optimum match depends on the forward (unmapped) part of the tree. Heuristic tree covering methods have been proposed for handling the general delay case (e.g., [38]). We will provide detailed descriptions of technology mapping targeting delay in Chapter 8.

7.8.9 Conclusions

The graph covering formulation of the problem of technology mapping, and the application of tree covering to solve it, is important for a number of reasons. From a theoretical standpoint the graph covering formulation provides a formal description of the problem, and the tree covering approach gives both an optimality criteria for evaluating solutions to the problem as well as an algorithm for achieving optimality.

The tree covering approach has been widely applied to the mapping of practical circuits capable of achieving aggressive area and timing requirements. This approach has been relatively easy to implement, as well as to adapt and extend, and the computational efficiency of this approach has made it amenable to use as the inner loop in iterative procedures.

The success of the graph covering formulation has helped to formulate the logic synthesis and optimization problem as a union of technology-independent and technology-dependent portions. The major limitation of graph covering, namely its dependence on the form of the subject graph is naturally compensated by the application technology-independent optimization. On the other hand, graph covering based technology mapping is able to address a morass of technology specific issues, such as technology libraries and their area and timing characterization, that would significantly complicate higher level optimizations.

7.9 Field Programmable Gate Arrays

In the late 1970s, most circuits were designed in the full-custom design style, in which the designers had complete control over the size and placement of every transistor on the chip. The fabrication of a full-custom *integrated circuit* (IC) requires that every step of the fabrication process be performed for each IC. This made the time for fabrication of an IC long and also increased its cost. In order to cut down the fabrication time and cost, *mask programmable gate arrays*

(MPGAs or simply gate arrays) were developed in the early 1980s. In MPGAs the transistors on the chip are prefabricated but the interconnection between the transistors can be customized. Therefore, only the last steps of fabrication have to be performed to obtain each IC. This makes fabrication of gate arrays fast and less expensive than for custom design.

For large and complex circuits, designers have always felt the need to quickly fabricate their design in order to verify its functionality. MPGAs, though better in this regard than full-custom ICs, do not satisfy this need because they are still too expensive to be used for prototyping. This prompted the development of another technology in the late 1980s, namely *field programmable gate arrays* (FPGAs). FPGAs correspond to an implementation medium that has two advantages over MPGAs — the manufacturing time is reduced from months to minutes and the cost of fabrication is reduced from tens of thousands to hundreds of dollars. FPGAs contain regular arrays of programmable logic blocks connected by a general-purpose programmable interconnection fabric. Therefore, any kind of digital circuit can be implemented. FPGAs were introduced in [18] and newer versions have been presented in [1, 5, 6, 19, 26, 33, 10, 4, 41, 34, 35].

FPGAs have become very important for the rapid prototyping of digital circuits and also for the implementation of low-end *application-specific integrated circuits* (ASICs). Since the structure of FPGAs is different from that of MPGAs, different logic synthesis algorithms are required to synthesize a design for FPGAs. In addition, the logic synthesis algorithms depend strongly on the architecture of the particular FPGA. In this section, we describe various FPGA architectures that are in use. Section 7.10 contains the description of various synthesis algorithms for FPGAs.

7.9.1 FPGA Architectures

A FPGA is an IC consisting of user-programmable logic blocks and interconnection fabric that can be used in the design of digital circuits. Each block in the circuit is user-programmable, meaning that the functionality implemented by the block can be determined by the user after the block has been fabricated. The blocks may consist of combinational or sequential logic elements. The blocks can be connected together after fabrication in a general way using a programmable routing fabric, as illustrated in Figure 7.15. This routing fabric consists of programmable switches that connect various hori-

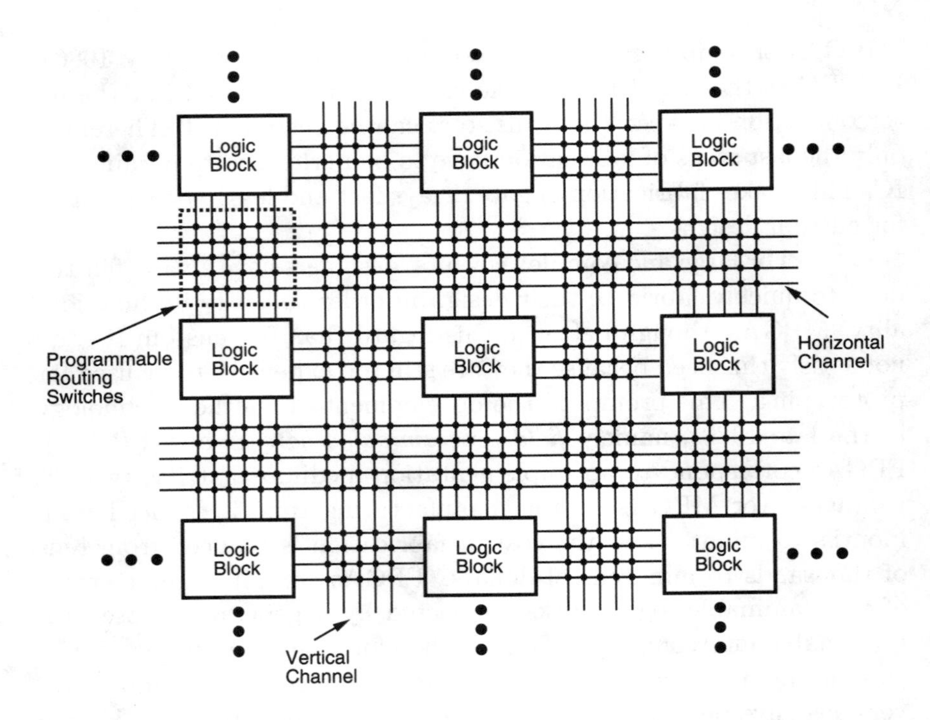

Figure 7.15: Basic FPGA architecture

zontal and vertical routing segments together with the pins from the logic blocks.

The user programmability of FPGAs makes it easy to fabricate a circuit in a matter of minutes (after logic synthesis and placement and routing). However, the programmable elements introduce two major drawbacks. The programmable switches require larger area than the metal wires used in MPGAs to make connections. Therefore, FPGAs will have a lower amount of logic per unit area than MPGAs. Current generation FPGAs are roughly a factor of 10 less dense than MPGAs in the same fabrication process technology. In addition to this, the programmable switches have higher series resistance and capacitance than metal wires, resulting in slower circuits. Current FPGAs are roughly a factor of 3 slower than MPGAs.

The effect of these drawbacks on circuit area and speed can be reduced by careful choice of the FPGA architecture. The architecture of choice is application dependent and there is no one architecture that is best suited for all applications. By choosing a logic block that minimizes the use of programmable elements, both logic density and speed can be improved. If a routing structure with few programmable elements is used, then density and speed can be improved at the expense of increased routing difficulty. In addition, the utility of an architecture is dependent on the existence of logic and layout synthesis tools to support it. For example, a logic block with high functionality will have low area efficiency if the logic synthesis tool cannot make good use of its functionality. Similarly, an intricate routing structure may be useless if there does not exist any automatic router that can achieve 100 percent routing completion.

7.9.2 FPGA Terminology

Each programmable element in an FPGA is programmed using a certain method called the *programming technology.* The properties of the programming technology dictate many of the tradeoffs in FPGA architecture. The static *random access memory* (RAM) programming technology uses static RAM bits to control pass transistors and multiplexors. For pass transistors, the static RAM output is connected to the gate input of the transistor. By setting the RAM bit to 1 or 0 the pass transistor can be switched on and off, respectively, thereby either enabling or disabling a connection. For multiplexors the static RAM bits feed the control inputs and therefore the data input that is selected by the multiplexor is user controllable. Since static RAM is volatile, the device must be loaded and configured at the time of power-up, which can be time consuming. Also, static RAM bits require larger area than other programming technologies. Despite these disadvantages, it remains the only high-speed reprogrammable technology.

Instead of static RAMs, *electrically programmable read-only memories* (EPROMs) or *electrically erasable and programmable read-only memories* (EEPROMs) can be used. However, special semiconductor processing steps are required to manufacture an FPGA with EPROM or EEPROM programmable elements. Other programming technologies use *antifuses.* Unlike a fuse, an antifuse normally presents a very high resistance between the two points it connects. When a high voltage is applied across it (the fuse is blown), a low resistance link between the two points is created. Attractive features of antifuses are their small size, low series resistance, and small parasitic capacitance. However, this is offset by the size of the transistors and wires required to route the high voltage and current to the antifuses for programming. Moreover, a chip that has been programmed once cannot be reprogrammed.

The programming technology is used to program both logic blocks and routing switches. A *logic block* is the basic unit of combinational and sequential logic in the FPGA. It may or may not contain programmable elements, but usually more than one function can be realized using a logic block. A *routing switch* is a programmable element that can optionally make an electrical connection between two wires. A *routing selector* is a programmable element that can selectively make a connection between one of n wires and another wire. A wire is broken up into several *segments* connected together by routing switches. The routing architecture of an FPGA consists of the placement and distribution of fixed wiring segments and the placement of connection points of the routing switches and selectors (that connect from the logic block to the routing fabric and between wiring segments). The architecture of the logic block together with the routing architecture determines the architecture of the FPGA. In the sequel, we discuss various logic blocks and routing architectures.

7.9.3 FPGA Logic Block Architectures

The architecture of logic block has significant impact on the density and performance of an FPGA. The primary factor in these considerations is the amount of functionality or the *granularity* of the block. Fine-grain blocks typically consist of one or two logic gates with a small number of inputs, typically between two and three. The active combinational logic, i.e., the hardware that computes the value of the function, has no more than 15 transistors and no more than two to three inputs per gate. Medium-grain logic blocks implement logic functions with four to 20 inputs to the combinational logic and use in the range of 15 to 100 transistors to implement the logic. Large-grain blocks have in excess of 20 inputs to the combinational logic and use more than 100 transistors to implement the logic.

One example of a fine-grain logic block is a two-input NAND gate introduced in [34]. A circuit is implemented in the usual way by connecting NAND gates to achieve the desired function. There are other examples of fine-grain logic blocks that use two input multiplexors [5] and two-input NAND and XOR gates [19]. The principal advantage of fine-grain architectures is that each logic block is highly utilized. However, they require a significant amount of general-purpose routing to connect all the blocks. Since general-purpose routing is expensive in terms of circuit delay as well as area, the logic density and speed of such FPGAs are poor.

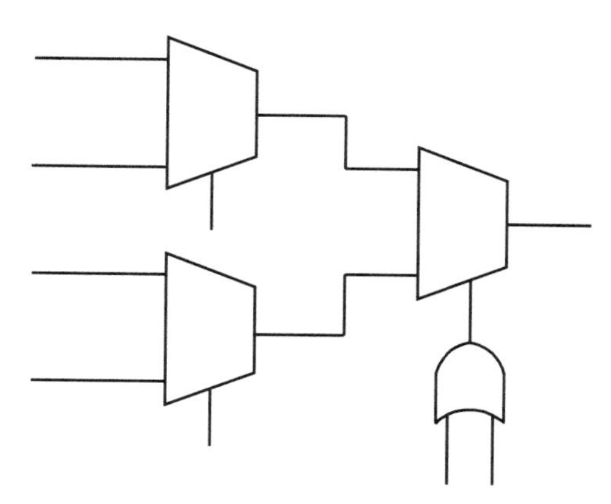

Figure 7.16: Actel Act-1 logic block

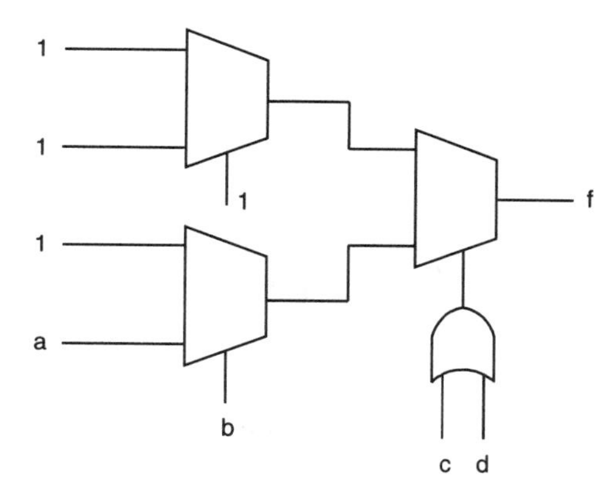

Figure 7.17: Implementation of $f = a + b + c + d$ using Act-1

Medium-grain logic blocks are the most popular ones in use. One example of such a block is the Actel *Act-1* logic block [25] shown in Figure 7.16. It consists of three multiplexors and one logic gate, having a total of eight inputs and one output. This logic block can implement 702 distinct logic functions. For example, the function $f = a + b + c + d$ is implemented in Figure 7.17. The *Act-2* logic block from Actel [4] is similar except that the two multiplexors at the input are connected to a two-input AND gate. These logic blocks provide a large amount of functionality for a relatively small number of transistors. Note that there are no programmable elements in the logic block.

Another nonprogrammable logic block with a larger number of inputs and increased functionality is from QuickLogic [10]. It is

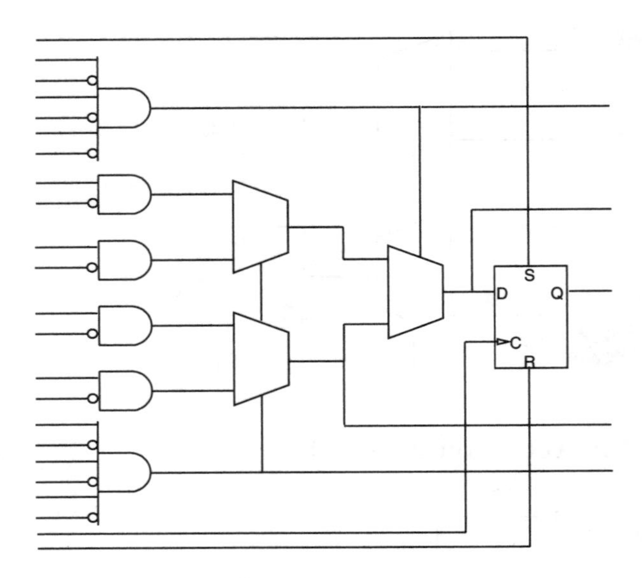

Figure 7.18: The QuickLogic logic block

similar to the Actel blocks because it has a four-to-one multiplexor as its basic element. In addition, this block has every input of the multiplexor connected to the output of an AND gate as shown in Figure 7.18. The AND gates together with the alternating inputs that are inverted provide increased functionality.

The logic blocks used in FPGAs from Xilinx contain programmable elements and use ROM-based *lookup tables* (LUTs) to implement functions. An LUT with k address lines and one output line can be used to implement any function of k variables by storing the value of the function for each input combination in the LUT. The Xilinx 3000 series logic block [27] contains a five-input single-output LUT, as illustrated in Figure 7.19. In addition, it contains multiplexors, some of which have programmable control inputs (these multiplexors are shown shaded in the figure). Each logic block also has set-reset D-type flip-flops. The LUT can be configured into two LUTs with four inputs each, as long as together they do not have more than five inputs. This reconfigurability provides flexibility that translates into better logic block utilization because many common logic functions do not require as much as five inputs. The programmable multiplexors allow the selection of particular functions generated by the LUTs. Note that the output of the combinational elements can be directly sent to the output or the output of the block can be taken

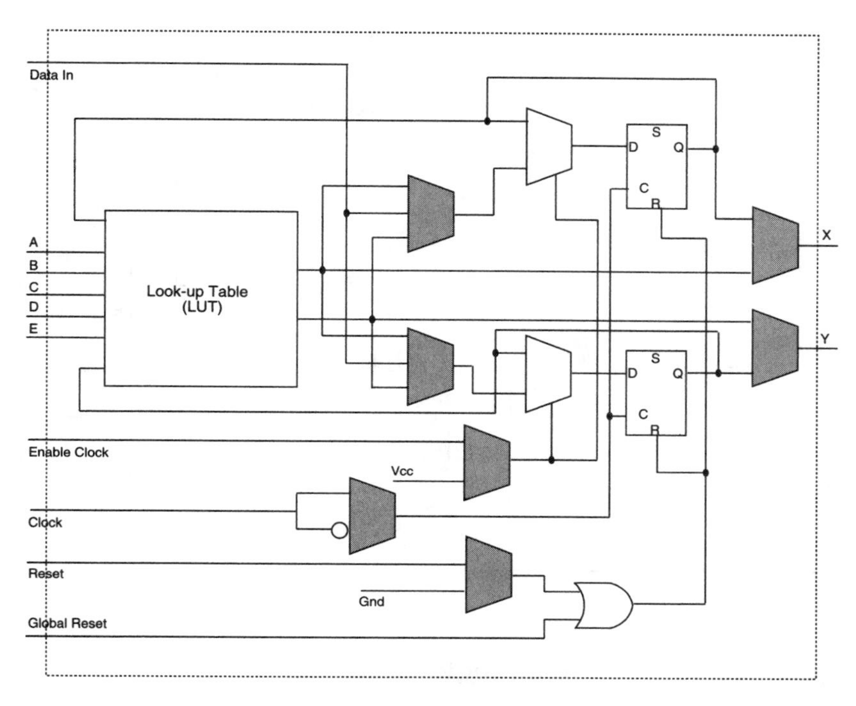

Figure 7.19: Xilinx 3000 logic block

from the D-type flip-flops.

The Xilinx 4000 series logic block [26] contains two four-input LUTs feeding into a three-input LUT as illustrated in Figure 7.20. LUTs with a large number of inputs are like high-fan-in gates while LUTs with a low number of inputs are like low-fan-in gates. Higher fan-in gates require fewer levels of logic but are usually less area efficient. A mixture of high-fan-in and low-fan-in gates provides the ability to optimize for speed in some portions of the circuit and logic density for other portions of the circuit. For this reason, the four- and three-input LUTs in the 4000 series block allow for a better tradeoff between speed and logic density. In addition, there is a non-programmable link between the four-input LUTs and the three-input LUT. Since this connection is significantly faster than any general-purpose interconnection link, proper use of this connection along the critical path may significantly speed up the circuit. However, because of the inflexibility of this link the three-input LUT may be unused in many blocks.

Commercial large-grain logic blocks have evolved from two-level *programmable logic devices* (PLDs). They typically have wide

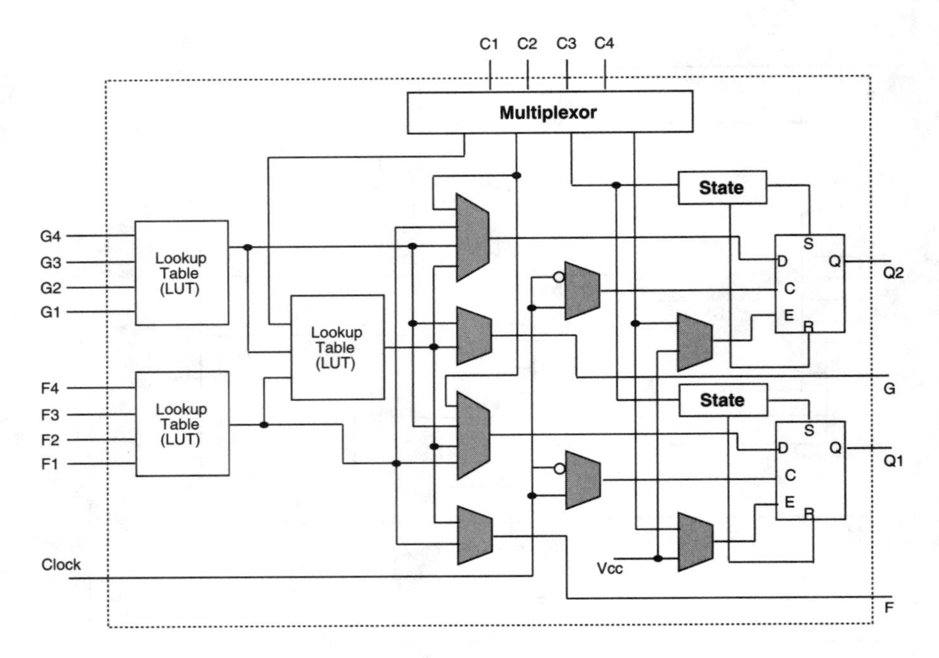

Figure 7.20: Xilinx 4000 logic block

fan-in AND gates (with 20 or more inputs) feeding into an OR gate with three to eight inputs. The output of the OR gate can be inverted by feeding it through an XOR gate, as in the Altera 5000 logic block. The advantage of this kind of block is that wide input AND gates can be used to make logic functions with very few levels of logic blocks, and hence the slow programmable interconnect can be avoided. However, it is difficult to make efficient use of all of the large gates, and so these devices will suffer in terms of logic density.

7.9.4 FPGA Routing Architecture

The routing architecture of an FPGA determines the manner in which the programming switches are positioned to allow the programmable interconnection of the logic blocks. The most important part of an FPGA is its routing architecture because the size, resistance, and capacitance of the programming technology result in routing structures that require large area and exhibit large delays.

Routing architectures for FPGAs consist of one or more logic blocks adjacent to one or more routing channels, as illustrated in Figure 7.15. There are a large number of alternatives to be selected from in determining the routing architecture, e.g., length of dedicated segments, symmetry, hierarchy, and the manner in which switches are placed between wires and between segments of the same wire.

As shown in Figure 7.15, there are horizontal and vertical routing channels in an FPGA. For the Xilinx 3000 architecture the number of tracks in the horizontal and vertical channels are more or less the same. In addition, there are connections from the logic block to both the horizontal and vertical channels. Such a routing architecture is said to be *symmetric.* In contrast, in the Actel *Act-1* architecture there are far fewer tracks in the vertical direction than there are in the horizontal direction. Also, the logic block pins connect only to the horizontal channel. This kind of architecture is called *asymmetric with horizontal bias.*

The general routing structure shown in Figure 7.15 illustrates a one-level hierarchy because all the routing and logic blocks are tiled identically across the chip. Several FPGAs, e.g., the Altera 5000 series, use two or more levels of hierarchy in the interconnection structure. The entire chip is divided into *logic array blocks* (LABs). Interconnection within a LAB is achieved using a local bus while LABs are connected together using a global bus. The local connections within one level of hierarchy can be made faster than the longer global connections. In this case, a performance advantage can be obtained if an entire critical path can be kept within one LAB. In some sense, the internal connectivity of the Xilinx 3000 and 4000 logic blocks represent hierarchical connectivity.

The number of tracks per channel is an important architectural factor as it determines whether 100 percent routing completion can be achieved or not. The Xilinx 3000 architecture has five tracks per channel. There are approximately seven pins entering each channel from the blocks adjacent to the channel. If a circuit uses a good portion of these pins, there are few tracks left for connections that pass by this block that neither originate nor terminate there. The Actel *Act-1* series uses 22 tracks per horizontal channel which gives a very good chance of routing completion.

There are several ways to connect logic blocks pins to the routing channels. There is a tradeoff between the flexibility and the cost of the switches in terms of area and speed penalty. For the Xilinx 3000 series, each pin of a logic block connects to some of the tracks in the adjacent channel. Such a connection scheme is called *partially populated.* For the Xilinx 4000 series, each pin can connect to all the tracks in the adjacent channel and this pin-to-channel connection is called *fully populated.* In partially populated connections, the choice of which tracks connect to which pins can affect routability.

The second issue in pin-to-channel connection is the number of channels each pin can access. In the Actel *Act-1* architecture, the logic block output pins can access a few channels above and below the adjacent channels. The input pins can connect only to the adjacent channels. In the Xilinx architectures, only the adjacent channels are accessible. The ability to connect to more than one channel increases the ease of routing at the expense of larger area.

A third issue in pin-to-channel connection is the use of routing selectors or multiplexors to connect from several tracks into the logic block inputs, rather than using individual switches. This has the advantage of using fewer programming technology elements to configure the multiplexor. This is especially advantageous when the programming technology is large, as in the case of static RAMs. Xilinx architectures that use RAM-based programming therefore use routing selectors. For smaller programming technologies, like the antifuse, a direct switch is more efficient.

Channel-to-channel connection is the next important issue. The point where horizontal wires are broken with switches and where they can connect to vertical tracks is called a *switch block*. There is a large degree of freedom in the design of this structure. At one extreme, every horizontal wire could be connected to every crossing vertical wire and every opposing horizontal wire. At the other extreme, there could be relatively few programmable switches and most connections would be made in a predetermined configuration. The Actel architecture provides complete flexibility when switching from horizontal to vertical tracks as every vertical track can connect to every horizontal track. In the horizontal direction, the tracks can connect only to the directly opposing tracks. In the Xilinx 3000 architecture the possible horizontal-horizontal, horizontal-vertical, and vertical-vertical connections are a small subset of the full connectivity case. Each wire entering the switch matrix from one direction can typically connect to one or two wires on each of all three opposite sides.

An important architectural issue is the choice of the distance each wiring track travels without being broken by a programmable switch. Each contiguous wire is referred to as a *segment*. The length of each segment is called the *segmentation length*. Longer segments have fewer switches and are thus faster and more dense. Shorter segments, on the other hand, provide more flexibility. Segmentation lengths are measured in terms of the number of logic blocks spanned.

The Actel *Act-1* and *Act-2* routing architectures provide a broad distribution of segment lengths. This makes it likely that a segment of the appropriate length can be found for each connection. Furthermore, each track contains segments of varying lengths. The Xilinx 4000 series architecture provides four segment lengths.

One problem with the high resistance and capacitance interconnect used in FPGAs is that the variable delays along different paths can cause large timing skews. For clock nets, this can severely affect the system performance. Therefore, almost all FPGAs provide dedicated global clock nets that connect only to the clock signals of the flip-flops.

Thus far we have described various FPGA architectures and have explored their advantages and disadvantages. It is clear that FPGAs are very different from MPGAs, and different FPGA architectures differ significantly. The techniques used for logic synthesis for full-custom or gate array-based designs cannot fully exploit the characteristics of an FPGA. In the following section, we explore the use of logic synthesis tools for designing circuits with FPGAs.

7.10 FPGA Synthesis Methods

The manual mapping of an optimized technology-independent design into an FPGA is difficult and time consuming due to the complexity of FPGA architectures. The reduction in turnaround time due to the user-programmability of a FPGA may be offset by the time spent in manual mapping. For this reason, algorithms have been developed for the automated mapping of designs to various FPGA architectures. These algorithms assume that a technology-independent optimization has been carried out on the design (using the methods described previously in this chapter).

A straightforward approach to synthesis for FPGAs is to adapt the synthesis methods developed for standard cell or mask-programmable gate array libraries (described in Section 7.7) to FPGAs. A design is first mapped into simple gates (such as two-input NAND gates), and subcircuits of simple gates are replaced by logic blocks of the target FPGA. This approach is called the *library mapping* approach. An alternative approach, termed the *direct mapping* approach is more ambitious – the design is mapped directly into logic blocks.

In this section, we describe both classes of approaches to

FPGA synthesis. In Section 7.10.1, we describe mapping approaches targeting FPGAs with lookup table (LUT) logic blocks. A similar description for FPGAs with multiplexor-based logic blocks follows in Section 7.10.2.

7.10.1 Lookup-Table-Based Architectures

Lookup-table-based logic blocks can implement any logic function of no more than a fixed number of variables. Additional functions can also be implemented depending on the details of the block. For example, in the Xilinx series 3000 architecture of Figure 7.19, any logic function F of up to five inputs can be implemented, or any two logic functions F and G with up to four inputs each and five total inputs can be implemented.

Existing approaches for synthesis for lookup-table-based FPGAs begin with a network that has been optimized using a technology-independent method. We describe first the most straightforward adaptation of technology mapping approaches to lookup-table-based FPGAs.

In a straightforward adaptation of the tree covering method to technology mapping, the technology-independent network is transformed into a network of two-input NAND gates and inverters, and a library of subcircuits also implemented as two-input NAND gates and inverters is created. A lookup table with k inputs can implement 2^{2^k} different Boolean functions. However, some of these functions can be obtained by permuting the inputs of other functions. While the unique functions are fewer than all possible functions, their number still grows super-exponentially. The number of unique functions for $k = 2$ is 10, for $k = 3$ is 78, and for $k = 4$ is 9014 [23]. In addition, some of these functions have a large number of possible two-input NAND gate representations resulting in a huge library. One may heuristically select a a relatively small subset of these functions and include them in the library, but restricting the library in such a manner results in a poor quality of mapping, especially for $k \geq 4$.

The alternative approach is the direct mapping approach in which the functionality of the logic block is taken into account directly, while mapping and a library of gates are not explicitly constructed. The tree covering algorithm can be modified to significantly reduce the time required by standard technology mapping algorithms [23].

In the direct approach, the network is first decomposed into a forest of trees as illustrated previously in Figure 7.8. The only difference is that the network does not have to be represented as two-input NAND gates and inverters, since we do not have an explicit library. An optimal mapping of each tree into the lookup table is performed using dynamic programming. The main difference in lookup table mapping is in the way the optimal mapping is done. It is not the structure of the logic function that matters in the matching, rather what matters is the the number of variables that the function depends on. Given a tree, every subtree that has at most k leaf nodes can be implemented by a single lookup table.

Consider the problem of finding a minimum area cover for a subject tree T. To find the minimum cost implementation of a tree rooted at node i we choose different subtrees T_{ij} rooted at i. For any given T_{ik} rooted at i, we combine the cost of T_{ik} with the minimum cost implementation of the subtrees rooted at the leaf nodes of T_{ik}. We choose the T_{ik} which results in the minimum overall cost. The technology mapping problem for a tree can thus be solved recursively starting at its leaf nodes and working towards its root.

In standard technology mapping, we choose different subtrees at node i that match the patterns in the library. In the case of lookup table mapping, all subtrees rooted at a node that have a number of leaf nodes less than or equal to k must be considered to make sure that all applicable solutions are searched. All these subtrees have a cost of 1.

Consider the circuit of Figure 7.21. Let us assume that we can implement any single-output function with up to five inputs using the lookup table. In Figure 7.21 a covering requiring four lookup tables is shown. However, the circuit can be mapped into three lookup tables as shown in Figure 7.22. In Figure 7.23 the dynamic programming approach to finding the optimal solution is illustrated. The subtrees rooted by the NAND gates 1 through 4 all have cost 1. At the NAND gate 5 we have several choices. We can merge gates 1 and 5 into a single lookup table, use a second table for gate 2, and a third table for gate 3. However, the best solution with cost 2 is to merge gates 2, 3, and 5 into a single table and use another table for gate 1. At the output of the final NAND gate we find a covering with cost 3 corresponding to the covering shown in Figure 7.22.

If a node in the tree has a number of inputs significantly larger than k, the number of subtrees to examine is very large. For example, if we have a node with 10 inputs and $k = 5$, we have to look at the subtrees corresponding to all possible subsets of cardinality

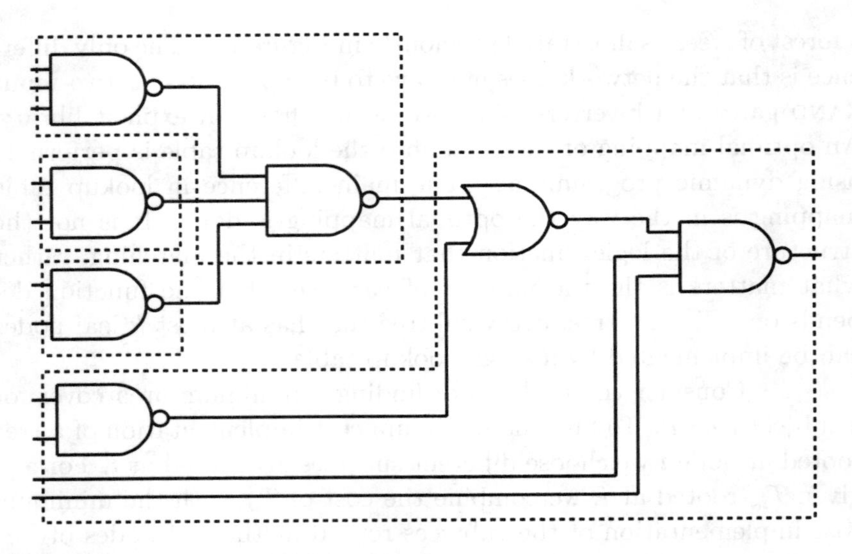

Figure 7.21: A possible covering

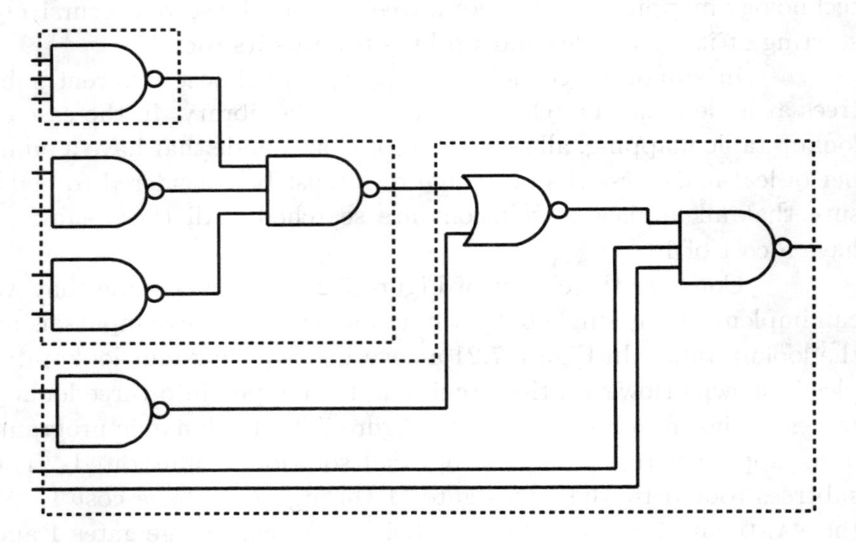

Figure 7.22: Better covering

5 among the 10 inputs. To avoid this explosion, the nodes of the network can be predecomposed so that each has a number of inputs less than l, where $l > k$. This can be done by splitting a large node into two nodes with nearly the same number of inputs. However, by doing so the optimality of the solution is no longer guaranteed. Improvements to the above approach that exploit node duplication

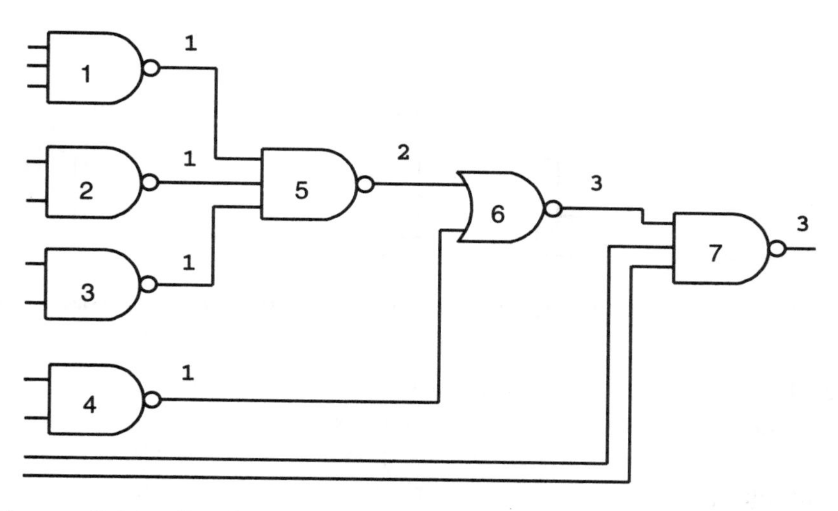

Figure 7.23: Optimum covering

to achieve better results are presented in [24].

7.10.2 Multiplexor-Based Architectures

Multiplexor-based architectures use logic blocks that combine a number of multiplexors and possibly additional AND or OR gates.

These logic blocks can implement a fairly large number of logic functions. For example, for the *Act-1* module of Figure 7.16, all two-input functions, most three-input functions, and several functions with four or more inputs can be implemented. However, some of these functions can be obtained by permuting the inputs of other functions. The *Act-1* module can implement 702 unique functions and the *Act-2* module can implement 766 unique functions.

As in the lookup-table-based architectures, there are two approaches to synthesis for multiplexor-based architectures, namely the *library mapping* approach and the *direct mapping* approach. We describe both these approaches in this section.

In the library mapping approach, a library is created which has gates representing all the functions obtained from the multiplexor-based logic block. For the Actel logic blocks, these libraries will contain approximately 700 gates. This is significantly smaller than the number of gates in lookup table architectures. Technology mapping algorithms based on dynamic programming are quite effective for libraries with 100 or 200 gates, but are considered too slow for

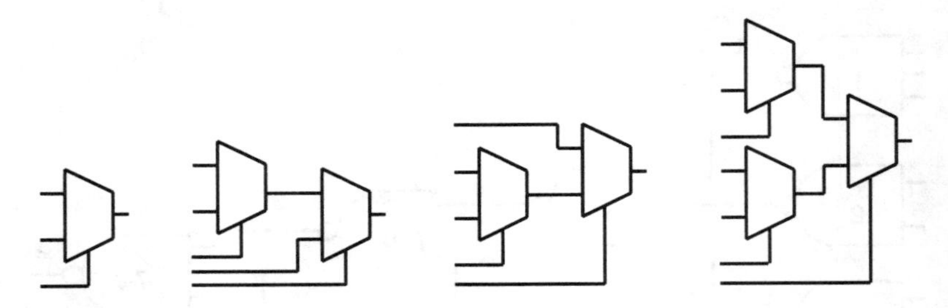

Figure 7.24: Pattern graphs for simplified Act-1 architecture

significantly larger libraries. Heuristic methods are therefore applied to reduce the size of the library. For example, the least frequently used gates are removed. Unlike lookup table architectures, the library mapping approach has proven fairly successful in multiplexor-based architectures.

Binary Decision Diagrams (BDDs) have been used as the basis for most existing direct approaches (e.g., [32, 31]). As described in Section 6.4.10, BDDs can be viewed as multiplexor-based networks.

The dynamic programming approach to technology mapping can be extended to pattern graphs and subject graphs described in terms of two-input multiplexors instead of two-input NAND gates. This is the approach taken in [32]. Since the structure of the Actel logic block corresponds to two-input multiplexors, only a few pattern graphs are needed to characterize all the nonequivalent gates corresponding to the logic block. For example, only four patterns suffice to describe the simplified structure of the *Act-1* logic block where the OR gate feeding the output multiplexor has been removed. These patterns are shown in Figure 7.24. Each multiplexor has two data inputs, and implements the function $i \cdot x + \overline{i} \cdot y$ where i is the control input, x is the top data input, and y is the bottom data input.

For the original block, the heterogenous structure yields a larger set of pattern graphs, but the number of pattern graphs required is substantially less than in the library mapping approach. The approach taken in [32] consists of three steps:

1. A BDD is built for the subject graph.

2. The BDD is converted into a forest of trees and dynamic programming is used to optimally cover each tree using the pattern graphs.

3. An iterative improvement phase consisting of transformations

such as partial collapsing and redecomposition is used to improve the result obtained.

In Step 1, a BDD is built for the subject graph. It is possible to build a BDD for the entire network in terms of its primary inputs, but the size of such a BDD may be very large. Furthermore, the multiplexor-based network corresponding to the BDD can be very inefficient as a starting point for implementation as an FPGA because the area-efficient structure obtained using technology-independent optimization has been lost. Therefore, BDDs are only built for the functions at each node in the technology-independent Boolean network. A decomposition step is performed to force all nodes to have at most k inputs. (k varies between 3 and 6 in [31, 32]). Given that k is relatively small, all possible orderings for ROBDDs representing the node function are tried and the smallest ROBDD is selected.

In Step 2, given a BDD for each node function, the BDD is decomposed into a forest of trees. Each of the subject trees is mapped using dynamic programming. A set of eight patterns is used in the case of the *Act-1* module, four of which were shown in Figure 7.24. The remaining patterns are shown in Figure 7.25. When one of the data inputs of a multiplexor is connected to a logical high value, the multiplexor effectively becomes an OR of the control input (or its complement) and the other data input. (If $f = i \cdot x + \overline{i} \cdot y$ and $x = 1$, then $f = i + \overline{i} \cdot y$, which simplifies to $f = i + y$.) Thus, the OR gate in the *Act-1* module can be modeled within the two-input multiplexor framework.

In Step 3, an iterative improvement phase is used to improve the final result. Partial collapsing of the network, decomposition and covering are iteratively applied till no improvements are found. The best result encountered during the phase is stored.

Problems

1. Given a Boolean network with N inputs and M intermediate nodes, the satisfiability don't-care (SDC) set is defined as

$$\sum_{i=1}^{M} \overline{y}_i F_i + y_i \overline{F}_i$$

If there are no external don't-cares in the space B^{N+M}, how many care points and how many don't-care points are there in SDC? Prove your result.

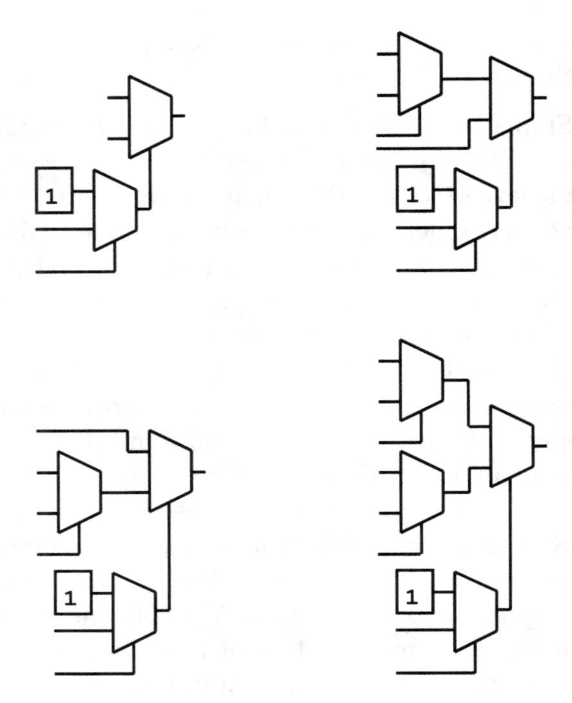

Figure 7.25: Additional pattern graphs for Act-1

2. Prove that the tree covering algorithm for technology mapping is optimal under a delay cost function where the delay of each library gate is a fixed quantity, irrespective of the fan-out of the gate. Give an example of a circuit where a suboptimal solution is attained by using the tree covering algorithm in the case where the delay of a library gate is proportional to its fan-out, i.e., the number of gates it drives in the mapped circuit.

3. Give a rigorous proof that the kernel generation algorithm of Figure 7.1 does not generate a cokernel more than once.

4. Modify the procedure of Figure 7.1, such that it generates only the level 0 kernels, i.e., kernels that do not contain any other kernels. Your modification should improve the efficiency of the procedure.

5. Generate all the kernels of

$$f = a \cdot c \cdot e + a \cdot c \cdot g + b \cdot c \cdot e + b \cdot c \cdot g + a \cdot d \cdot e + a \cdot d \cdot g + b \cdot d \cdot e + b \cdot d \cdot g$$

by using the kernel generation algorithm of Figure 7.1 and alternately by constructing the cube-literal matrix and generating all the prime rectangles using the procedure of Figure 7.3.

6. Use a lexicographic ordering heuristic in choosing literals as divisors to factor f in Problem 4. Choose kernels as divisors to factor f, and compare the literal counts of the two factored functions.

7. Give a $O(|f| + |g|) \log(|f| + |g|)$ algorithm for computing the quotient f/g.
Hint: Assign different bit positions for each literal and each complemented literal, and encode each of the cubes in f and g.

8. Perform Boolean division and compute $f//g$ given the logic functions f and g below.

$$f = a \cdot b + \overline{a} \cdot \overline{c} \cdot d + a \cdot c \cdot d + \overline{a} \cdot \overline{b} \cdot c \cdot \overline{d}$$

$$g = c \cdot \overline{d} + \overline{c} \cdot d$$

9. Given

$$f = a \cdot \overline{b} \cdot d \cdot \overline{e} + c \cdot d \cdot \overline{e} + a \cdot \overline{b} \cdot f + c \cdot f + \overline{a} \cdot \overline{c} \cdot e + \overline{a} \cdot \overline{c} \cdot g + b \cdot \overline{c} \cdot e + b \cdot \overline{c} \cdot g$$

$$g = a \cdot \overline{b} + c$$

Algebraically factor f using g and $\overline{g}$ if possible.

10. Prove that c is a cokernel of a Boolean function f if and only if it is the cube corresponding to a prime rectangle of the cube-literal matrix of f with at least two rows.

11. The formulation of algebraic decomposition as a rectangle covering problem where the rectangles are allowed to overlap allows for valid decompositions that are nonalgebraic. Give examples of cube and kernel extraction where nonalgebraic decompositions are produced by rectangle covering.

12. Find an optimum covering using the library of Figure 7.6 for the circuit of Figure 7.10 with an additional inverter at the output.

13. Map the circuit of Figure 5.14 into two-input NAND gates and inverters, and determine an optimum covering using the library of Figure 7.6.

14. Augment the library of Figure 7.6 with a cascade of two inverters that can be covered by a wire with cost 0. Replace each inverter-free connection between primary inputs and NAND gates and between NAND gates in Figure 7.10 with a pair of inverters. Find an optimum covering for the modified circuit under the augmented library. Is the cost of the covering smaller than the cost of the optimum covering for the original circuit under the original library?

    (a) If the answer is no, then give an example of a circuit for which the above modification of the circuit and augmentation of the library gives a better covering?

    (b) Can you prove that the cost of the optimum covering for an arbitrary circuit modified in the above fashion under an arbitrary library augmented in the above fashion is no greater than the optimum covering for the original circuit under the original library?

15. Verify that the number of functions that can be implemented with the Actel *Act-1* logic block is 702.

16. How many different functions can be implemented with the Xilinx 3000 logic block?

17. Optimally map the circuit of Figure 5.14 into a lookup table architecture where all functions of up to four inputs can be

implemented by a single lookup table. Perform the mapping for minimum area and minimum delay.

REFERENCES

[1] Advanced Micro Devices. *Mach Devices High Density EE Programmable Logic Data Book*, 1990.

[2] A. Aho and M. Corasick. Efficient String Matching: An Aid to Bibliographic Search. *Communications of the ACM*, 18(6):333–340, June 1975.

[3] A. Aho and S. Johnson. Optimal Code Generation for Expression Trees. *Journal of the ACM*, 23(2):488–501, July 1976.

[4] M. Ahrens, A. El Gamal, D. Galbraith, J. Greene, S. Kaptanoglu, K. R. Dharmarajan, L. Hutchings, S. Ku, P. McGibney, J. McGowan, A. Samie, K. Shaw, N. Stiawalt, J. Whitney, T. Wong, W. Wong, and B. Wu. An FPGA Family Optimized for High Densities and Reduced Routing Delay. In *Proceedings of the Custom Integrated Circuits Conference*, pages 31.5.1–31.5.4, May 1990.

[5] Algotronix Ltd, Edinburgh, Scotland. *CAL 1024 Datasheet*, 1989.

[6] S. Baker. Lattice Field FPGA. *Electronic Engineering Times*, June 1991.

[7] K. Bartlett, R. K. Brayton, G. D. Hachtel, R. M. Jacoby, C. R. Morrison, R. L. Rudell, A. Sangiovanni-Vincentelli, and A. R. Wang. Multilevel Logic Minimization Using Implicit Don't-Cares. *IEEE Transactions on Computer-Aided Design of Integrated Circuits*, 7(6):723–740, June 1988.

[8] K. Bartlett, W. Cohen, A. J. De Geus, and G. D. Hachtel. Synthesis of Multilevel Logic under Timing Constraints. *IEEE Transactions on Computer-Aided Design of Integrated Circuits*, CAD-5(4):582–595, October 1986.

[9] L. Berman and L. Trevillyan. Improved Logic Optimization Using Global Flow Analysis. In *International Conference on Computer-Aided Design*, pages 102–105, November 1988.

[10] J. Birkner. A Very High Speed Field Programmable Gate Array Using Metal-to-Metal Antifuse Programmable Elements, 1991. New product introduction at Custom Integrated Circuits Conference.

[11] D. Bostick, G. D. Hachtel, R. Jacoby, M. R. Lightner, P. Moceyunas, C. R. Morrison, and D. Ravenscroft. The Boulder Optimal Logic Design System. In *Proceedings of the International Conference on Computer-Aided Design*, pages 62–65, November 1987.

[12] R. Boyer and J. Moore. A Fast String Searching Algorithm. *Communications of the ACM*, 20(10):762–772, October 1977.

[13] R. Brayton, R. Rudell, A. Sangiovanni-Vincentelli, and A. Wang. MIS: A Multiple-Level Logic Optimization System. *IEEE Transactions on Computer-Aided Design of Integrated Circuits*, CAD-6(6):1062–1081, November 1987.

[14] R. K. Brayton, G. D. Hachtel, and A. L. Sangiovanni-Vincentelli. Multilevel Logic Synthesis. *Proceedings of the IEEE*, 78(2):264–300, February 1990.

[15] R. K. Brayton and C. McMullen. The Decomposition and Factorization of Boolean Expressions. In *Proceedings of the International Symposium on Circuits and Systems*, pages 49–54, Rome, May 1982.

[16] R. K. Brayton and C. McMullen. Synthesis and Optimization of Multistage Logic. In *Proceedings of the International Conference on Computer Design*, pages 23–28, October 1984.

[17] J. R. Burch, E. M. Clarke, K. L. McMillan, and D. Dill. Sequential Circuit Verification Using Symbolic Model Checking. In *Proceedings of the* 27th *Design Automation Conference*, pages 46–51, June 1990.

[18] W. Carter, K. Duong, R. H. Freeman, H-C. Hsieh, J. Y. Ja, J. E. Mahoney, L. T. Ngo, and S. L. Sze. A User Programmable Reconfigurable Gate Array. In *Proceedings of the Custom Integrated Circuits Conference*, pages 233–235, May 1986.

[19] Concurrent Logic, Sunnyvale, California. *Concurrent Logic CFA6006 Field Programmable Gate Array Data Sheet*, 1991.

[20] J. Darringer, D. Brand, J. Gerbi, W. Joyner, and L. Trevillyan. LSS: A System for Production Logic Synthesis. IBM *Journal of Research and Development*, 28(5):537–545, September 1984.

[21] J. Darringer, W. Joyner, L. Berman, and L. Trevillyan. Logic Synthesis through Local Transformations. IBM *Journal of Research and Development*, 25(4):272–280, July 1981.

[22] S. Devadas, A. R. Wang, A. R. Newton, and A. Sangiovanni-Vincentelli. Boolean Decomposition in Multilevel Logic Optimization. *IEEE Journal of Solid State Circuits*, 24(2):399–408, April 1989.

[23] R. J. Francis, J. Rose, and K. Chung. A Technology Mapping Program for Lookup-Table-Based Field Programmable Gate Arrays. In *Proceedings of the* 27th *Design Automation Conference*, pages 613–619, June 1990.

[24] R. J. Francis, J. Rose, and Z. Vranesic. Chortle-crf: Fast Technology Mapping for Lookup-Table-Based FPGAs. In *Proceedings of the* 28th *Design Automation Conference*, pages 227–233, June 1991.

[25] A. El Gamal, J. Greene, J. Reyneri, E. Rogoyski, K. A. El-Ayat, and A. Mohsen. An Architecture for Electrically Configurable Gate Arrays. *IEEE Journal of Solid State Circuits*, 24(2):394–398, April 1989.

[26] H-C. Hsieh, W. S. Carter, J. Ja, E. Cheung, S. Schreifels, C. Erickson, P. Freidin, L. Tinkey, and R. Kanazawa. Third Generation Architecture Boosts Speed and Density of Field Programmable Gate Arrays. In *Proceedings of the Custom Integrated Circuits Conference*, pages 31.2.1 – 31.2.7, May 1990.

[27] H-C. Hsieh, K. Dong, J. Y. Ja, R. Kanazawa, L. T. Ngo, L. G. Tinkey, W. S. Carter, and R. H. Freeman. A 9000-Gate User-Programmable Gate Array. In *Proceedings of the Custom Integrated Circuits Conference*, pages 15.3.1–15.3.7, May 1988.

[28] K. Keutzer. DAGON: Technology Mapping and Local Optimization. In *Proceedings of the* 24th *Design Automation Conference*, pages 341–347, June 1987.

[29] D. Knuth, J. Morris, and V. Pratt. Fast Pattern Matching in Strings. *Siam Journal of Computing*, 6(2):323–350, 1977.

[30] P. C. McGeer and R. K. Brayton. Efficient, Stable Algebraic Operations on Logic Expressions. In *Proceedings of The International Conference on Very Large Scale Integration*, August 1987.

[31] R. Murgai, R. K. Brayton, and A. Sangiovanni-Vincentelli. An Improved Synthesis Algorithm for Multiplexor-Based PGAs. In *Proceedings of the* 29th *Design Automation Conference*, pages 380–387, June 1992.

[32] R. Murgai, Y. Nishizaki, N. Shenoy, R. K. Brayton, and A. Sangiovanni-Vincentelli. Logic Synthesis for Programmable Gate Arrays. In *Proceedings of the* 27th *Design Automation Conference*, pages 620–625, June 1990.

[33] H. Muroga, H. Murata, Y. Saeki, T. Hibi, Y. Ohashi, T. Noguchi, and T. Nishimura. A Large Scale FPGA with 10K Core Cells with CMOS 0.8μm 3-Layered Metal Process. In *Proceedings of the Custom Integrated Circuits Conference*, pages 6.4.1–6.4.4, May 1991.

[34] Plessey Semiconductor, Swindon, England. *ERA60100 Preliminary Data Sheet*, 1989.

[35] Plus Logic, San Jose, California. *FPGA2040 Field Programmable Gate Array Data Sheet*, 1989.

[36] R. Rudell. *Logic Synthesis for VLSI Design.* PhD thesis, University of California at Berkeley, April, 1989. ERL Memo 89/49.

[37] H. Savoj, R. K. Brayton, and H. J. Touati. Use of Image Computation Techniques in Extracting Local Don't-Cares and Network Optimization. In *International Workshop on Logic Synthesis*, May 1991.

[38] H. Touati, C. W. Moon, R. K. Brayton, and A. Wang. Performance-Oriented Technology Mapping. In *Sixth MIT Conference on Advanced Research on VLSI*, pages 79–97, April 1990.

[39] H. Touati, H. Savoj, B. Lin, R. Brayton, and A. Sangiovanni Vincentelli. Implicit State Enumeration of Finite State Machines Using BDDs. In *Proceedings of the International Conference on Computer-Aided Design*, pages 130–133, November 1990.

[40] L. Trevillyan, W. Joyner, and L. Berman. Global Flow Analysis in Automatic Logic Design. *IEEE Transactions on Computers*, C-35(1):77–81, January 1986.

[41] S. C. Wong, H. C. So, J. H. Ou, and J. Costello. A 5000-Gate CMOS EPLD with Multiple Logic and Interconnect Arrays. In *Proceedings of the Custom Integrated Circuits Conference*, pages 5.8.1–5.8.4, May 1989.

Chapter 8

Delay of Multilevel Circuits

8.1 Component and Circuit Delay

After correct logical functioning the speed of an integrated circuit is the single most important design characteristic. For this reason timing optimization is the single most important aspect of logic optimization. Any optimization system is only as good as the estimators that guide it, and as a result good timing optimization is entirely dependent on accurate timing estimation. For these reasons we will spend a good deal of attention on techniques for accurate timing estimation. In our discussion of timing estimation we will focus entirely on synchronous sequential circuits.

Accurate timing estimation relies on *component delay calculation* and *circuit delay calculation.* We have found no standard use of terminology in this area, so we will define these terms for our purposes.

Component delay calculation is the method used for actually calculating the delay of individual components within a circuit, and the method uses precalculated timing data in this calculation. Data used in component delay calculation consist of information such as the *inertial delays* and *transport delays* of components and the *wiring delays* due to capacitances of wires. These data are typically gathered from extensive transistor-level and/or device-level simulation of the circuit components. *Wiring delays* can be estimated through simulation or can be back-annotated from the final circuit layout.

If we view a circuit as a graph, then the method used for

delay calculation at the vertices of the graph is component delay calculation and *circuit delay calculation* is the model used for calculating delay over the entire graph. While accurate delay calculation depends on each of these parts, each of them can be presented independently.

Calculation of delay data is very dependent on the circuit implementation method, and we will not be discussing it further. We first present a simple component delay calculation method and then spend several sections on the topic of circuit delay calculation, which is the most challenging and relevant problem in timing estimation for a developer of a logic optimization system.

8.1.1 Component Delay Calculation

The component delay model used in the MIS logic optimization system [1] is a simple linear model. The delay across a pin i of a gate is given by the equation

$$delay(i) = transport(i) \; + \; resistance(i) \times load \tag{8.1}$$

The parameter *load* is the total capacitive load that is driven by this gate. This includes gates that are in the fan-out of the output of this gate plus any interconnect capacitance. This collected capacitance is modeled here as though it were simply lumped at the output of the gate. The delay *transport(i)* is the transport delay for the gate at pin i. This models the time required for transistors associated with this pin to switch on or off. It does not consider more refined details such as the effect of slow rising transitions on the transistors associated with this pin. The parameter $resistance(i)$ is the inverse of the drive capability of the gate. The drive capability of a gate is correlated to the ratios of the width and length of the transistors in the gate. For any given transition at a pin a charge or a discharge path causes the gate output to rise or fall. The width over length ratios of the transistors on this path can be used to compute $resistance(i)$.

Component delay calculations are performed extensively in timing analysis and logic optimization, and as a result tradeoffs have evolved between the accuracy of estimate and the time allowed for calculation. Equation 8.1 is a very simple approximation. More sophisticated models are often required for accuracy. A more detailed discussion of component delay calculation can be found in Chapter 3 of [10].

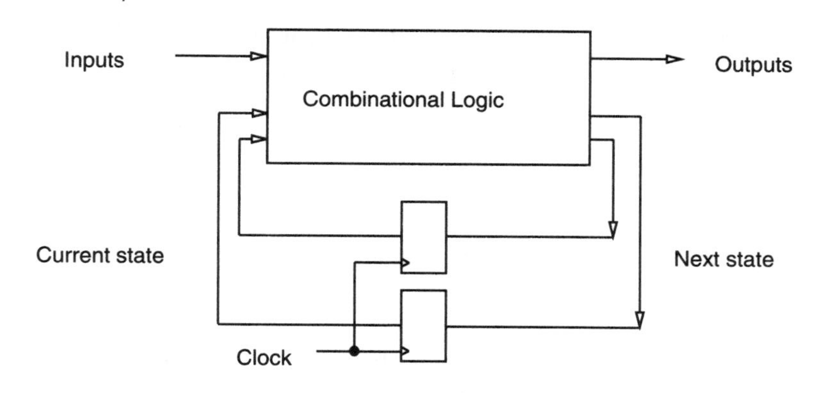

Figure 8.1: Clocked model for a sequential circuit

8.1.2 Circuit Delay Calculation

We now proceed to explain how to use component delay calculation to compute the delay of an entire synchronous circuit. A simple implementation model of a clocked or synchronous sequential circuit is shown in Figure 8.1. The approach is to use a clocked memory device, such as a register composed of edge-triggered flip-flops. At each active clock edge the next state is loaded into the flip-flops and becomes the current state.

Memory elements have a *propagation delay* associated with the interval between a clock edge and valid outputs. In order to guarantee that an input is not sampled when invalid a period of validity extending slightly before and after the active edge is specified. Specification of a *setup time* t_s and *hold time* t_h dictates that the memory element inputs must be valid and stable during a period that begins t_s before the active clock edge and ends t_h after the edge.

Given a sufficiently long clock period and appropriate constraints on the timing of transitions on the inputs, the inputs to the flip-flops can be guaranteed to be stable at each active clock edge, ensuring correct operation. Correct operation depends on the assumptions that:

1. The clock period is longer than the sum of the maximum propagation delay through the combinational logic, the setup time of the memory, and the maximum propagation delay through the memory.

2. The circuit's input lines are stable and valid for a sufficient period surrounding each active clock edge to accommodate both

the maximum propagation delay through the combinational logic and the setup time of the memory.

3. The minimum propagation delay around the cycle exceeds the hold time requirement of the flip-flop.

The most important constraint above is Condition 1. It is clear that the length of the clock period of a sequential circuit is directly related to the maximum propagation delay through the combinational logic of the circuit.

Given that the delay calculation of the sequential circuit primarily depends on the delay of the combinational subcircuit, in this chapter we will focus on the problem of correctly computing the maximum propagation delay of a multilevel combinational circuit. We will then show how to optimize a circuit so as to minimize the delay through the circuit.

For some time the most common approach to estimating or verifying the delay of a circuit was *timing simulation*. In timing simulation component delay calculations are driven by input stimuli and the result is the delay of a circuit on a set of input patterns. There are three problems with using timing simulation to calculate the delay of the circuit. The first is the effort required to build a set of input patterns. In a large circuit this effort can be considerable. The second problem is ensuring that the set of input patterns comprehensively exercises the circuit enough to accurately determine the delay. The third problem is the computational expense involved in simulation. These three problems are practically insurmountable, and as a result use of timing simulation for verifying the timing of synchronous circuits is diminishing. Instead, *timing verification* is being used for validating the timing of circuits, and we will focus exclusively on using timing verification for estimating and validating the timing of a synchronous circuit.

In Section 8.2 we will precisely characterize the delay of a multilevel logic circuit. We will see in Section 8.2 that the delay of a multilevel circuit depends on various assumptions relating to the mode of operation of the circuit and the delay model chosen. Furthermore, the correct circuit delay cannot be calculated by simply treating the circuit as a *directed acyclic graph* (DAG); circuit delay depends on the Boolean functionality of every node in the circuit. In Section 8.3 we will present methods to calculate the true delay of a circuit.

We will describe how the logic transformation methods described in Chapter 7 can be applied to minimize circuit delay. The application of technology-independent logic transformational methods to minimize the delay of a circuit is the subject of Sections 8.4 and 8.5. In Section 8.6, we treat technology mapping methods that target a delay cost function.

8.2 Timing Analysis and Verification

Timing optimization systems such as those described in Sections 8.4 and 8.5 assume a fixed delay model for gate and wire delays. Most timing analyzers fall into the *topological timing analysis* category, where the topologically longest path in the circuit is assumed to dictate the critical delay of the circuit.

In Section 8.2.1 we describe a topological timing analyzer that determines the longest path in the circuit without regard to the Boolean functionality of the circuit. We introduce false paths in Section 8.2.2 by using a well-known example of a carry bypass adder. We describe the different modes of operation of a circuit and different delay models in Sections 8.2.3 through 8.2.5. Delay-independent sensitization conditions are the subject of Sections 8.2.6 and 8.2.7. Finally, in Section 8.2.8 we describe necessary and sufficient conditions for a path to be responsible for the delay of a circuit.

8.2.1 Topological Timing Analysis

Circuit speed is measured by most optimization systems using a fixed delay model (see Section 5.5), where each gate and wire in the network has a given and fixed delay. As described in Section 8.1.1 the delay information for each gate is dependent on many parameters, e.g., the layout of the gate, the fabrication technology, and the transistor-level structure. Typically, a worst-case design methodology is followed, where the given delay for the gate is an upper bound on the actual delay of the fabricated gate.

The *arrival time* of a signal s, denoted A_s, is the time at which the signal settles to its steady state value. The *required time* of a signal s, denoted R_s, is the time at which the signal is required to be stable. The *slack time* of a signal is the difference between its required time and arrival time.

$$S_s = R_s - A_s$$

The *topologically longest path* in the circuit is a path where each signal has the minimum slack.

It is clear that the slack value of a signal measures its criticality. Static timing analyzers assume that the *critical delay* of the circuit is the delay of the topologically longest path. Under this assumption the longest path is also called the *critical path*. However, as we will see in Section 8.2.2, this can be a pessimistic assumption.

Starting with the primary input arrival times the arrival time for each of the signals is computed, and using the required times at the outputs we compute the required times for all the signals. The slack at each node is also calculated. This computation of arrival times, required times, and slack at each node is termed a *delay trace* through the network.

An example of a delay trace is shown in Figure 8.2. In the figure the arrival time, required time, and slack of each signal has been shown as a 3-tuple. We are given the arrival times for the four primary inputs and the required time for the output. The delay of each node is indicated within the node. Using the arrival times of the primary inputs we compute the arrival times for each node in the network. The arrival time at the signal corresponding to the node output is the maximum of the arrival times of the signals, corresponding to the inputs to the node plus the delay of the node. In Figure 8.2 the arrival time of signal e is the maximum of the arrival times of primary inputs a and b $(= 1)$ plus the delay of the node $(= 1)$, equaling 2. Similarly the arrival times of the other signals can be calculated.

The required time at the output can be used to compute the required time at the other signals in the network. The required times of the signals at the inputs to the node are obtained by subtracting the delay of the node from the required time at the output of the node. Given a required time of 8 at output h the required times for signals f and g can be computed as 8 minus the delay of the output node $(= 2)$, equaling 6. However, given the required time of 6 at f the required times at signals e and g are calculated to be 4. The required time for signal g is the minimum of the computed required times, namely 4. This is intuitive because if g does not stabilize by time 4 f will not stabilize by time 6 and the output h will not stabilize by time 8. Similarly, the required times at the other signals can be calculated.

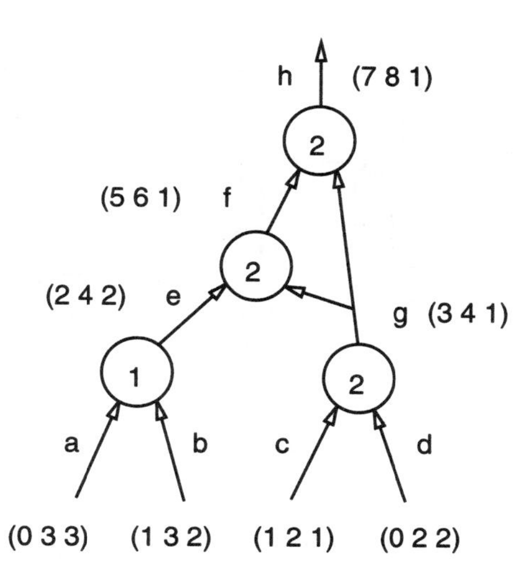

Figure 8.2: An example of a delay trace

8.2.2 False Paths in an Adder

The problem with topological analysis of a circuit is that not all paths in a circuit need be responsible for the delay. Paths in a circuit can be *false*; false paths are not responsible for the delay of a circuit. The *critical delay* of a circuit is defined as the delay of the longest *true* path in the circuit. Thus, if the topologically longest path in a circuit is false, then the critical delay of the circuit will be less than the delay of the longest path. In this section we will show that the critical delay of a combinational logic circuit is dependent on not only the topological interconnection of gates and wires, but also the Boolean functionality of each node in the circuit. The earliest reference to the existence of false paths is made by Hrapcenko [11] who demonstrated the existence of false paths on a parametric circuit that he constructed.

Assume the fixed delay model, and consider the carry bypass circuit of Figure 8.3.

The circuit of Figure 8.3 uses a conventional ripple-carry adder (the output of gate `11` is the ripple-carry output) with an extra AND gate (gate `10`) and an additional multiplexor. If the propagate signals `p0` and `p1` (the outputs of gates `1` and `3`, respectively) are high, then the carry-out of the block `c2` is equal to the carry-in of the block `c0`. Otherwise it is equal to the output of the ripple-carry adder. The multiplexor thus allows the carry to skip the ripple-carry

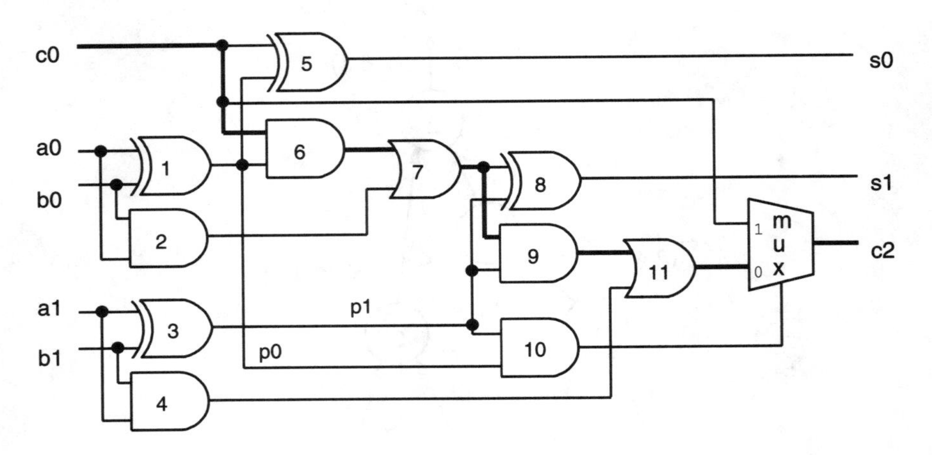

Figure 8.3: 2-bit carry-bypass adder

chain when all the propagate bits are high. A carry-bypass adder of arbitrary size can be constructed by cascading a set of individual carry-bypass adder blocks, such as those of Figure 8.3.

Assume the primary input `c0` arrives at time $t = 5$ and all the other primary inputs arrive at time $t = 0$. Let us assign a gate delay of 1 for AND and OR gates and gate delays of 2 for the XOR gates and the multiplexor. The longest path including the late arriving input in the circuit is the path shown in bold, call it P, from `c0` to `c2` through gates `6`, `7`, `9`, `11`, and the multiplexor (the delay of this path is 11). A transition can never propagate down this path to the output because in order for that to happen the propagate signals have to be high, in which case the transition propagates along the bypass path from `c0` through the multiplexor to the output. This path is false since it cannot be responsible for the delay of the circuit.

For this circuit the path that determines the worst-case delay of `c2` is the path from `a0` to `c2` through gates `1`, `6`, `7`, `9`, `11`, and the multiplexor. The output of this critical path is available after 8 gate delays. The critical delay of the circuit is 8 and is less than the longest path delay of 11.

8.2.3 Delay Models and Modes of Operation

Recall our definition of a false path — a path is false if it is not responsible for the delay of a circuit. This is a general definition that has to be augmented with the assumptions regarding the delay model and the mode of operation of the circuit.

Consider the operation of a circuit over the period of application of two consecutive input vectors v_1 and v_2. In the *transition* mode of operation the circuit nodes are assumed to be ideal capacitors and retain their value set by v_1 until v_2 forces the voltage to change. Thus, the timing response for v_2 is also a function of v_1 (and possibly other previously applied vectors). In the *floating* mode (this term was originally coined in [3]) of operation the nodes are not assumed to be ideal capacitors, and hence their state is unknown until it is set by v_2. Thus, the timing behavior for v_2 is independent of v_1.

As mentioned earlier, the most common delay model for a circuit component is one in which the delay is assumed to be a fixed number d. This is referred to as the fixed delay model. This number is typically either an upper bound or lower bound on the delay of the component in the fabricated circuit. The monotone speedup delay model takes into account the fact that the delay of each component can vary. It specifies the delays as a pair of numbers, $[0, d]$, where d is the upper bound on the actual delay and 0 is the lower bound.

For the purpose of developing the ideas in this section it is sufficient that we restrict the delays in a circuit to gates. This is general enough to accommodate other delay quantities such as wire delays and pin-to-pin delays by introducing buffers with appropriate delays in the circuit.

There is thus a spectrum of possibilities in proposing sensitization criteria — we can choose either the floating or transition mode of operation, and either of the two delay models. We will consider both transition mode and floating mode operation in the sequel.

8.2.4 Transition Mode and Monotone Speedup

In our analysis of the carry-bypass adder we assumed fixed delays for the different gates in the circuit and applied a vector pair to the primary inputs. It was clear that an event could not propagate down the longest path in the circuit. A precise characterization is that the path cannot be sensitized under the transition mode of operation and under (the given and) fixed gate delays. The path is therefore false under transition mode and (the given) fixed gate delays. Varying the gate delays in Figure 8.3 does not change the sensitizability of the path shown in bold.

Consider the circuit of Figure 8.4(a), taken from [12]. The delays of each of the gates are given inside the gates. In order to determine the critical delay of the circuit we will have to simulate

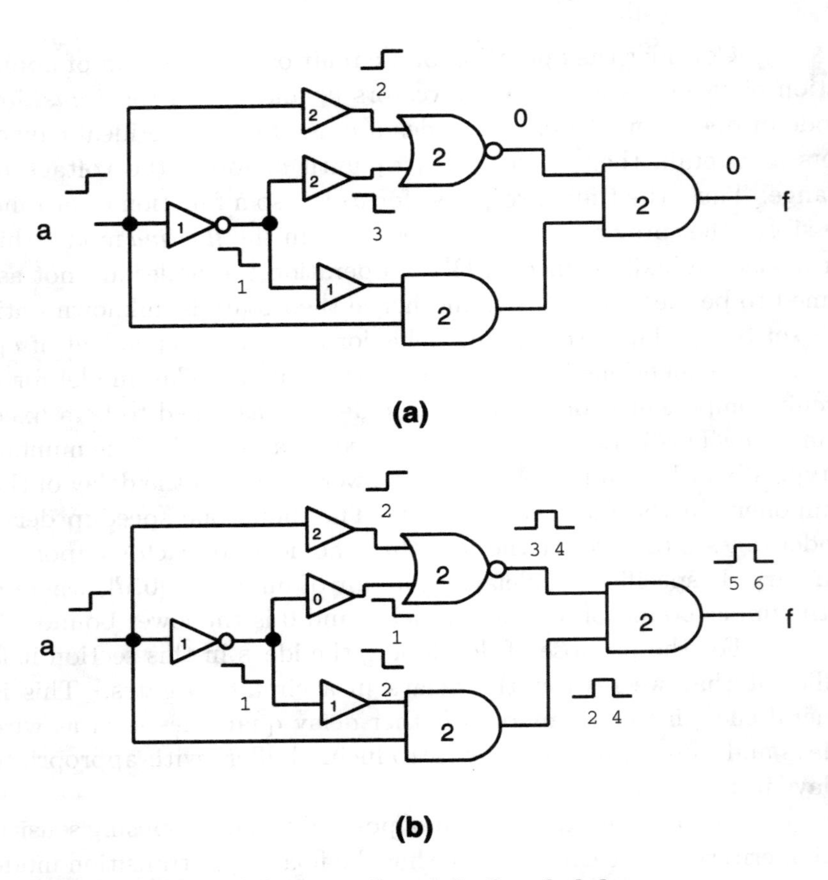

Figure 8.4: Transition mode with fixed delays

the two vector pairs corresponding to a, making a $0 \to 1$ transition and a $1 \to 0$ transition. Applying $0 \to 1$ and $1 \to 0$ transitions on a does not change the output f from 0. Thus, one can conclude that the circuit has critical delay 0 under the transition mode of operation for the given fixed gate delays.

Now consider the circuit of Figure 8.4(b) which is identical to the circuit of Figure 8.4(a) except that the buffer at the input to the NOR gate has been sped up from 2 to 0. We might expect that speeding up a component in a circuit would not increase the critical delay of a circuit. However, for the $0 \to 1$ transition on a the output f switches both at time 5 and time 6, and the critical delay of the circuit is 6.

The above example shows that a sensitization condition based on transition mode and fixed gate delays is unacceptable in

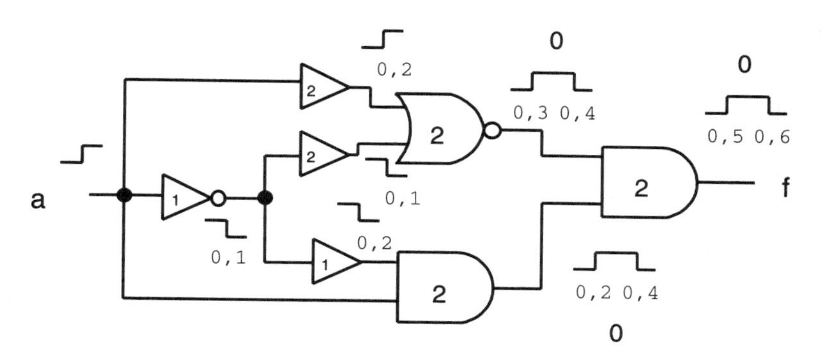

Figure 8.5: Transition mode with monotone speedup

a worst-case design methodology. Recall that in a worst-case design methodology, we are given the upper bounds on the gate delays and are required to report the (worst-case) critical path in the circuit. The gate delays of the circuit of Figure 8.4(a) could correspond to the upper bounds. Unfortunately, if we use only the upper bounds in the transition mode of operation, an erroneous critical delay may be computed as illustrated with the example above.

In order to obtain a useful sensitization condition one strategy is to use the transition mode of operation and monotone speedup. We will illustrate this on our example. Consider the circuit of Figure 8.5, which is identical to the circuit of Figure 8.4(a), except that each gate delay can vary from 0 to its given upper bound. As before, in order to determine the critical delay of the circuit we will have to simulate the two vector pairs corresponding to a making a $0 \rightarrow 1$ transition and a $1 \rightarrow 0$ transition. However, the process of simulating the circuit is much more complicated since the transitions at the internal gates may occur at varying times. In the figure, the possible combinations of waveforms that appear at the outputs of each gate are given for the $0 \rightarrow 1$ transition on a. For instance, the NOR gate can either stay at 0 or make a $0 \rightarrow 1 \rightarrow 0$ transition, where the transitions can occur between $[0, 3]$ and $[0, 4]$, respectively. In order to determine the critical delay of the circuit, we scan all the possible waveforms at output f and find the time at which the last transition occurs over all the waveforms. This provides us with a critical delay of 6. For a detailed description of a bounded delay simulation method see [7].

Timing analysis for a worst-case design methodology can use the above strategy of monotone speedup delay simulation under the transition mode of operation. However, this has several disadvantages. Firstly, the search space is 2^{2n} where n is the number of primary inputs to the circuit, since we may have to simulate each possible vector pair. Secondly, monotone speedup delay simulation is significantly more complicated than fixed delay simulation.

8.2.5 Floating Mode and Monotone Speedup

The above difficulties have motivated delay computation under the floating mode of operation. Under floating mode the delay is determined by a single vector. As compared to transition mode critical delay under floating mode is significantly easier to compute for the fixed or monotone speedup delay model – large sets of possible waveforms do not need to be stored at each gate.

Single-vector analysis and floating mode operation, by definition, make pessimistic assumptions regarding the previous state of nodes in the circuit. The assumptions made in floating mode operation make the fixed delay model and the monotone speedup delay model equivalent. To understand this consider a circuit C with fixed values on its component delays. Let π be a path through C and v_2 be a vector applied to C. In order to determine if π is responsible for the delay of C on v_2, we inspect the side-inputs of π. At any gate g on π the side-inputs have to be at noncontrolling values when the controlling or noncontrolling value propagates along π through g. If the value at a side-input i to g is noncontrolling on v_2, monotone speedup (under transition or floating mode) allows us to disregard the time that the noncontrolling value arrives, since we can always assume that it arrives before the value along π. Let the delay of all paths from the primary inputs to i be greater than the delay of the subpath corresponding to π ending at g. Under monotone speedup, we can speed up all the paths to i, ensuring that the noncontrolling value arrives in time. Under floating mode with fixed delays we cannot change the delays of the paths to i, but we can assume that v_1, the vector applied before v_2, was providing a noncontrolling value! We do not have to wait for v_2 to provide the noncontrolling value. In either case, the arrival time of noncontrolling values on side-inputs does not matter.

Because of this apparent equivalence between the floating mode of operation and monotone speedup, it has been conjectured that transition mode delay under the monotone speedup model can be reduced to a single-vector condition. This conjecture has been

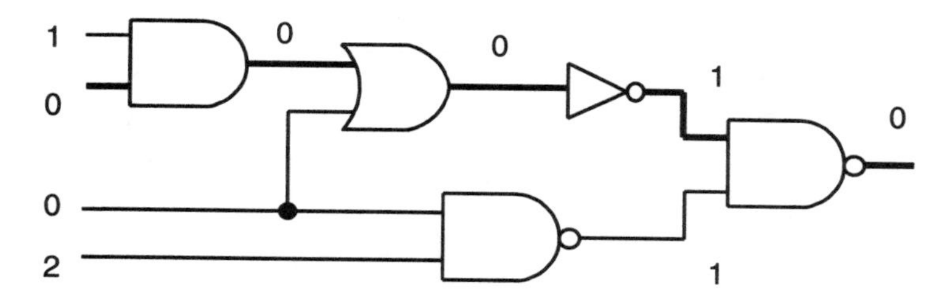

Figure 8.6: Static sensitization of a path

disproved [5]. However, for the purposes of the discussion here it is worthwhile to target the definition of a "correct" single-vector sensitization criterion such that an unsensitizable path is not responsible for the delay of a circuit under the transition mode of operation and monotone speedup, and a sensitizable path is responsible for the delay of the circuit under the transition mode of operation and monotone speedup.

There has been significant research done in an effort to arrive at the correct sensitization criterion. A detailed history may be found in [13]. Here, we will first look at delay-independent sensitization conditions, since they are the easiest to understand, and augment them to arrive at the correct sensitization criterion.

8.2.6 Static Sensitization

Recall that a path π is statically sensitized by a vector if all the side-inputs along π settle to noncontrolling values. This sensitization condition is delay-independent.

Consider the circuit of Figure 8.6. The path shown in bold has been statically sensitized by the vector 1002, where 2 stands for the unknown or undefined value. We can propagate a transition down the path by applying the vector pair $\langle 1102, 1002 \rangle$. Next, consider the circuit of Figure 8.3. We cannot find an input vector that statically sensitizes the path shown in bold. This is because in order to sensitize the bold path we require a 0 at the multiplexor input, which implies that either `p0` or `p1` have to be 0 or both. If `p0` or `p1` is 0, then gate `6` or gate `9`, respectively, will have a controlling value at the side-input to the bold path. We also know that we cannot find an input vector pair under the given gate delays that propagates a transition down the path.

The above observations may lead us into thinking that static

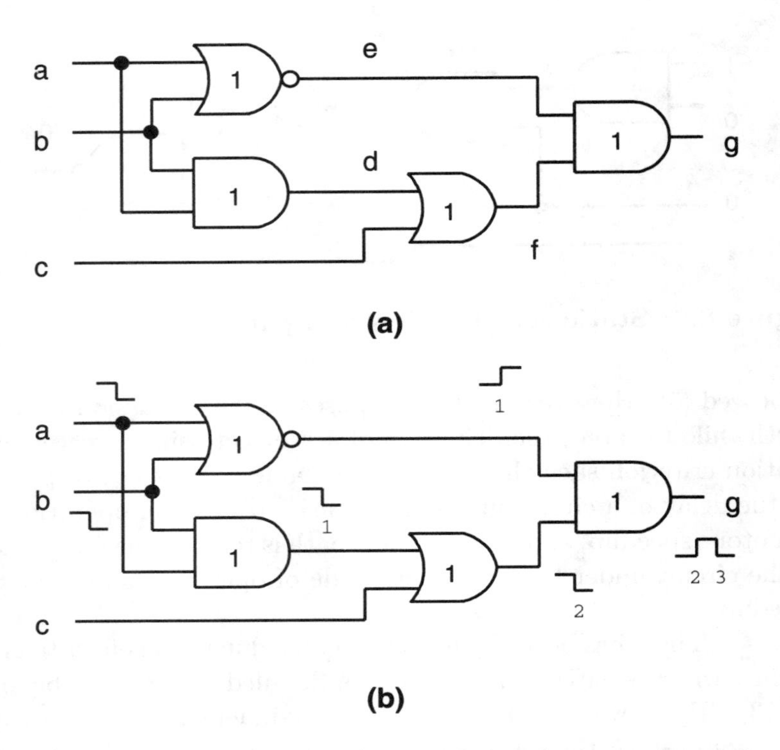

Figure 8.7: Static sensitization underestimates the delay

sensitization is the sensitization condition that we are looking for. Consider the circuit of Figure 8.7, taken from [12]. Assume that all the gate delays are 1. The paths $\{a, d, f, g\}$ and $\{b, d, f, g\}$ are not statically sensitizable. This is because to statically sensitize path $\{a, d, f, g\}$ we require a 1 on e, implying that both a and b have to be 0, but this requires that we have two controlling values at the AND gate with output d. A similar analysis can be done for the path $\{b, d, f, g\}$. This may lead us to conclude that the critical delay of the circuit is 2. However, keeping c at 0 and toggling a and b from $1 \rightarrow 0$ propagates a transition at time 3 to the output of the circuit. Thus, a path may not be statically sensitizable but can be responsible for the delay of a circuit.

It can be shown that static sensitizability is a sufficient condition for a path to be responsible for the delay of a circuit under the floating mode of operation. To understand this consider a vector v_2 being applied to a circuit C where v_2 statically sensitizes path π to a 1. On v_2 all the side-inputs to π are at noncontrolling values.

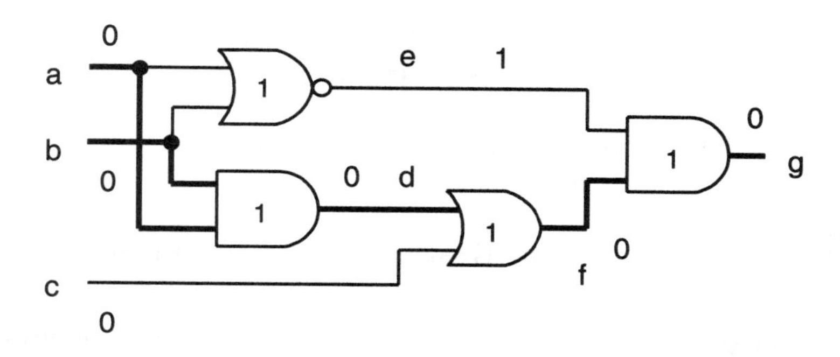

Figure 8.8: Static cosensitization of paths

Under the floating mode of operation, on any particular gate g on π, the values of the side-inputs of g on the previously applied vector v_1 can be assumed to be at noncontrolling values. This implies that we have steady noncontrolling values at each side-input. Thus, the value propagating down π "sees" noncontrolling values on all side-inputs and propagates all the way to the primary output. This of course implies that π is responsible for the delay of C.

8.2.7 Static Cosensitization

Let $\pi = \{v_0, e_0, ..., v_n, e_n, v_{n+1}\}$ be a path in a circuit C. We say that an input vector w *statically cosensitizes to a 1* path π in C if and only if the value of v_{n+1} is 1, and for each v_i, $1 \leq i \leq n+1$, if v_i has a controlled value, then the edge e_{i-1} presents a controlling value [4].

Static cosensitization is a delay-independent condition similar to static sensitization but is weaker than static sensitization. In Figure 8.8 the application of the vector 000 statically cosensitizes the paths shown in bold $\{a, d, f, g\}$ and $\{b, d, f, g\}$. Thus, if we used static cosensitization as our delay criterion, we would report the correct delay of 3 for this circuit.

Static cosensitization can be pessimistic. Consider the trivial circuit of Figure 8.9. The path of length 6 is statically cosensitized by the vector $a = 0$. However, the critical delay of the circuit is no greater than 5. (To see this apply the $0 \rightarrow 1$ and $1 \rightarrow 0$ transitions on a.)

It can be shown that static cosensitization is a necessary condition for a path to be responsible for the delay of a circuit under the floating mode of operation. To understand this consider a path π

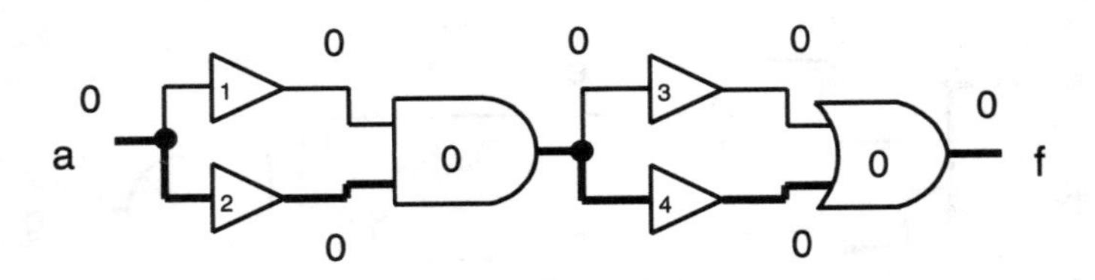

Figure 8.9: Static cosensitization can be pessimistic

that is not statically cosensitizable. This means that for any applied vector v_2 there is at least one gate g on π at which the input to g corresponding to π has a noncontrolling value and some other side-input to g has a controlling value. The controlling value at this side-input will always control the output of g. If the noncontrolling value arrives before the controlling value, the gate output will go to the controlling value after the controlling value arrives. Alternately, if the controlling value arrives before the noncontrolling value, the gate output will be at the controlling value before the noncontrolling value arrives. In either case, π is not responsible for the delay of the circuit on v_2.

8.2.8 True Floating Mode Delay

We now have two delay-independent conditions, static sensitization and static cosensitization that are sufficient and necessary, respectively, for a path to be responsible for the delay of a circuit. The necessary and sufficient condition for a path to be responsible for circuit delay under the floating mode of operation is a delay-dependent condition that is stronger than static cosensitization but weaker than static sensitization.

The fundamental assumptions made in single-vector delay-dependent analysis are illustrated in Figure 8.10. Consider the AND gate of Figure 8.10(a). Assume that the AND gate has delay d and is embedded in a larger circuit, and a vector pair $\langle v_1, v_2 \rangle$ is applied to the circuit inputs, resulting in a rising transition occurring at time t_1 on the first input to the AND gate and a rising transition at time t_2 on the second input. The output of the gate rises at a time given by $MAX(t_1, t_2) + d$. The abstraction under floating mode of operation only shows the value of v_2. In this case a 1 arrives at the first and second inputs to the AND gate at times t_1 and t_2, respectively, and a 1 appears at the output at time $MAX(t_1, t_2) + d$. Similarly, in

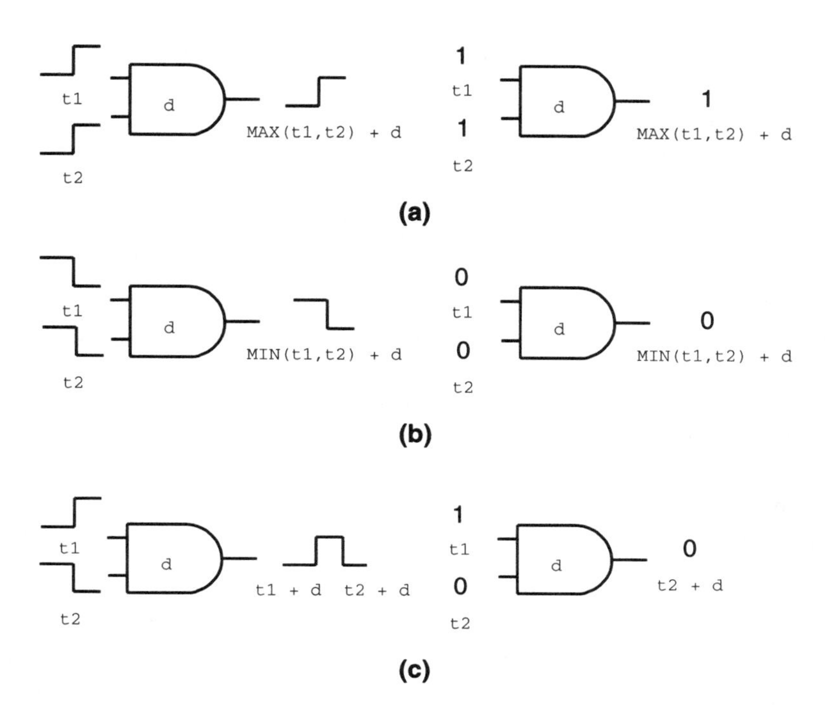

Figure 8.10: Fundamental assumptions made in floating mode operation

Figure 8.10(b) two falling transitions at the AND gate inputs result in a falling transition at the output at a time that is the minimum of the input arrival times plus the delay of the gate.

Now consider Figure 8.10(c), where a rising transition occurs at time t_1 on the first input to the AND gate and a falling transition occurs at time t_2 on the second input. Depending on the relationship between t_1 and t_2 the output will either stay at 0 (for $t_1 \geq t_2$) or glitch to a 1 (for $t_1 < t_2$). It is possible to accurately determine whether the AND gate output is going to glitch or not if a simulation is carried out to determine the range of values that t_1 and t_2 can have on $\langle v_1, v_2 \rangle$. (This was illustrated in Figure 8.5.) However, under the floating mode of operation we only have the vector v_2. The 1 at the first input to the AND gate arrives at time t_1, and the 0 at the second input arrives at time t_2. The output of the AND on v_2 obviously settles to 0 on v_2, but at what time does it settle? If $t_1 \geq t_2$, then the output of the gate is always 0, and the 0 effectively arrives at time 0. If $t_1 < t_2$, then the gate output becomes 0 at $t_2 + d$. In order not to underestimate the critical delay of a circuit all single-vector sensitization conditions *have* to assume that the 1 (the noncontrolling value for the AND gate) arrives before the 0 (the controlling value for the AND gate), i.e., that $t_1 < t_2$. Under the floating mode of operation this corresponds to assuming that the values on the previous vector v_1 were noncontrolling. (The above assumption also captures the essence of transition mode delay under the monotone speedup delay model. Given that the AND gate is embedded in a circuit, under the monotone speedup model the subcircuit that is driving the first input can be sped up to cause the rising transition to arrive before the falling transition.)

The rules in Figure 8.10 represent a timed calculus for single-vector simulation with delay values that can be used to determine the correct floating mode delay of a circuit under an applied vector v_2 (assuming pessimistic unknown values for v_1) and the paths that are responsible for the delay under v_2. The rules can be generalized as follows:

1. If the gate output is at a controlling value, pick the minimum among the delays of the controlling values at the gate inputs. (There has to be at least one input with a controlling value. The noncontrolling values are ignored.) Add the gate delay to the chosen value to obtain the delay at the gate output.

2. If the gate output is at a noncontrolling value, pick the maximum of all the delays at the gate inputs. (All the gate inputs have to be at noncontrolling values.) Add the gate delay to the chosen value to obtain the delay at the gate output.

To determine whether a path is responsible for floating mode delay under a vector v_2, we simulate v_2 on the circuit using the timed calculus. A path is responsible for the floating mode delay of a circuit on v_2 if and only if for each gate along the path:

1. If the gate output is at a controlling value, then the input to the gate corresponding to the path has to be at a controlling value and furthermore has to have a delay no greater than the delays of the other inputs with controlling values.

2. If the gate output is at a noncontrolling value, then the input to the gate corresponding to the path has to have a delay no smaller than the delays at the other inputs.

These conditions were first presented in [3]. They were related to static sensitization and static cosensitization in [4].

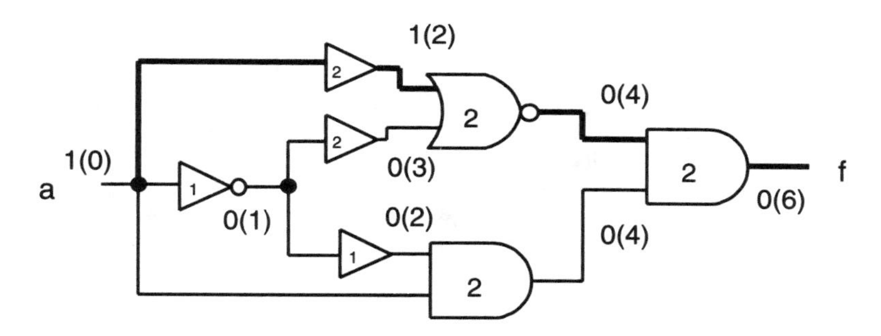

Figure 8.11: First example of floating mode delay computation on a circuit

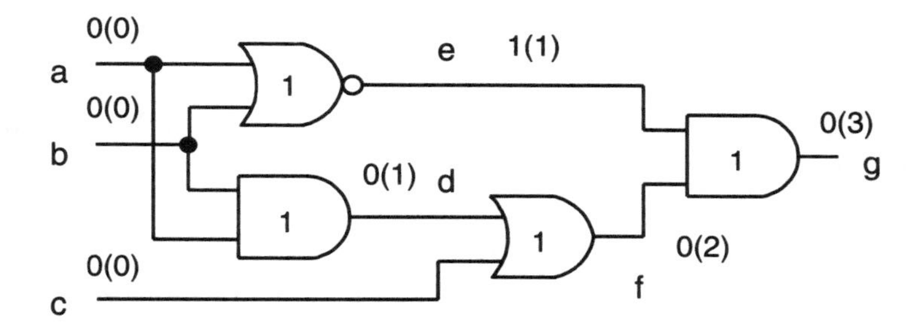

Figure 8.12: Second example of floating mode delay computation on a circuit

Let us apply the above conditions to determine the delay of the circuits we have encountered previously. Consider the circuit of Figure 8.4(a) reproduced in Figure 8.11. Applying the vector $a = 1$ sensitizes the path of length 6 shown in bold, illustrating that the sensitization condition takes into account monotone speedup (unlike transition mode fixed delay simulation). Each wire has both a logical value and a delay value under the applied vector.

Next, consider the circuit of Figure 8.7 reproduced in Figure 8.12. Applying the vector 000 gives a floating mode delay of 3, illustrating that the sensitization condition is weaker than static sensitization. The paths $\{a, d, f, g\}$ and $\{b, d, f, g\}$ can be seen to be responsible for the delay of the circuit.

Finally, consider the circuit of Figure 8.9 reproduced in Figure 8.13. Applying $a = 0$ and $a = 1$ results in a floating mode delay of 5, illustrating that the sensitization condition is stronger than static cosensitization. Recall that static cosensitization reports a delay of 6 for the circuit.

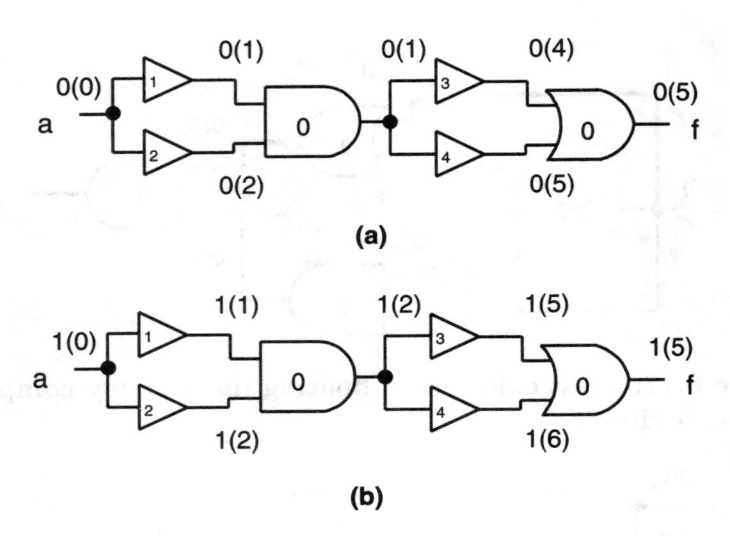

Figure 8.13: Third example of floating mode delay computation on a circuit

We presented informal arguments justifying the single-vector abstractions of Figure 8.10 to show that the derived sensitization condition is necessary and sufficient for a path to be responsible for the delay of the circuit under the floating mode of operation. For a topologically oriented formal proof of the necessity and sufficiency of the derived condition, see [3]. For a proof utilizing equivalent normal form analysis, see [4].

8.3 Floating Mode Delay Computation

In Section 8.2 we derived a sensitization condition that represents true floating mode delay. We now turn our attention to methods that can efficiently compute the true critical delay of a circuit.

Most methods to compute critical delay operate on a per-path basis, i.e., the longest path in the circuit is found and the method searches for a v_2 that sensitizes the path according to the chosen sensitization conditions. If the search fails, then the next longest path is picked and the process is iterated until the longest true path is found. The delay of this path is reported as the critical delay of the circuit. Searching for a v_2 for a particular path could take $O(2^n)$ time where n is the number of primary inputs to the circuit. Furthermore, this search has to be carried out for each chosen path. Many circuits such as parallel multipliers have thousands of longest paths, some of which are true and the others false. Per-path delay computation methods (reviewed in [13]) cannot be used on large circuits.

An alternate strategy pioneered in [4] is to directly answer the question of what the true critical delay of the circuit is and operate on sets of paths rather than a single path at a time. The question posed is: Is the delay of the circuit $\geq \delta$? δ can initially be set to the length of the longest path in the circuit and can be progressively decreased till the answer to the posed question is yes.

A straightforward $O(2^n)$ algorithm to find the true critical delay of a circuit that does not require path enumeration is to simulate each of the 2^n input vectors or minterms using the timed calculus and determine the longest delay seen at the circuit output. This process can be speeded up considerably by using cube simulation rather than minterm simulation. To this end, we describe the implicit enumeration algorithm **PODEM** [9] in Section 8.3.1. We generalize the timed calculus to allow unknown (or 2) values in Section 8.3.2 so we are in a position to use the **PODEM** search strategy in a timed test generation procedure (originally presented in [6]) that calculates the true floating mode delay. The timed test generation procedure is described in Section 8.3.3.

8.3.1 The PODEM Algorithm

PODEM is an implicit enumeration algorithm that implicitly and exhaustively enumerates the input space using cubes rather than minterms in an effort to satisfy a given objective. We will describe the **PODEM** algorithm when it is used to justify a particular value (0 or 1) at the primary output of a circuit[1]. The procedure **PODEM** is shown in Figure 8.14. It simply places the given primary output line `po` on a justification list `jlist` with an appropriate value, say 1, and calls a search procedure **SEARCH_1**.

The search procedures are described in Figures 8.15 and 8.16. The procedure **SEARCH_1** calls a **BACKTRACE** procedure to find a primary input whose logical value is currently unknown beginning from the primary output `po`. The primary input is set to a 1 value and the **IMPLY** procedure is called. This procedure in **PODEM** corresponds to standard three-valued cube simulation (without any delay information). The implication procedure may produce a logical conflict. If no conflict occurs, **SEARCH_1** is called

[1]This can viewed as detecting a primary output stuck-at fault.

```
PODEM(po, lvalue) {

    jlist = po with logical value lvalue ;

    status = SEARCH_1(jlist) ;
    return(status) ;
}
```

Figure 8.14: The PODEM algorithm

```
SEARCH_1(jlist)
{
    if (length of jlist is zero) return SUCCEED ;

    if (BACKTRACE(po, po_value, &pi, &pi_value) == FALSE)
        return(FAILED) ;

    if (IMPLY(pi, pi_value, jlist) != IMPLY_CONFLICT) {
        search_status = SEARCH_1(jlist) ;
        if (search_status == FAILED) {
            restore the state of the network to what it was
            prior to the most recent primary input assignment ;
            search_status = SEARCH_2(jlist, pi, 1 - pi_value) ;
        }
    } else {
        restore the state of the network ;
        search_status = SEARCH_2(jlist, pi, 1 - pi_value) ;
    }
    return(search_status) ;
}
```

Figure 8.15: First search procedure

recursively. The procedure terminates successfully in **SEARCH_1** if the justification list is empty. In the case of a logical conflict in **SEARCH_1** we backtrack to the most recent primary input setting. The primary input is set to the 0 value, and the **SEARCH_2** proce-

```
SEARCH_2(jlist, pi, pi_value)
{
    backtracks = backtracks + 1 ;
    if (backtracks > BACKTRACK_LIMIT) return(ABORTED) ;
        if (IMPLY(pi, pi_value, jlist) != IMPLY_CONFLICT) {
        search_status = SEARCH_1(jlist) ;
        if (search_status == FAILED)
            restore the state of the network ;
    } else {
        search_status = FAILED ;
        restore the state of the network ;
    }
    return(search_status) ;
}
```

Figure 8.16: Second search procedure

dure shown in Figure 8.16 is called. Failure results if either the backtrace procedure is unable to find a primary input to set or if the space has been completely enumerated without success in **SEARCH_2**. In the case of conflict or failure, we have to restore the state of the network to what it was immediately prior to the primary input setting that caused this failure.

We will illustrate the **PODEM** algorithm with a simple example. We wish to find an input vector that sets the logical function of Figure 8.17 to a 1. We begin with $a = 2$, $b = 2$, and $c = 2$. The simulation or implication of these input values results in a 2 on all wires including the output. Backtracing sets input a to 1. Implication immediately sets the output to a 0 as shown in Figure 8.17(a). The value of 0 at the output corresponds to a conflict since we wish a 1 value at the output. We therefore backtrack to the most recent primary input setting and change it. In this case we set a to 0. The output remains at a 2 for the vector $a = 0$, $b = 2$, and $c = 2$. We therefore backtrace to another primary input b and set it to a 1. Implication sets the output to a 0 as illustrated in Figure 8.17(b), and we again have a conflict. The conflict results in backtracking to the most recently set primary input, in this case b. The input value is changed to 0. The vector $a = 0$, $b = 0$, and $c = 2$ results in an output of 2 as shown in Figure 8.18(a). Setting the remaining

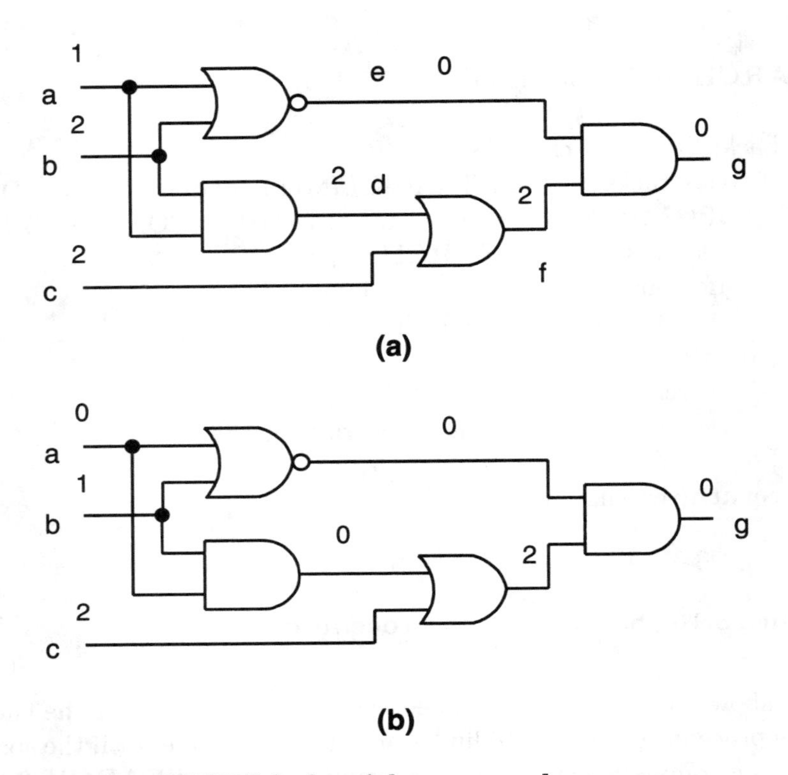

Figure 8.17: PODEM algorithm example

unknown input c to 1 sets the output to the required value of 1, as shown in Figure 8.18(b). The search procedure ends in success.

The binary decision tree for the above example is shown in Figure 8.19. The tree is similar to the binary decision diagram described in Chapter 6. The nodes of the tree correspond to primary inputs, and the left and right edges from each node correspond to input settings of 1 and 0, respectively. The leaves of the tree correspond to primary output settings.

8.3.2 Cube Simulation

In order to use **PODEM** for delay computation, we are interested in arriving at a timed calculus over three logical values, 0, 1, and 2 (unknown), that has the following properties:

1. It is equivalent to the timed calculus of Section 8.2.8 in the case where inputs are completely specified.

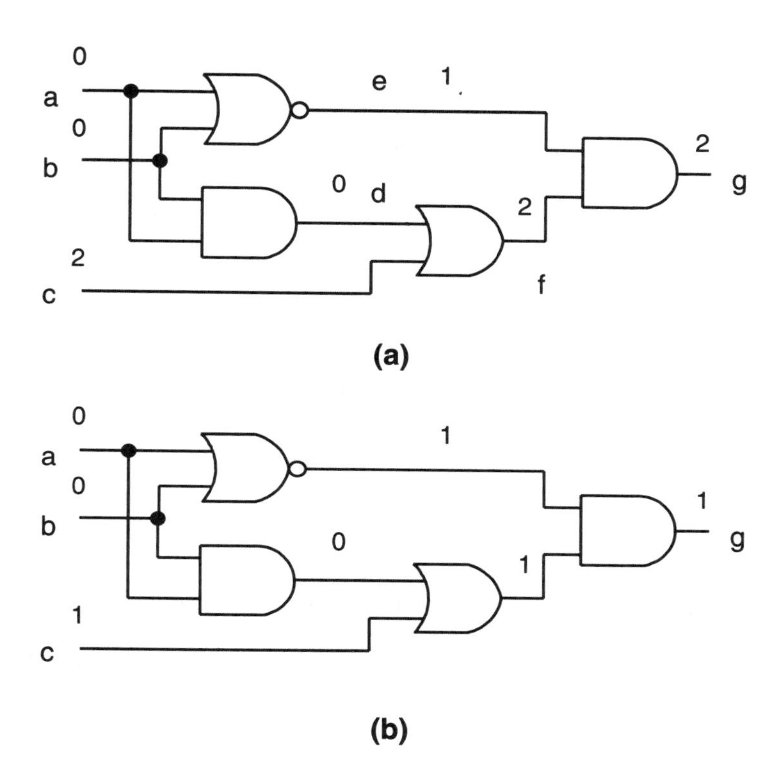

Figure 8.18: PODEM algorithm example (continued)

2. Given an incompletely specified vector, it produces an upper bound on the achievable delay over any of the minterms in the vector.

3. Given an incompletely specified vector, it produces a lower bound on the achievable delay over any of the minterms in the vector.

The timed calculus for cube simulation is given in Table 8.1 for a two-input AND gate. The calculus for an OR gate would be similar, except that the role of controlling and noncontrolling values is interchanged. For each entry, we have the logical value at the AND gate output and the lower and upper bounds on the achievable delay. (d is the delay of the AND gate.)

Each input to the AND gate has an associated logical value in $\{0, 1, 2\}$ and a lower and upper bound on the delay corresponding to the value. For input i_1 the lower bound is l_1 and the upper bound is u_1. Similarly for i_2. The logical values in the table correspond to

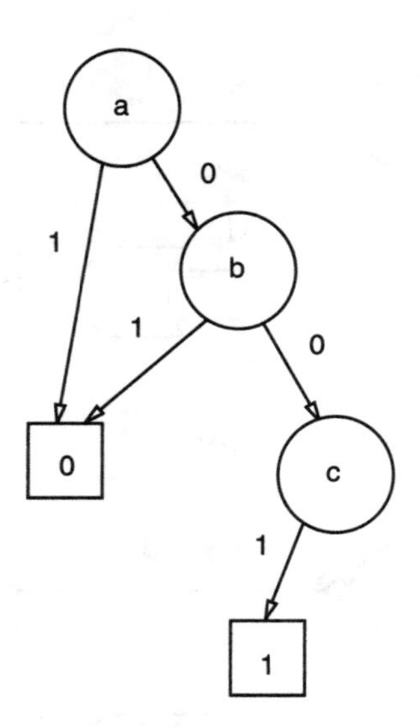

Figure 8.19: Binary decision tree for PODEM

$i_1 \rightarrow$ $i_2 \downarrow$	0	1	2
0	0 $MIN(l_1, l_2) + d$ $MIN(u_1, u_2) + d$	0 $l_2 + d$ $u_2 + d$	0 $MIN(l_1, l_2) + d$ $u_2 + d$
1	0 $l_1 + d$ $u_1 + d$	1 $MAX(l_1, l_2) + d$ $MAX(u_1, u_2) + d$	2 $l_1 + d$ $MAX(u_1, u_2) + d$
2	0 $MIN(l_1, l_2) + d$ $u_1 + d$	2 $l_2 + d$ $MAX(u_1, u_2) + d$	2 $MIN(l_1, l_2) + d$ $MAX(u_1, u_2) + d$

Table 8.1: Timed Calculus with unknown Values

standard three-valued simulation with 0, 1, and 2 values. We will explain the calculation of the lower and upper bounds.

When the inputs are in $\{0, 1\}$, we follow the rules of Figure 8.10. When we have two 0s at the AND gate inputs, we choose the minimum of the lower (upper) bounds to calculate the lower (upper) bound at the output. When we have two 1s, we choose the maximum of lower (upper) bounds to calculate the lower (upper) bound at the output. When we have a 0 and a 1, we simply use the lower and upper bounds of the 0 input.

When a 2 is at the input to an AND gate, the calculus gives the lower and upper bounds on the achievable delay with either (0 or 1) setting of the 2 value. For example, consider the entry in the table when i_1 is a 1 and i_2 is a 2. The logical value at the AND gate output is a 2. If i_2 is set to 0, the lower (upper) bound on the delay will be $l_2 + d$ ($u_2 + d$). However, if i_2 is set to 1, the lower (upper) bound on the delay will be $MAX(l_1, l_2) + d$ ($MAX(u_1, u_2) + d$). Since $l_2 \leq MAX(l_1, l_2)$, l_2 represents the lower bound on the achievable delay, and since $MAX(u_1, u_2) \geq u_2$, $MAX(u_1, u_2)$ represents the upper bound on the achievable delay.

Now consider another entry where i_1 is a 0 and i_2 is at a 2. In this case the value and the upper bound of i_1 are passed through to the output. If i_2 is set to 1, its delay will not matter. If i_2 is set to 0, then the delay at the output will be $MIN(u_1, u_2) + d$. However $u_1 \geq MIN(u_1, u_2)$, and therefore we pick $u_1 + d$. By the same reasoning, the lower bound will be $MIN(l_1, l_2) + d$.

When both inputs to the AND gate are 2s, the lower bound on the achievable delay at the output is clearly $MIN(l_1, l_2) + d$. The upper bound on the achievable delay at the output is $MAX(u_1, u_2) + d$. Similarly for the other entries.

An example of cube simulation using the timed calculus is given in Figure 8.20. Each wire has a logical value and two delay values. For example, $0(1, 2)$ implies that the logical value of the wire is 0, the lower bound on the achievable delay is 1 and the upper bound on the achievable delay is 2. In Figure 8.20(a) we have a vector with two 2 entries being simulated on the circuit. The output of the circuit is 2, and for the minterms that are contained in 00212, the maximum achievable delay is 4. When c is set to 1, the maximum achievable delay at the output is reduced to 3.

We emphasize that the computed delays using the timed calculus merely give the range of the achievable delays over all the minterms contained in the partial input setting (input cube). Even if a wire is at a *known* value under a partial input setting v, the delay of the wire may be a range rather than a constant. For instance, consider the case where the first input to a two-input AND gate with zero delay is at 0 with $l_1 = u_1 = 4$. If the second input is at a 2 with $l_2 = u_2 = 3$, then the lower bound on the output is achieved when the second input is set to a 0; we obtain a delay of 3 at the output. The upper bound of 4 is achieved when the second input is set to a

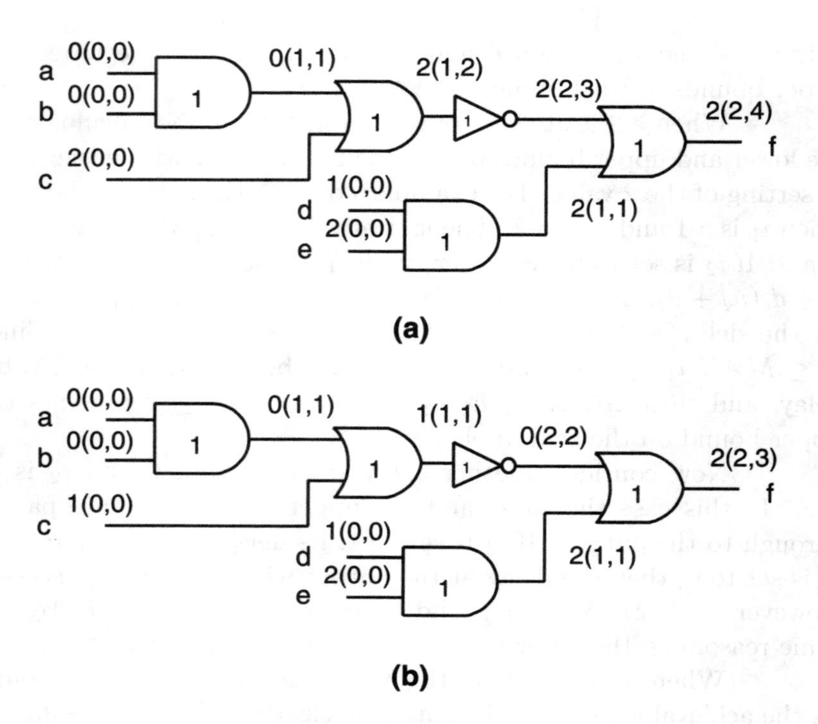

Figure 8.20: Cube simulation using timed calculus

logical 1. The output of the AND gate is a 0 for both cases.

8.3.3 Timed Test Generation

We will now describe the application of the **PODEM** algorithm to the problem of floating mode delay computation. In order to use the **PODEM** algorithm for floating mode delay computation, we have to use the timed calculus of Table 8.1. We will be justifying both logical values on wires as well as delay values, and therefore the procedure is termed *timed test generation* [6].

The procedure is given a logical value $L \in \{0, 1\}$ and a number δ, and asked to find an input vector, if such a vector exists, which sets the output of the circuit to L and which results in a floating mode delay $\geq \delta$.

The procedure follows the same steps as the **PODEM** algorithm. However, there are some important differences. The implication procedure uses the timed calculus of Table 8.1 rather than a purely logical calculus. Conflicts occurring during implication may be logical conflicts or time conflicts. Logical conflicts as before cor-

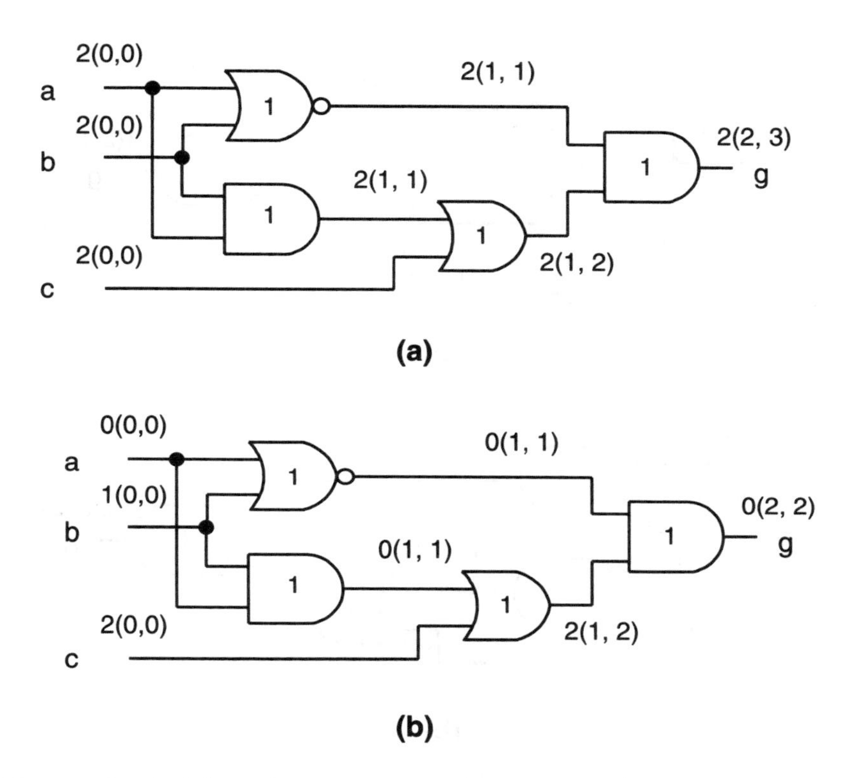

Figure 8.21: Timed test generation example

respond to the case when the output is set to the value $\overline{L}$. A time conflict occurs when the output is set to L but the upper bound on the delay at the output is strictly less than δ. In this case too we have to backtrack since we cannot find an input vector within the cube corresponding to the current settings of the primary inputs that has a delay $\geq \delta$. The procedure ends successfully if the output has been set to L and the lower bound on the computed delay at the output is $\geq \delta$.

Consider the example of Figure 8.21. The delay of each gate is indicated inside the gate. We wish to justify a 1(3) at the output of the circuit. Initially, all the inputs are set to 2 values with zero delay. Implication results in a 2 logical value at the output with a delay range $(2, 3)$ as shown in Figure 8.21(a). Backtracing sets a to 1. Implication immediately sets the output to 0, which is a logical conflict. We backtrack to the most recently set primary input and change its value. Therefore, a is set to 0. Implication results in a 2 at the output with a delay range of $(2, 3)$. Backtrace sets b to 1,

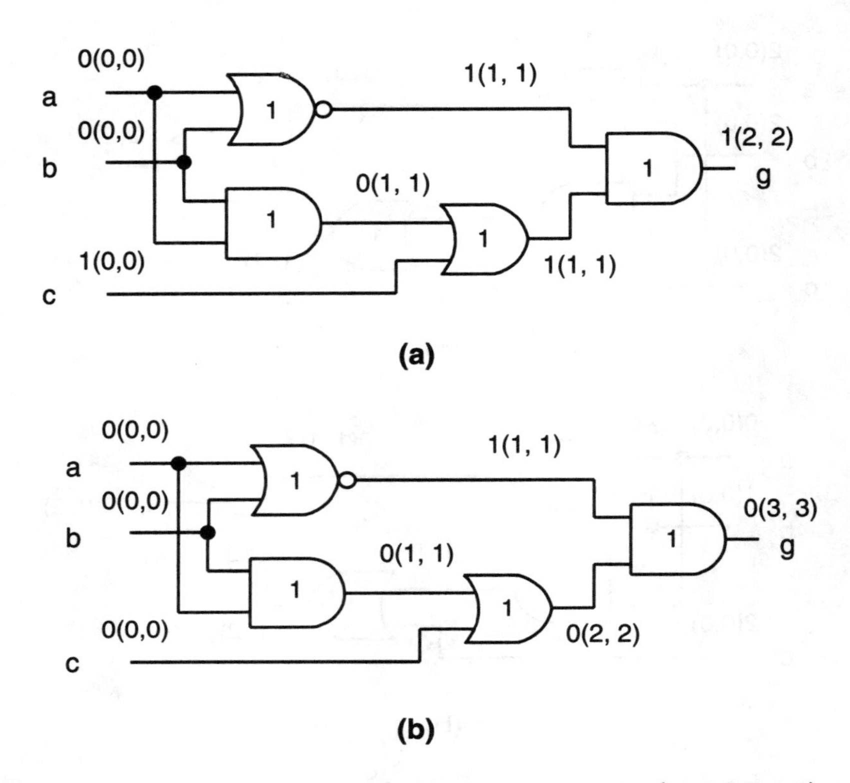

Figure 8.22: Timed test generation example (continued)

resulting in a 0 at the output, a logical conflict, as shown in Figure 8.21(b). We backtrack to set b to 0. The output is unknown with the input setting corresponding to $a = 0$, $b = 0$, and $c = 2$. Setting c to 1 results in the output being set to 1, but with a delay of $(2, 2)$, as shown in Figure 8.22(a). This is a time conflict since the upper bound on the delay is strictly less than the required delay. Finally, setting c to 0 results in a 0 at the output, a logical conflict, as shown in Figure 8.22(b). We have failed to find an input vector that results in a 1 at the output of the given circuit with floating mode delay 3. Note that we did not have to simulate all of the 2^3 different input minterms.

It is possible to find an input vector that results in a delay of 3 that produces a 0 at the output, as shown in Figure 8.22(b).

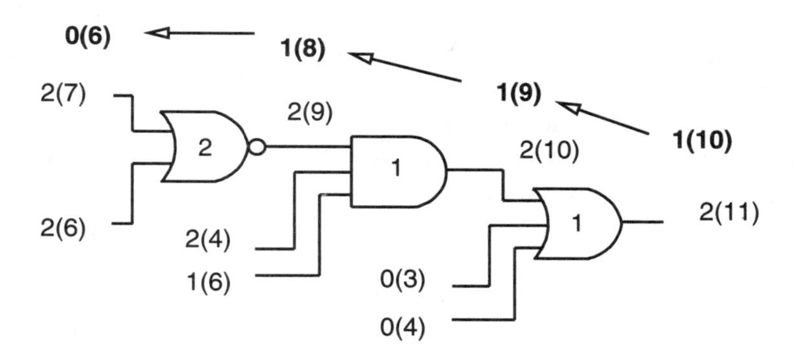

Figure 8.23: Backtrace example

8.3.4 Backtrace

We have not described the backtrace procedure thus far. In **PODEM** the backtrace procedure is called when the primary output of the circuit is at the unknown value for the current primary input settings. It begins from the primary output of the circuit and traces back through the gates in the circuit to a primary input that is still at the unknown value. The gates traversed all have the property that their outputs are at unknown values. The required value at the output may dictate the required value at these gates, but in general there will be choices as to which path the backtrace follows. The backtrace procedure uses heuristics in following paths that begin from a primary input with an unknown value which when set will likely set the output to the desired value. The particular value that the primary input is set to, either 0 or 1, is also decided heuristically.

The backtrace procedure in timed test generation is similar to the purely logical backtrace of **PODEM** except that it uses both the logical and desired delay value at the output to choose what path to follow. The backtrace procedure is called when the primary output is at the 2 value, or if the lower and upper bounds on the computed delay at the output are not equal. This is illustrated in Figure 8.23, where we have a fragment of a circuit with a partial setting of its primary inputs. The logical value and the upper bound on the delay value for each wire is shown in the figure. The desired value at the primary output is 1(10) and is shown in bold. The current value of the output is unknown with an upper bound on the delay being 11.

Backtracing begins at the OR gate connected to the primary output. Since we require a 1 at the primary output and the other two inputs to the OR gate are 0, we can infer that we require a 1 at the first input to the OR gate. Furthermore, the delay value required is 9. We now move to the AND gate and note that we require a 1(9) at its output. (This desired value is shown in bold in the figure.) This means that all of its inputs have to be a 1. The first and second inputs to the AND gate are at 2s. We will choose to follow the path corresponding to the first input because that is the only path that can satisfy the delay value of 9 at the AND gate output. Next, we move to the NOR gate with a required output value of 1(8). Both its inputs are at unknown values, and it is possible for either input to provide a 0(6) value. The backtrace procedure randomly selects one of the inputs and continues until it reaches a primary input.

8.4 Technology-Independent Optimization

Being able to meet speed requirements is absolutely essential in synthesizing logic circuits. Timing optimization of combinational circuits is performed both at the technology-independent level and during technology mapping. In this section we will consider the restructuring operations used in current logic synthesis systems to improve circuit speed. We will give an overview of a basic restructuring method that takes into account timing constraints specified as input-arrival times of the primary inputs and output-required times of the primary outputs. Then, in Section 8.5, we will describe an implementation of this method in greater detail. The goal is to meet the timing constraints while keeping the area increase to its minimum. The methods described here all use topological timing analysis, i.e., the delay tracing mechanism described in Section 8.2.1, to compute arrival times, required times, and slack.

8.4.1 Circuit Restructuring

The *critical section* of a Boolean network is composed of all the critical paths from primary inputs to primary outputs. Given a critical path, the total delay on the path can be reduced if any section of the path is sped up. For example, in Figure 8.24(a) we have a critical path $\{a, x, y\}$. The critical path can be reduced by first collapsing x and y and then redecomposing y in a different way to minimize the critical path as shown in Figure 8.24(b). This method is the basic step taken in restructuring. The nodes along the critical paths chosen to be collapsed and redecomposed form the *redecomposition region*.

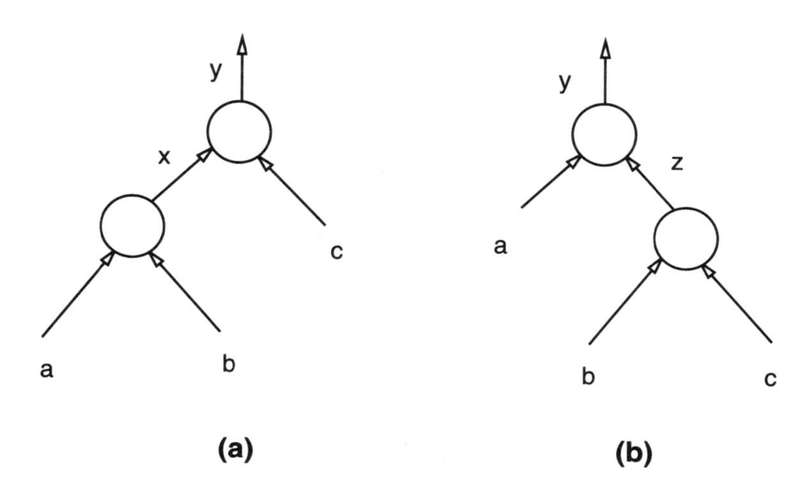

Figure 8.24: Collapsing and redecomposition

Since a critical section usually consists of several overlapping critical paths, the algorithm selects a minimum set of subsections, called redecomposition points, which when sped up will reduce the delays on all of the critical paths. (Note that it is not always possible to do so.) A weight is assigned to each candidate redecomposition point to account for possible area increase and for the total number of redecomposition points required. The goal is to select a set of points which cut all the critical paths and have the minimum total weight.

Once the redecomposition points are chosen, they are sped up by the collapsing-decomposing procedure. The new critical section of the network is then found. The algorithm proceeds iteratively until the requirement is satisfied or no improvement in delay can be made. A detailed exposition of speed optimization algorithms can be found in [16]. We will describe one such method in the following section.

8.5 The Speedup Algorithm

The algorithm we will describe takes as input a network of 2-input NAND gates and inverters. Timing constraints are specified as the arrival times at the primary inputs and required times at the primary outputs. The algorithm manipulates the network topology to achieve improved speed until the timing constraints are satisfied or it becomes apparent that no further decrease in the delay is possible. The final network produced by the algorithm is also a network composed of

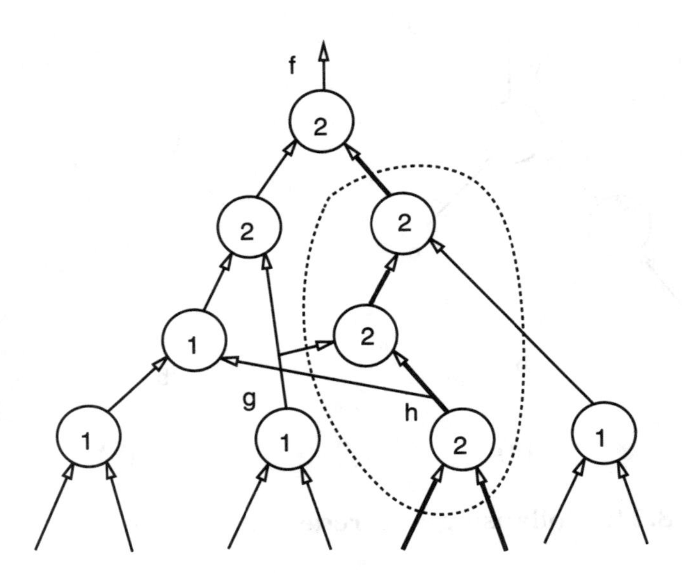

Figure 8.25: Example of the d_critical_fan-in of a node

2-input NAND gates and inverters.

8.5.1 Definitions

An ϵ-network is defined as a subnetwork in which all the signals have a slack within ϵ of the most negative slack.

The *distance* between two nodes f and g is the minimum number of nodes that have to be traversed from g to reach f, including g. For each node in the network we define a *d_critical_fan-in* section as the set of nodes that (1) are in the transitive fan-in of the node, (2) are at most distance d away from the node, and (3) are part of the ϵ-network.

The above definitions are explained pictorially in Figure 8.25. In the figure the delays of each node in the network is given inside each node. Assume zero arrival times for each of the primary inputs and a required time at the output of 8. The critical path of the circuit is marked in bold. The nodes on the critical path have zero slack. The 3_critical_fan-in of node f is shown encircled in the figure, assuming that $\epsilon = 0$. If $\epsilon = 1$, then we would include node g in the 3_critical_fan-in of node f, because the slack at node g is 1.

Partial collapsing of a node collapses all the nodes in the d_critical_fan-in of the node into two levels of logic. The nodes in the d_critical_fan-in that fan-out to some node not in the d_critical_fan-in

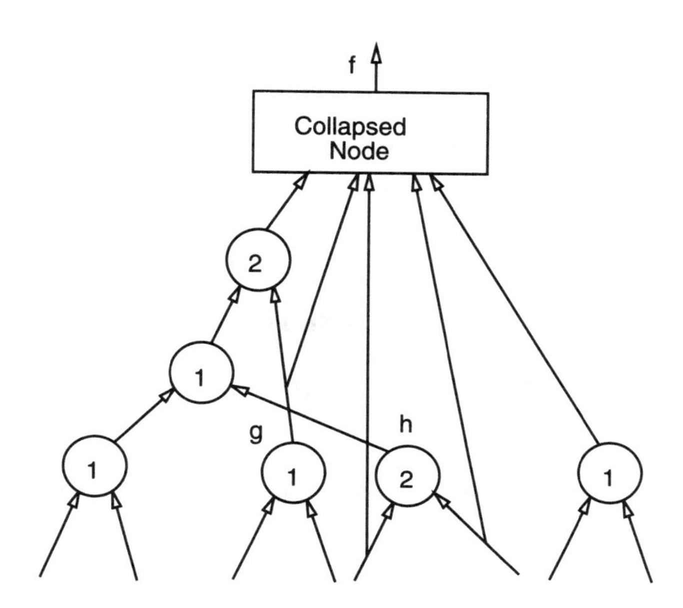

Figure 8.26: Example of partial collapsing

of the node are duplicated. Partial collapsing is illustrated in Figure 8.26. The 3_critical_fan-in of node f of Figure 8.25 is collapsed. Note that the logic corresponding to node h has been collapsed into f but is also required to be duplicated since the signal h serves as an input to other nodes in the network.

8.5.2 Outline of the Algorithm

After a delay trace through the Boolean network that computes arrival times, required times, and the slack at each node we create an ϵ-critical network from the Boolean network as shown in Section 8.5.1. We then assign weights to the nodes in the ϵ-network and find a minimum weighted cutset of nodes of the ϵ-network. A *cutset* of nodes in a network has the property that if the nodes in the cutset are removed from the network, the network breaks down into two disconnected subnetworks. After partially collapsing each node in the cutset, we do a timing decomposition of the collapsed nodes to yield a new network with a smaller delay. This process of delay tracing, cutset analysis, partial collapsing, and timing decomposition is iterated until the user specified constraints are met or no further speedup can be obtained.

The algorithm uses the user provided delay values or library

```
SPEED_UP( network, d, ε) {

    /* d is the distance up to which the fan-ins are collapsed */
    /* ε is the threshold for generating the ε-critical network */
    while (delay decreases or timing constraints not satisfied) {
        DELAY_TRACE() ;
        GENERATE( ε-network) ;
        node_list = NODE_CUTSET( ε-network) ;
        foreach( node ∈ node_list ) {
            PARTIAL_COLLAPSE( node, dist) ;
        }
        foreach( node ∈ node_list) {
            SPEEDUP_NODE( node) ;
        }
    }
}
```

Figure 8.27: Outline of the speedup algorithm

gate delay values at each stage. This is possible since we use a bottom-up decomposition at each stage, which enables us to do a delay trace on the decomposed network before moving on to a part of the network that is not yet resynthesized. The overall algorithm for speedup is shown in Figure 8.27.

8.5.3 Weight of the Critical Nodes

The function chosen to assign weights to the nodes in the ϵ-network is crucial since it determines the nodes that will be selected for speedup. Associated with a node we define two components for its weight — an area penalty ($\mathcal{W}_a$) and a potential for speedup ($\mathcal{W}_t$). The necessity for an area penalty is due to the duplication of some logic during the partial collapsing of a node. The potential for speedup ($\mathcal{W}_t$) is determined by various heuristics which try to identify the possibility of reducing the arrival time of the node after partial collapsing and resynthesis. These components are weighted depending on the area and delay tradeoff desired.

$$\mathcal{W} = \mathcal{W}_t + \alpha \times \mathcal{W}_a$$

where $\mathcal{W}_a$ = number of literals in the duplicated logic

$\mathcal{W}_t$ = potential for speedup
α = coefficient controlling the area-delay tradeoff

In order to arrive at one specific definition of the "potential" for speedup we look at the relation between the delays and arrival times in the given network. Consider a node n in the ϵ-network. For each node, i, that fans into the d_critical_fan-in of n, we know its arrival time, (A_i), and the delay, (D_i), from it to the node n. Consider the following situations:

1. The standard deviation (σ) of the vectors (A_i, D_i) is small. This implies a near balanced decomposition already exists in the transitive fan-in of the node when the inputs arrive at similar times. Hence, there is not much scope of improving the existing decomposition.

2. Let $D = \beta \times A + \delta$ be the least square error straight line that fits the data points (A_i, D_i). A negative value of β indicates that early arriving signals (small A_i) pass through a larger delay. This too suggests that the current decomposition is skewed in the right direction, reducing the "potential" for speedup.

With the above in mind, the larger σ is and the closer β is to 1, the smaller is $\mathcal{W}$ (favored for selection).

8.5.4 Minimum Weighted Cutset

After assigning the node weights, the maxflow-mincut algorithm is applied to generate a minimum weighted node cutset (also called separator set) of the ϵ-network. Maxflow-mincut is a very well-studied problem and several standard algorithms have been developed. (See [14] for descriptions of these algorithms.) Since critical paths extend from the primary inputs to the primary outputs, speeding up all the nodes on the node cutset of the ϵ-network by a given amount, say γ $(\gamma < \epsilon)$, would, in most cases, reduce the delay through the ϵ-network by at least γ. However, resynthesis may increase the delay through other parts of the circuit, resulting in a speedup of less than γ. The minimum weighted cutset provides us with a minimal area increase when the nodes are resynthesized.

8.5.5 Partial Collapsing

The network used for the cutset analysis is in the form of 2-input NAND gates. We collapse all the nodes in the d_critical_fan-in of the node to generate a large node with an associated sum-of-products expression to decompose later. The choice of the distance *d* parameter in the **PARTIAL_COLLAPSE** procedure influences the algorithm and is explained in Section 8.5.9. Decomposition of the collapsed node takes into consideration the arrival times and is explained in the next subsection.

8.5.6 Timing Decomposition

The general idea in the timing decomposition of a node is to place the critical signals closer to the output, thus making them pass through a smaller number of gates. This is illustrated in Figure 8.28. The critical paths in the original network are shown in bold and begin from signals *c* and *d*. Node *f* is collapsed, and a divisor *k* is selected which has the desired property that substituting *k* into *f*, places the critical signals *c* and *d* closer to the output. Note that the critical paths in the decomposed network may have changed.

Since in a multilevel network we can reduce the area by sharing common functions, we first attempt to extract area saving divisors that do not contain critical signals. After all such divisors have been extracted, we decompose the node into a NAND-NAND tree using the same heuristic, placing late arriving signals nearer the output. Figure 8.29 describes the algorithm for the decomposition of the node, taking into account the arrival times at the inputs.

The **SUBSTITUTE** procedure is a call to algebraic resubstitution described in Section 7.3. An important aspect of the procedure **SPEEDUP_NODE** is that the decomposition of any node is based on the correct updated arrival times of its input. This is a result of the bottom-up approach adopted for the decomposition of a node. During the decomposition we decompose first the parts that will eventually be closer to the input (e.g., the divisors which are extracted) and update their arrival times. This allows the decomposition to adjust dynamically to the updated delay of the extracted divisor which is an input to that node.

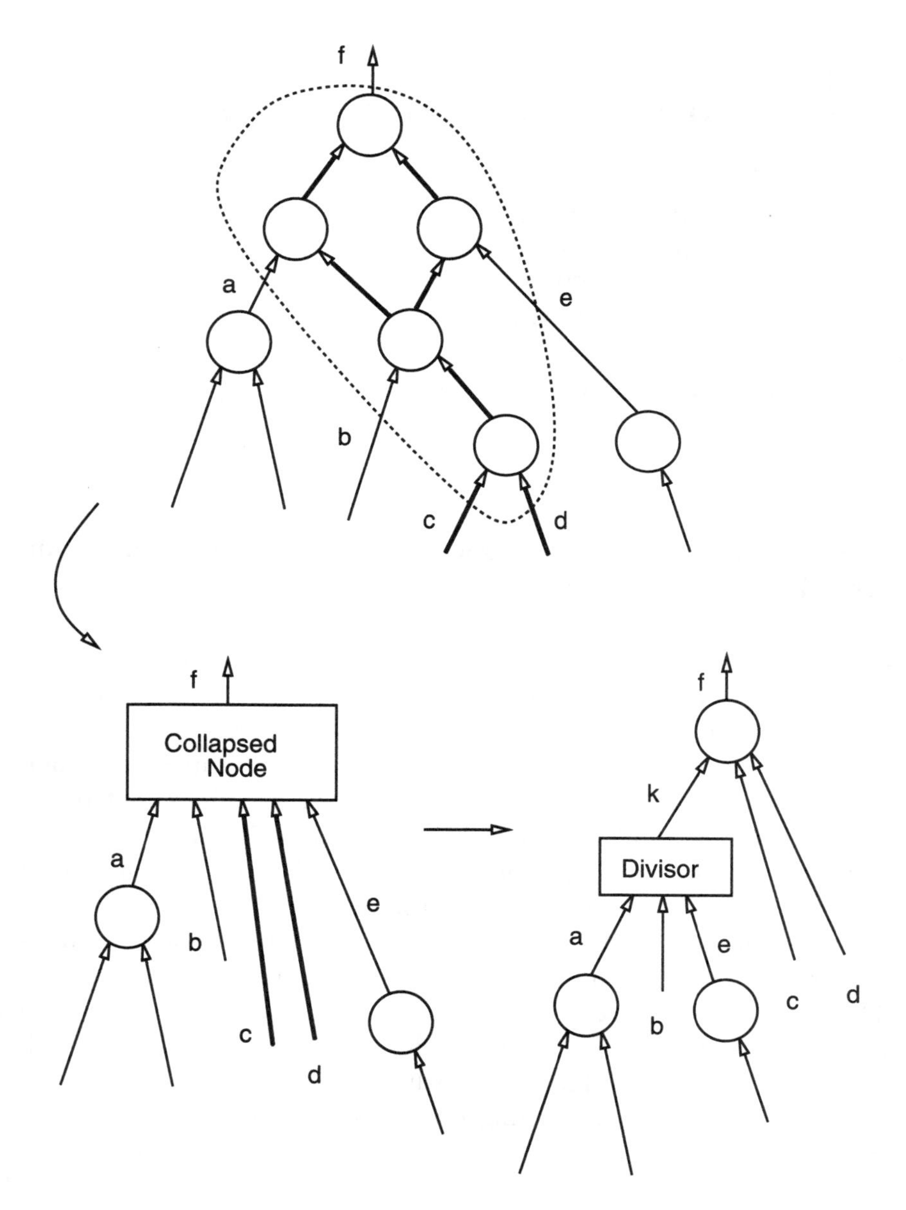

Figure 8.28: Basic idea of timing decomposition

8.5.7 Kernel-Based Decomposition

We will now describe the kernel-based recursive decomposition procedure in **SPEEDUP_NODE**, i.e., the case where k is nonnull.

After collapsing the node the notion of a critical path no

```
SPEEDUP_NODE( f ) {
    k = CHOOSE_BEST_TIMING_DIVISOR(f) ;
    if ( k != NULL ) {
        SUBSTITUTE ( f, k ) ;
        SPEEDUP_NODE( k ) ;
        /* Update the arrival time at inputs of f */
        DELAY_TRACE() ;
        SPEEDUP_NODE( f ) ;
    } else {
        AND_OR_DECOMP( f ) ;
    }
}
```

Figure 8.29: Procedure to speed up a node based on input arrival times

longer exists since the critical path depends on the decomposition of the node. With this in mind the objective is, given the arrival times at the inputs of a node, to decompose it in such a manner as to reduce the arrival time at its output. In decomposing a node to reduce its delay we want to preserve the area savings that result from extracting good divisors. The search for divisors is typically restricted, for efficiency reasons, to the set $\mathcal{K}$ consisting of level 0 kernels and level 0 kernel intersections.

The weight of a divisor is a linear sum of an area component and a timing component, depending on the tradeoff desired. The area component reflects the literals saved if the divisor were extracted. The timing component is designed to prefer divisors with early arriving signals. Thus, the earlier the latest input of a divisor arrives, the lower is the timing cost of that divisor. The detailed procedure for weighting the divisors is given in Figure 8.30.

8.5.8 AND-OR Decomposition

After all divisors containing early arriving signals are exhausted, i.e., when k becomes null in procedure **SPEEDUP_NODE**, a NAND-NAND decomposition is carried out, again based on input arrival times. The procedure **AND_OR_DECOMP** (Figure 8.31) decomposes a function, say $\mathcal{F}$, into a NAND-NAND tree. Each cube of the

```
CHOOSE_BEST_TIMING_DIVISOR( f ) {

    /* K ={level 0 kernels} ∪ {level 0 kernel intersections} */
    D = K ; /* D is the set of divisors */
    ρ = 0.1 ; /* Determined experimentally */
    for ( n ∈ K) {
        f = q · n + r ;
        D = q ∪ D ;
    }
    for ( n ∈ D) {
        F_in = Signals that fan into n ;
        C_t = ρ × MIN_{i∈F_in} A_i + (1 − ρ) × MAX_{i∈F_in} A_i ;
        C_a = Literals saved if n is extracted ;
        C( n ) = C_t + α × C_a ;
    }
    return (j  s.t. C( j ) is minimum) ;
}
```

Figure 8.30: Selection of divisors for timing resynthesis

function is decomposed (using the procedure **AND_DECOMP**) into a tree of 2-input NAND gates. A delay trace updates the arrival times at the output (x_i) of the NAND-trees, representing the cubes $c_i \in \mathcal{F}$. The cube $\prod \overline{x_i}$ represents the function $\overline{\mathcal{F}}$. Since the arrival times at the inputs of this new cube are known, it can be decomposed by the procedure **AND_DECOMP**.

The procedure **AND_DECOMP** reduces a cube to an AND tree. Again, decomposition is guided by the arrival times at the inputs to the node. This is a recursive procedure (Figure 8.32) which creates a tree of AND gates. We combine the two earliest inputs in an AND function and update the arrival time at the function output. This new function is substituted in the original node, and the resulting node is passed to the **AND_DECOMP** procedure. The recursion stops when a cube with one or two literals is encountered. Here, too, the decomposition of a cube is based on the actual arrival times of its inputs. Since the decomposition creates a tree, this procedure guarantees that the resulting 2-input AND decomposition has a minimum arrival time.

An example of a 4-input cube being decomposed is illus-

```
AND_OR_DECOMP( F ) {

    /* F is a multiple-cube function */
    foreach( cube c_i ∈ F ) {
        AND_DECOMP( c_i ) ;
    }
    DELAY_TRACE() ;
    F̄ = ∏_i x̄_i ; /* x_i represents the cube c_i */
    AND_DECOMP( F̄ ) ;
}
```

Figure 8.31: The AND_OR decomposition procedure

```
AND_DECOMP( F ) {

    /* F is a cube */
    if ( | F | > 2 ){
        l_1 = Earliest arriving input of F ;
        l_2 = Next earliest arriving input ;
        c = l_1 · l_2 ;
        SUBSTITUTE( F, c) ;
        DELAY_TRACE() ;
        AND_DECOMP( F ) ;
    }
}
```

Figure 8.32: The AND_DECOMP procedure to decompose a cube

trated in Figure 8.33. The arrival times of the inputs a through d are indicated in the networks. A cube with 2 inputs is factored out of f. The inputs a and b are selected because they arrive fastest. Assume that the delay of a 2-input AND gate is 2 units. Given the AND decomposition using e, we now have a cube with three inputs whose arrival times are 2, 1, and 3. We again select the two fastest inputs to arrive at the final AND decomposition.

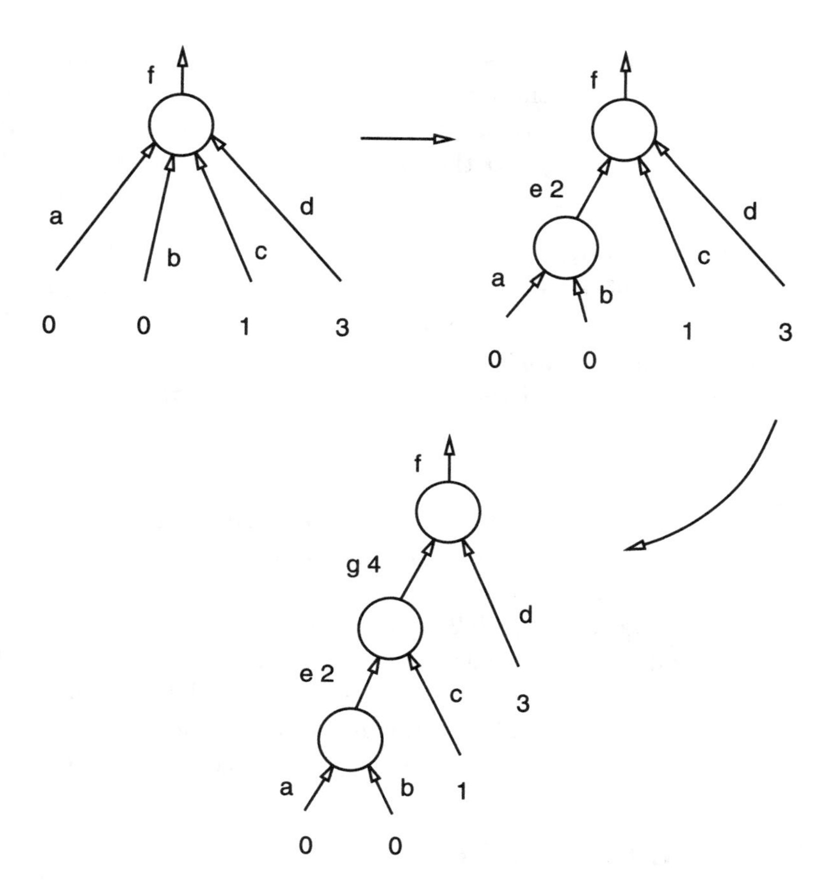

Figure 8.33: Example of the use of the AND_DECOMP procedure

8.5.9 Controlling the Algorithm

As is evident from the description a variety of parameters govern the running time of the algorithm and the quality of the results. We discuss the influence of each of these in this subsection.

1. ϵ: The size of the ϵ-network is governed by this parameter. Using a large ϵ might result in selecting nodes for speedup from a region where speeding up does not reduce the critical delay. Thus, area is wasted. Too small an ϵ results in a slow algorithm.

2. d: Since critical inputs are collapsed into a node in the node cutset of the ϵ-network up to a distance d, this determines the quality of the results and the running time of the algorithm.

A large value of d is useful in making relatively large changes in the delay since the larger nodes provide a greater degree of flexibility in restructuring the logic. However, due to the time spent in collapsing and the large number of divisors available for large nodes, the run time increases rapidly as d is increased.

3. α: This is the parameter that controls the tradeoff of area and speed. The larger is the value of α (the coefficient multiplying the area component of the weight), the more we want to avoid the duplication of logic during partial collapsing. In cases when we want a speedup irrespective of the increase in area, we can set $\alpha = 0$, thereby ignoring the duplication of logic.

4. model: The delay trace performed on the circuit can use a variety of delay models. The most primitive is the unit delay model which assigns a delay of 1 unit to a gate irrespective of its size and the loading. The unit fan-out delay model incorporates an additional delay of 0.2 units for each fan-out. The library delay model uses the delay data in the library cells to provide more accurate delay values. By using a crude delay model (e.g., unit fan-out delay model) in the initial stages and a more refined delay model later we can significantly reduce the run time.

8.6 Technology Mapping for Delay

Technology-independent delay optimization algorithms cannot estimate the delay of a circuit accurately, largely due to the lack of accurate technology-independent delay models. Therefore, such optimizations are not guaranteed to produce faster circuits, when circuit speed is measured after technology mapping and layout design. In this section, technology mapping algorithms that optimize the delay of the circuit will be presented. The two main technology-dependent delay optimization techniques are *tree covering* and *fan-out optimization.* Tree covering algorithms in the context of technology mapping for minimum area were introduced in Section 7.7. Modifications to these methods to target circuit speed are presented in the first half of this section. Tree covering alone does not generate good quality solutions because most circuits are not trees but DAGs. In such circuits, there is fan-out, i.e., a signal from a source is fed to two or more destinations. Due to the large amount of capacitance that has to be driven, the delay through the gate that drives this signal could be large. The optimization of this delay is called fan-out optimization and is the subject of the second half of this section.

8.6.1 Delay Model

The most accurate estimate of the delay of a gate in a circuit can only be obtained after the entire circuit has been placed and routed. Since technology mapping has to be performed before placement and routing, an approximate delay model with reasonable accuracy has to be used. The delay model used in this section is a simple linear delay model. The delay from an input pin i to the output of a gate g is characterized using an equation of the form

$$\delta_{i,g} = \alpha_{i,g} + \beta_{i,g} \times \gamma \tag{8.2}$$

In this equation $\delta_{i,g}$ is the delay between input pin i and the output; $\alpha_{i,g}$ denotes the intrinsic (load independent) delay from the input pin i to the output; the coefficient $\beta_{i,g}$ models the load dependent delay; and γ denotes the capacitive load at the output of the gate. The value of γ is determined by looking at the output connections of the gate. In many cases the values of $\alpha_{i,g}$ and $\beta_{i,g}$ are the same for all input pins of the gate. In such cases the subscript i will be dropped. In cases where the gate under consideration is clear from the context, the other subscript will be dropped too.

8.6.2 Delay Optimization Using Tree Covering

The tree covering algorithm presented in Section 7.7 can only be used if the cost of a match at a gate can be determined by examining the cost of the match and the cost of the inputs to the match (for which the cost has already been determined). For area optimization the cost of a gate depends on the area cost of the match and the area cost of the inputs of the match. For delay optimization the cost is signal arrival time at the output of the match. Therefore, the cost of a match for delay optimization depends not only on the structure of the tree beneath the gate but also on the capacitive load seen by the match. This load cannot be determined at the time of the selection of the match as it depends on the unmapped portion of the tree. Several attempts have been made to generalize tree covering to produce minimum delay implementations [2, 15, 17]. We will describe some of these approaches here.

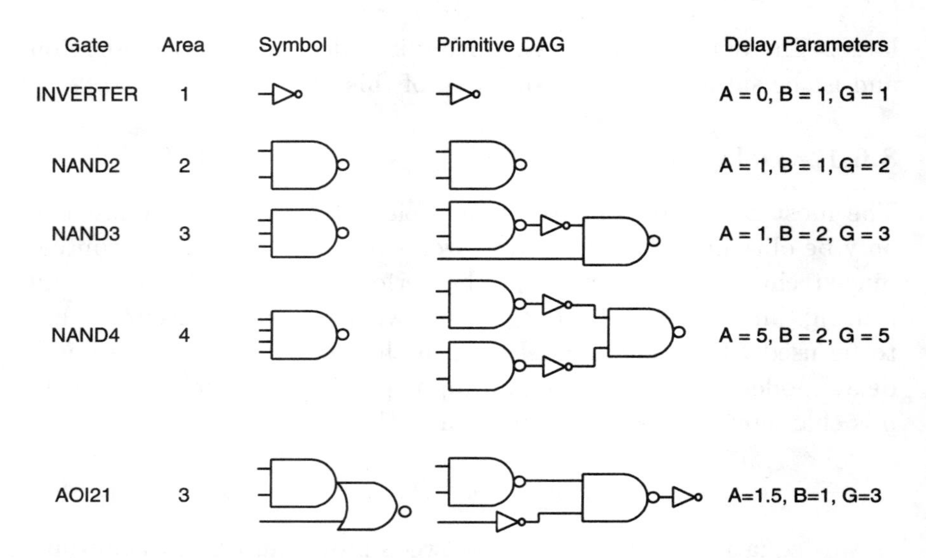

Figure 8.34: Gate library

The tree covering algorithm of Section 7.7 can be used to produce a minimum delay implementation of a circuit provided the load of all the gates in the network are the same. Consider the technology library shown in Figure 8.34 and the circuit shown in Figure 8.35(a). For each gate in the library, its name, area, symbol, and primitive DAG are presented. In addition, the delay parameters for our delay model are shown. The intrinsic delay, α, is denoted by A, the load dependent coefficient is denoted by B, and the load presented by the gate to any input gate is denoted by G. Note that in order to calculate the delay of a gate using Equation 8.2, we will use A and B for the gate and sum up the G values for all its fan-out gates.

If the load of each gate in the circuit is considered to be 1, then the perfect match at each gate can be determined in one bottom-up pass, as in Section 7.7. For gate **1** this corresponds to a two-input NAND gate with a delay of 2. The best match at gate **2** is a three-input NAND gate with a delay of 3. The best covering for this circuit under the fixed load assumption is shown in Figure 8.35(b).

The above algorithm does not necessarily produce the optimal solution because the load of all gates is not the same. As can be seen from the library in Figure 8.34 different gates provide different load values to their inputs. An algorithm, originally presented in [15], can be used to take into account the effect of different loads. This

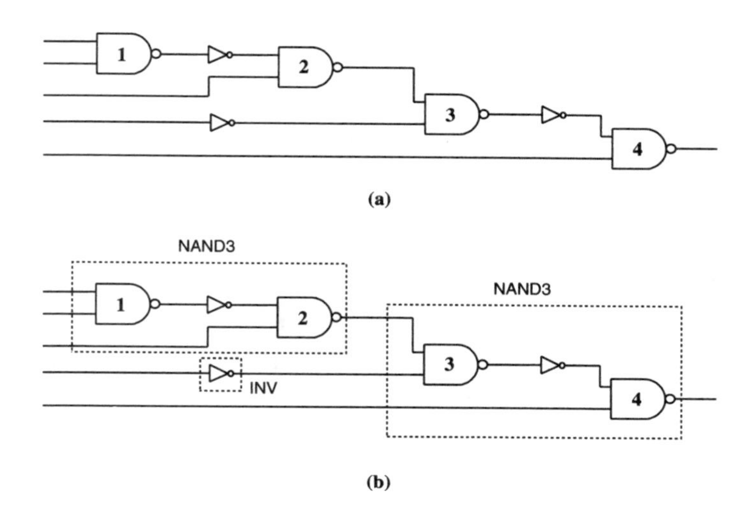

Figure 8.35: Circuit and its mapped implementation

algorithm will be illustrated using the circuit of Figure 8.35(a) and the library of Figure 8.34.

The first step of the algorithm is a preprocessing step over the technology library in order to create n load bins and quantize the load values for all the pins in the library. For each load bin a representative load value is selected, and the remaining load values are mapped to their closest value in the chosen set. The value of n determines the accuracy and the run time of the algorithm. If n is equal to the number of distinct loads in the library, then the algorithm is most accurate. However, the larger the value of n, the more computation will be required. Since the number of distinct load values in our example is only four, four bins are considered.

For a match at a gate an array of costs (one for each load value) is calculated. The cost is the arrival time of the signal at the output of the gate. For each bin or load value the match that gives the minimum arrival time is stored. For each input i of the match the optimum match for driving the pin load of pin i of the match is assumed, and the arrival time for that match is used. This is calculated by traversing the tree once from the leaves of the tree to its root. This is called postorder traversal. Following this the tree is traversed from the root to the leaves whereby the load values are propagated down, and for each gate the best match at the gate is selected, depending on the value of the load seen at the gate. This is called preorder traversal.

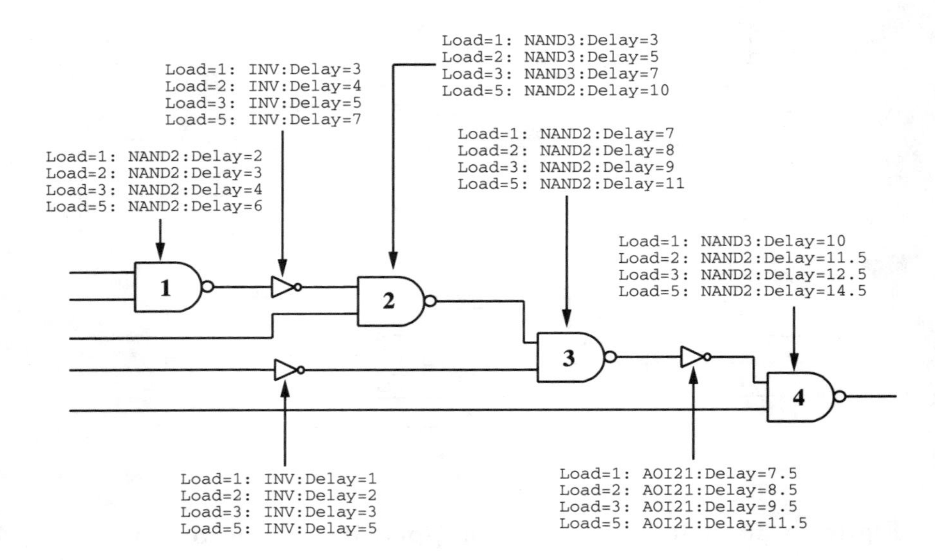

Figure 8.36: Technology mapping using load values

To illustrate this, consider the best matches shown in Figure 8.36. For gate 1 the only match is a two-input NAND gate. For each load value the delay of this gate then gives the arrival times at the output of the match (assuming zero arrival time at the inputs). For the inverter at the output of this NAND gate the only match is that of an inverter. Since the inverter presents a load of 1 to the NAND gate, the arrival time at the input of the inverter is the arrival time corresponding to the first bin of the NAND gate. Using this arrival time the arrival times at the output of the inverter for all possible load values are computed and are shown in the figure.

At gate 2 there are two possible matches corresponding to two-input and three-input NAND gates. If we consider the two-input NAND gate, the two arrival times at the inputs of the match are 0 (corresponding to the primary input connection to gate 2) and 4 (corresponding to the inverter connection to gate 2 seeing a load of 2). The maximum arrival time at the inputs is 4. The arrival times at the output of the gate for the four different load values are 6, 7, 8, and 10. For example, for a load value of 5 a two-input NAND gate has a delay $1 + 1 \times 5 = 6$. This delay added to the arrival time of 4 at the input of the NAND gate produces an arrival time of 10 at the output. For the three-input NAND gate the arrival times of all inputs are 0, and therefore the arrival times at the output are 3, 5, 7, and

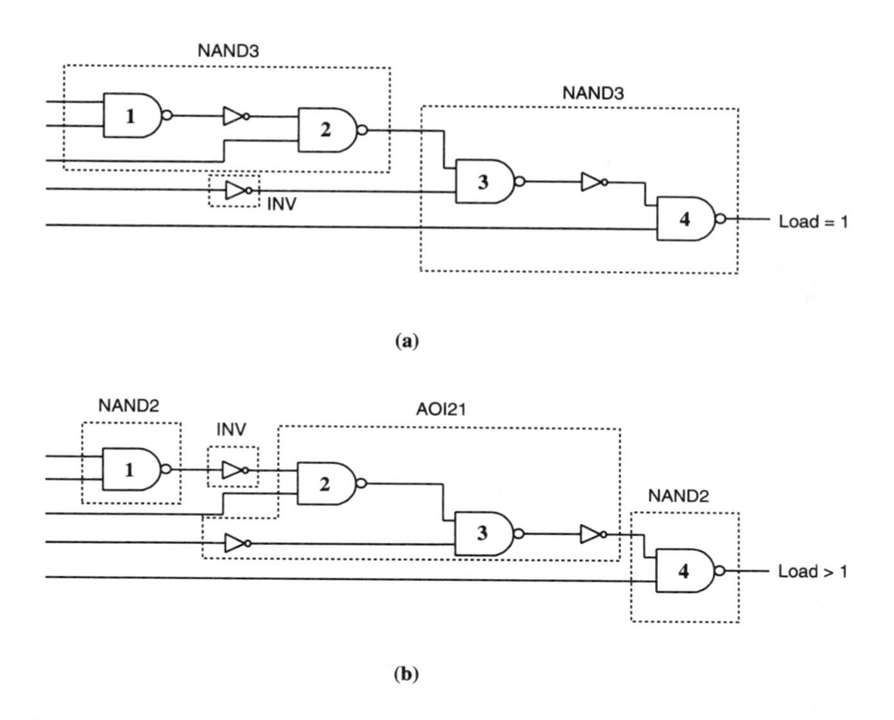

Figure 8.37: Two different implementations of the circuit depending on load value

11. Therefore, for the first three load values, the three-input NAND is a better choice, while for the last load value the two-input NAND is a better choice.

The final mapping is determined during preorder traversal and depends on the load seen by gate 4. Assuming a load of 1, the best match at gate 4 is a three-input NAND gate. This gate presents a load of 3 to its inputs, implying that the best match for a load value of 3 at gate 2 has to be chosen. This match is another three-input NAND gate. The resulting mapping is shown in Figure 8.37(a), which is coincidentally the same mapping obtained assuming constant load (Figure 8.35(b)). However, if the load is greater than 1, then the mapping of Figure 8.37(b) is better.

The above approach quantizes load values a priori based on the library information. Unless the quantization intervals are fairly small, this approach is not accurate. However, if the quantization intervals are small, the computation time increases. A better approach is to adapt the quantization intervals to each gate. In one precomputation phase, we can determine all the possible load values at a gate by examining all the possible matches at the gate. These load values can then be used to determine the values of the quantization intervals. For example, for gate 1 in the circuit of Figure 8.36 only a load value of 1 has to be considered because all possible matches at the inverter consist of only an inverter. On the other hand, for gate 2, load values of 2 and 3 have to be considered. This type of adaptive quantization of the load value produces results close to the optimum within reasonable amounts of computation time.

8.6.3 Minimizing the Area under a Delay Constraint

The tree covering algorithm used above can be generalized to minimize the area under a delay constraint. It may not be necessary to obtain the fastest circuit, but instead we may want to obtain a circuit that meets certain timing constraints and has the minimum possible area. This timing constraint is expressed as a required time at the root of the tree and can be propagated down the tree together with load values during preorder traversal. In this case the cost of a match at a gate includes not only the arrival time but also the area of a match. During preorder traversal the minimum area solution that meets the required timing constraint is chosen. If no such solution is available, then the minimum delay solution is chosen. It might appear that if the arrival time and the required time at the root of the tree are the same, no further area minimization is possible. This is not true because minimum delay technology mapping, even with the propagation of the required times, makes each subtree maximally fast. However, each subtree is not required to be maximally fast, and therefore the area of the circuit can be minimized, as will be shown in the following example.

Consider the mapping shown in Figure 8.38(a). The circuit has been mapped for minimum delay, and the arrival time at the output of gate 7 is 7. However, the required time at the output of this gate is 9, and the other match at gate 7 has an arrival time of 9 but a smaller area. Selecting this match gives us a circuit with the same delay but a smaller area, as shown in Figure 8.38(b).

There are other approaches to minimizing area based on recovering area at the expense of increased delay on noncritical paths. The most obvious candidates for area recovery are inverters. They are the most frequently used gates in circuits, and selecting the best inverter, among several inverters with varying areas and delays, to minimize the delay at any given point can be performed by enumer-

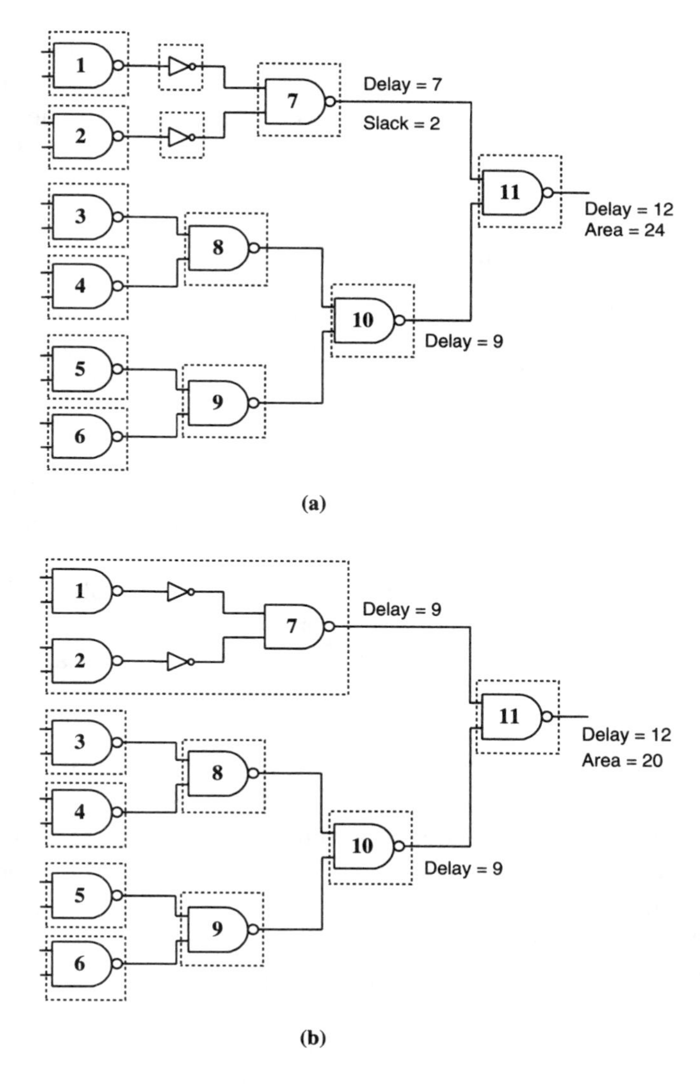

Figure 8.38: Example illustrating area recovery

ating all the possible choices and selecting the best one. There is a simple and optimal algorithm that can be used to select the best inverter after the circuit has been mapped. The algorithm for inverter selection proceeds as follows. Gates are visited during a preorder traversal, and loads and required times are propagated down. Each time an inverter is reached, if there is a smaller inverter that meets the timing constraint imposed by the required time, that inverter is selected.

There is no reason to restrict the above algorithm to inverters only. The inverter selection algorithm can be used for the selection of all gates that come in different sizes and strengths in the library. However, this algorithm can only replace a gate with a similar one with smaller area.

8.6.4 Optimality of Tree Covering

Given a tree decomposed into primitive elements, tree covering can produce an optimal implementation of the tree according to certain cost functions such as minimum area and minimum arrival time. Does this mean that for a *technology-independent* tree circuit, tree covering will produce the optimal solution, either in terms of delay or area? There are some factors that affect the final outcome of technology mapping for trees, and in this section we investigate these factors.

An initial decomposition of a tree into primitive gates is required before the tree covering algorithm can be applied. The results of technology mapping depend on the initial decomposition of the tree. Unfortunately, it is not practical to enumerate all possible decompositions of a tree into simple primitives, perform technology mapping, and choose the best one. Typically, one decomposition is chosen, and the final result depends on the quality of this decomposition. Consider the example shown in Figure 8.39. A technology-independent circuit consisting of a four-input NAND gate driving an inverter which in turn drives a two-input NAND gate has been decomposed in two different ways. The decomposition shown in Figure 8.39(a), when mapped for minimum delay using the library of Figure 8.34 yields a circuit shown in the figure with a delay of 8 units but an area of 9 units. The decomposition shown in Figure 8.39(b) on the other hand yields a slower circuit with a delay of 9 units but has a smaller area of 7 units. If area were important, the second decomposition would be better, while if speed were important, the first one would be chosen. The choice for decomposition when the delay of the circuit is the target should be made as a function of the arrival times at the leaves of the tree. Since such information is not available before technology mapping, a balanced tree decomposition of the nodes of the technology-independent tree circuit is made. There is no guarantee that this will produce the optimal solution.

Assignments of pins of a match to the inputs of the match is another reason why the tree covering algorithm can be suboptimal. During tree matching the symmetry of gate inputs is exploited to

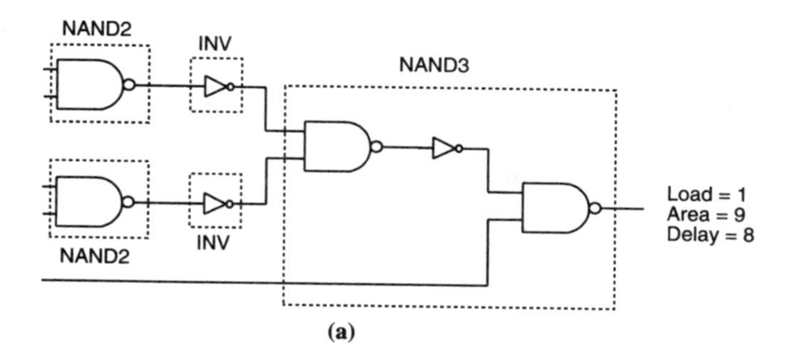

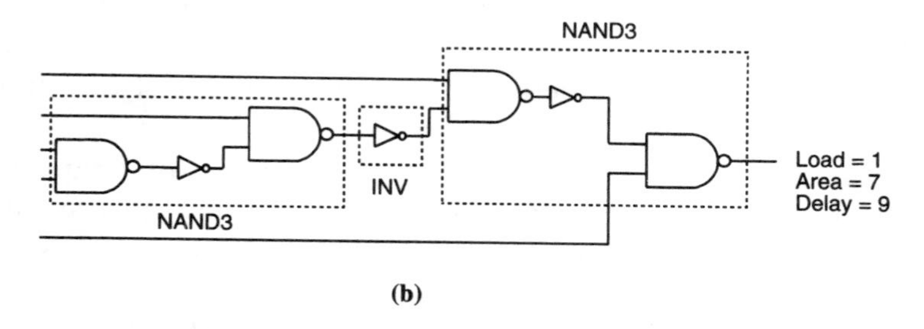

Figure 8.39: Effect of initial decomposition on final result

reduce the computational overhead of the algorithm. For area optimization, where the cost does not depend upon pin assignment, this does not matter. However, for delay optimization this is a factor because delay through a gate is pin dependent in general. Therefore, assigning the slowest arriving signal to the pin with the largest delay can increase the arrival time at the output of the gate. For example, if every NAND gate in the circuit of Figure 8.38(a) has an intrinsic delay of 1 unit between the top input pin and the output while it has an intrinsic delay of 1.5 units between the bottom pin and the output, then the circuit will have a delay of 14 units. If the inputs to the NAND gate 11 are interchanged, then the circuit will have a delay of 13.5 units.

In our delay model we did not distinguish between rise and fall times. In reality, rise and fall times are different, and therefore arrival times at the output of the gate will depend on whether the signal is rising or falling. In practice, arrival times are characterized by a pair of real numbers (a_r, a_f), where a_r is the arrival time of the rising signal while a_f is the arrival time of the falling signal, instead of a single number. To decide which of two solutions is better, we need to decide which of two pairs is faster. One criterion is as follows:

$$(a_r, a_f) < (b_r, b_f) \;\; if \;\; max(a_r, a_f) < max(b_r, b_f)$$

This selection is not guaranteed to be optimal in general.

8.6.5 Fan-Out Optimization

A circuit is usually not a tree but a DAG, and there are gates that fan-out to more than one other gate. A circuit can be divided into tree and fan-out regions as shown in Figure 8.40. Tree covering techniques can be used to obtain the best mapping for the shaded tree regions. The distribution of the signals in the unshaded fan-out regions is an important factor in the determination of the delay of the circuit. For example, consider a signal to be fed from a source gate A to the destinations shown in Figure 8.41(a). Given the total capacitive loads as shown in the figure, the arrival time of the signal at all its destinations is 9 units. However, if a simple buffering tree as shown in Figure 8.41(b) is used, then at the expense of extra area and power dissipation, the speed of the circuit can be improved. The objective of fan-out optimization is to build such fan-out trees that do not compute any function but simply distribute a signal to one or more destinations at a minimum cost.

Fan-out optimization is important for reducing circuit delay. If the output of a gate is connected to n destinations, then the delay through the gate is, to a first approximation, $O(n)$. By building a simple buffer tree this delay can be reduced to $O(log\ n)$. Fan-out optimization can also be used to impose fan-out considerations imposed by a technology.

The problem of fan-out optimization can be formulated as follows. Given a library of buffers and inverters with known delay parameters, a source signal s with a drive capability β_s, and required times r_i, loads γ_i, and the polarity p_i at all the sinks, find a tree of buffers and inverters that distributes the signal from the source to all the sinks and maximizes the required time at the source.

This problem is the dual of the tree covering problem. For delay minimization tree covering aims at minimizing the arrival time at the root of a tree given arrival times at its leaves, while fan-out optimization aims at maximizing the required time at the root of the fan-out tree given the required times at the leaves. In terms of complexity, fan-out optimization is *nondeterministic polynomial-time*

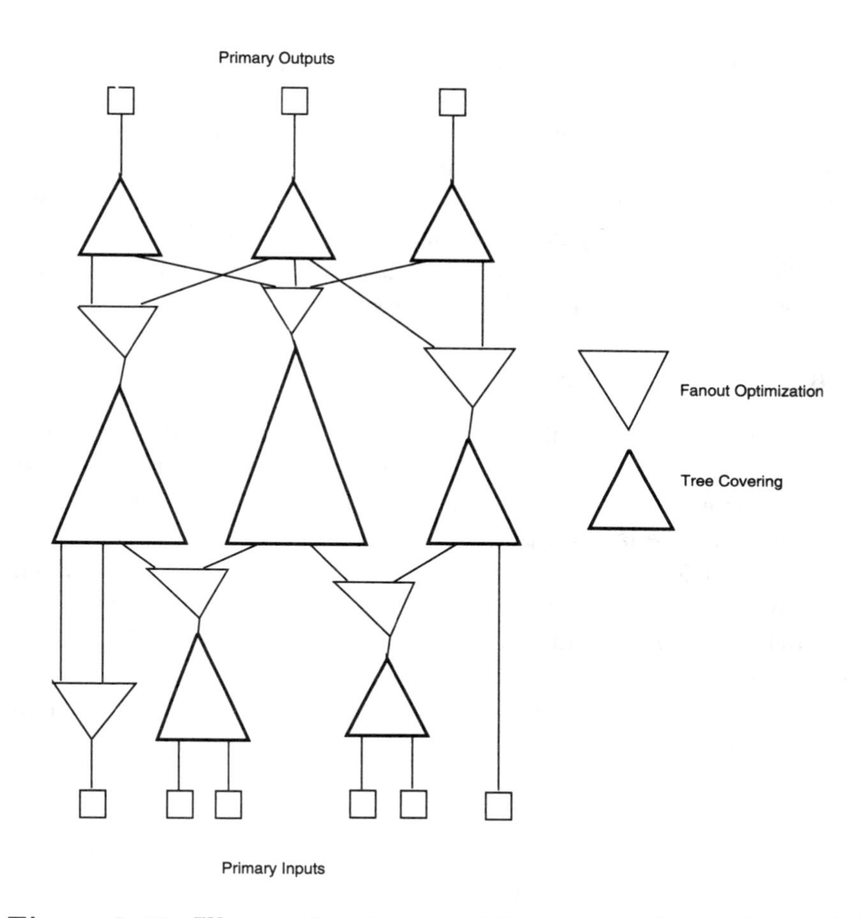

Figure 8.40: Illustrating tree and fan-out regions of a circuit

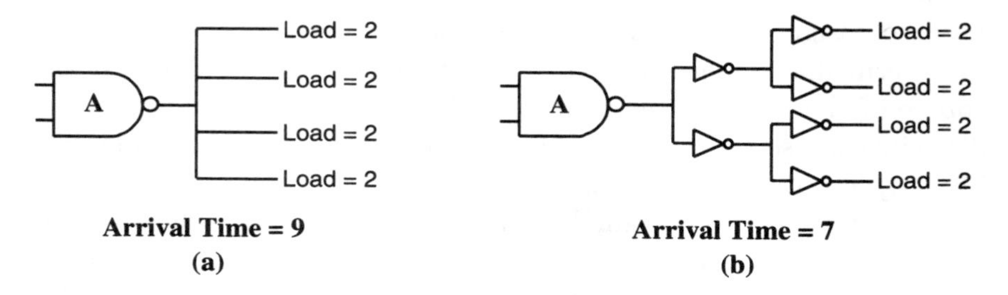

Figure 8.41: Effect of fan-out optimization

(NP)-complete for all but the simplest approximations. In particular, most strategies optimize each fan-out tree (each inverted triangle in Figure 8.41) separately. In contrast, tree covering itself is of linear complexity, but extending it to a globally optimum algorithm is a difficult problem as the global algorithm requires optimum DAG covering which is known to be NP-complete.

We will describe three different heuristic methods for fan-out optimization that all share the characteristic that they operate on a single fan-out tree of the given circuit. The first method we describe (Section 8.6.6) is based on restricting oneself to two-level trees. The second and third methods described in Sections 8.6.7 and 8.6.8 generate a richer set of fan-out-free structures.

8.6.6 Two-level Trees

A tree is a two-level tree if any leaf of the tree is separated from the root of the tree by exactly one intermediate node. One of the simplest ways to perform fan-out optimization is to insert a two-level tree of buffers at multiple fan-out points in the circuit.

Let the capacitive loads of the sinks be $(\gamma_1, \ldots, \gamma_n)$. Then we can precompute the quantities $\gamma_{i,n} = \sum_{i \leq k \leq n} \gamma_k$. Assume that β_b is the drive capability and γ_b the load of a buffer b. Assuming that a fan-out tree of only one level consisting of the same kind of buffer is used and the load is distributed evenly amongst these buffers, then for any buffer b, the optimum number of buffers to be chosen is given as $N_b = \lfloor \sqrt{\frac{\beta_b \gamma_{1,n}}{\beta_s \gamma_b}} \rfloor$.

Though this calculation is performed assuming that loads are equally distributed among all the intermediate buffers, this will not be generally true. Unfortunately, the assignments of sinks to the buffers is equivalent to the multiprocessor scheduling problem and is known to be NP-complete [8]. In practice, a greedy heuristic is used to assign sinks to intermediate buffers. A sink is assigned to an intermediate buffer in such a way that the required time at the source of the fan-out tree is decreased the least by this assignment. The sinks are sorted in order of increasing required times, and in case of a tie, in order of decreasing loads.

Consider the example of Figure 8.42(a), where the required times of the sinks of a fan-out source are shown. The loads of all sinks are the same, namely 1 and $\beta_s = 4$. Let us assume a buffer b with $\alpha_b = 1, \beta_b = 1$, and $\gamma_b = 1$. Then, according to the formula for the optimal number of buffers, only 2 buffers are required. ($N_b = \lfloor \sqrt{\frac{1(6+6+6+7+7)}{4(1)}} \rfloor$.) The sink with the earliest required time is picked first and assigned to buffer 1. The second sink is now picked. Assigning it to buffer 1 would make the required time at the source 3,

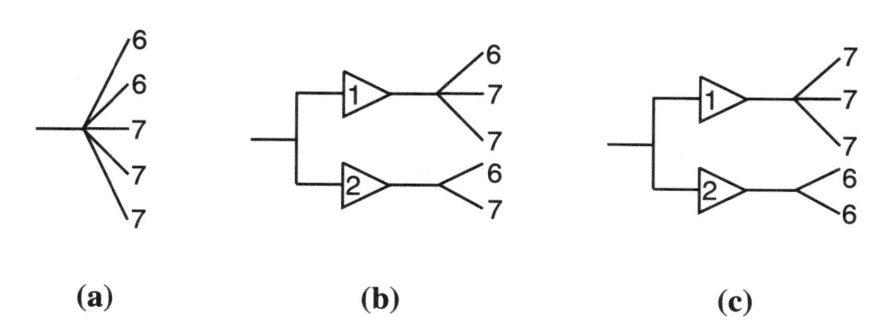

Figure 8.42: Two-level fan-out tree

while assigning it to buffer 2 would make the required time 4. Therefore, it is assigned to buffer 2. The third sink is now assigned to buffer 1 (buffer 2 would be equally good). The fourth sink would maximize the required time at the source only if it is assigned to buffer 2. The fifth sink can be assigned either to buffer 1 or 2. The final assignment is shown in Figure 8.42(b). The required time at the source is 3 since inverter 1 sees a load of 3 and has to produce valid outputs by time 6. Note that this greedy algorithm is not optimal, as the assignment shown in Figure 8.42(c) produces a larger required time of 4.

The selection of the best two-level fan-out tree is performed for each buffer in the library. The buffer that gives the maximum required time is then chosen. The greedy algorithm is of polynomial complexity.

8.6.7 Combinational Merging

Combinational merging is a simple algorithm that can be used to generate a rich set of fan-out tree structures, as opposed to just two-level trees. The algorithm begins by sorting the sinks (or the leaves of the fan-out tree) in order of increasing required times $(r_1, r_2, \ldots, r_n)$. A group of sinks with the largest required times $(r_k, \ldots, r_n)$ is chosen, and the sinks in the group are made the children of a new buffer node. The required time at the input of the buffer node is then computed and is merged with the sorted list of sinks. This transformation is applied until all nodes in the list can be selected at once (i.e., $k = 1$). In that case, the source is used to drive all the sink nodes, unless inserting a buffer between the source and the sinks yields a faster circuit.

A library typically has a set of buffers, each with a different drive capability and maximum allowable fan-out. In the algorithm

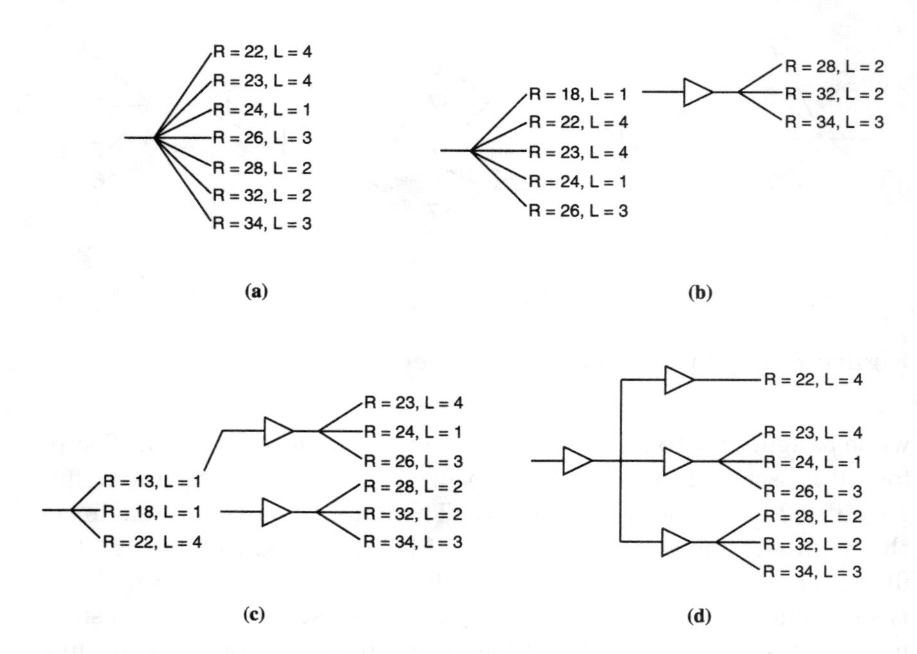

Figure 8.43: Combinational merging

above, two questions have to be answered before combinational merging can be used. The first question is which buffer to select, and the second is what is the value of k for the chosen buffer. Given that the capacitive loads of the sinks are $(\gamma_1, \ldots, \gamma_n)$, we have already seen in Section 8.6.6 that the optimum number of buffers for a two-level fan-out tree is given as $N_b = \sqrt{\frac{\beta_b \gamma_{1,n}}{\beta_s \gamma_b}}$, if the two-level tree exclusively contains buffer b. The value of the index k for this buffer, denoted as k_b, is selected heuristically to be such that $\gamma_{k_b,n}$ just exceeds $\frac{\gamma_{1,n}}{N_b}$. Having selected the value of k_b for a particular buffer the required time at the input of the buffer, r_b, is computed using the equation $r_b = r_{k_b} - \beta_b \gamma_{k_b,n} - \alpha_b - \beta_s \gamma_b$. For each buffer in the library the value of r_b is computed. The buffer that gives the largest required time r_b is selected, and the corresponding number k_b is used as the value of k.

Consider a fan-out tree in Figure 8.43(a). For each sink, the required time (R) and the load (L) are shown and the sinks are sorted according to increasing required times. Assume that the library has only two buffers. The first buffer has $\alpha_1 = 1, \beta_1 = 1$, and $\gamma_1 = 1$, and the second one has $\alpha_2 = 1, \beta_2 = 2$, and $\gamma_2 = 2$. The value of β_s is assumed to be 2. For both types of buffers the value of k is

6, but the first buffer gives a larger required time of 18. A new set of sinks are formed, as shown in Figure 8.43(b), and another pass of the algorithm is initiated. For this pass too the first buffer is chosen, and the fan-out tree of Figure 8.43(c) results. The final fan-out tree is shown in Figure 8.43(d). All buffers are of the first type.

The complexity of the combinational merging algorithm for n sinks is $O(n \; log \; n)$. Unfortunately, this algorithm relies on a simple but suboptimal heuristic to determine which type of buffer to use and the number of fan-outs to be grouped under one buffer.

8.6.8 LT-Trees

The LT-tree algorithm, unlike combinational merging, considers only a subset of the set of all possible fan-out trees. This subset is kept small enough to make the algorithm practically useful, yet powerful enough to perform buffering of large capacitive loads. This approach is not restricted to two-level trees. Moreover, this approach uses dynamic programming to select not only the shape of the tree but also the types of buffers to be used in the tree.

An LT-tree can be defined recursively as follows:

1. A leaf is an LT-tree.
2. A two-level tree is an LT-tree.
3. Let T be a tree rooted at r such that one child of r is an LT-tree and all the other children of r are leaves. Then T is an LT-tree.

In an LT-tree, there is at most one intermediate node that has more than one intermediate node as its children. If there is no such node, then the LT-tree is said to be of type 1. If there is such a node, the node is the root of a two-level tree that terminates the LT-tree. In such cases, the LT-tree is said to be of type 2. Examples of type 1 and type 2 LT-trees are shown in Figure 8.44. It can be shown for n sinks and d types of buffers in the library that the maximum number of distinct LT-trees of type 1 is $(d+1)^{n-2}$.

The LT-tree based algorithm selects the best LT-tree for a fan-out problem using dynamic programming. However, the following restrictions are imposed.

1. Two-level subtrees of LT-trees of type 2 are generated by the two-level fan-out algorithm.

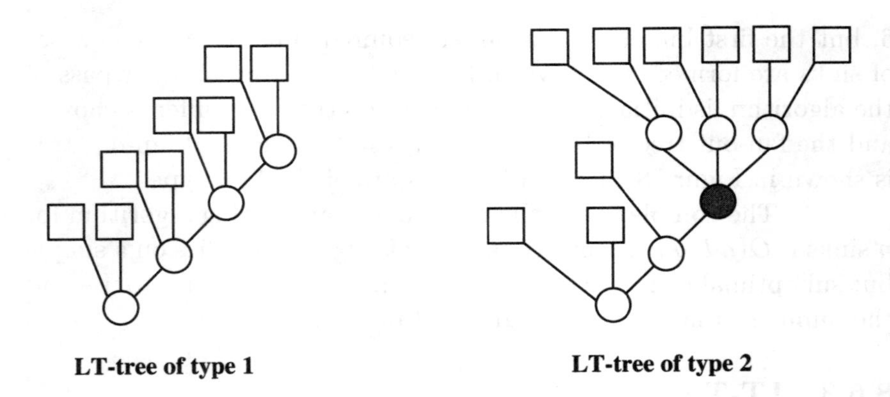

Figure 8.44: Examples of LT-trees

2. The sinks appear in the tree in order of increasing required times.

As in the previous case the sinks are sorted in order of increasing required times, and the quantities $\gamma_{i,n} = \sum_{i \leq j \leq n} \gamma_j$ are calculated. After that, all possible two-level trees that can be subtrees of an LT-tree are calculated. This is performed by calling the two-level fan-out algorithm of Section 8.6.6 with a fan-out problem composed of a source and $n - p + 1$ sinks. The source is either a buffer b from the library, or, if $p = 1$, the source s of the original fan-out problem. The sinks are the $n - p + 1$ sinks with the largest required times, $(r_p, \ldots, r_n)$. This computation is performed for all values of p, and for each $p > 1$, for all buffers in the library. For $p = 1$ the source directly drives all the sinks. For each value of p the best two-level tree is chosen using the algorithm of Section 8.6.6. For each value of p and a buffer g the required time is stored in a matrix $required_{two\_level}[p, g]$. Note that each pair (p, g) specifies a fan-out subproblem of source g and sinks with required times $(r_p, \ldots, r_n)$.

The algorithm relies on dynamic programming to compute by induction on p, with p varying from n to 1, an LT-tree that achieves the maximum required time for each fan-out subproblem (p, g). For a given pair (p, g), an optimal LT-tree, $T_{(p,g)}$, can be obtained by selecting the best of the following $(n - p)d + 1$ configurations, where d is the number of buffers in the library.

1. For some sink with index $l > p$ and buffer type b the root of the optimal LT-tree $T_{(p,g)}$ is directly connected to sinks with arrival times $(r_p, \ldots, r_{l-1})$ and to a buffer of type b. The subtree

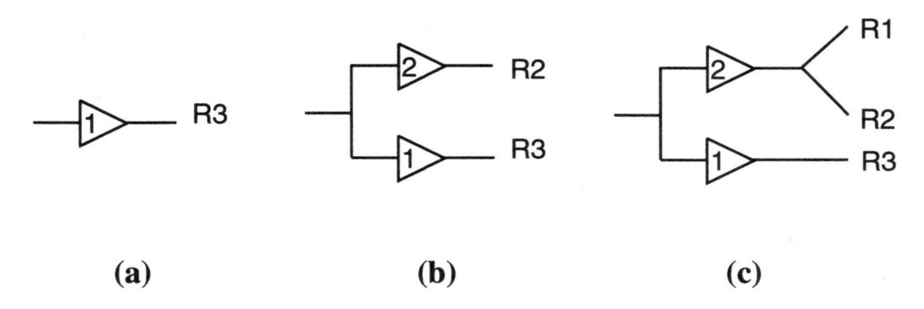

Figure 8.45: Precomputation of best two-level trees

connected to the buffer b is an optimal LT-tree $T_{(l,b)}$ for the subproblem (l, b). Since the algorithm proceeds by induction from n to 1, $T_{(l,b)}$ has already been computed and is available.

2. $T_{(p,g)}$ is a two-level tree precomputed for pair (p, g).

Consider an example where the number of sinks is $n = 3$ and the number of buffers in the library is $d = 2$. We first compute the best two-level tree for the three subproblems as shown in Figure 8.45. For instance in the case of three sinks we use buffer 2 in the library for driving the first and second sinks and buffer 1 for driving the third sink.

We begin with $p = n = 3$. This implies that the number of sinks we are considering is 1. We have a single sink with a required time of r_3. Our choice is the best two-level tree we have computed for the single sink case shown in Figure 8.45(a) denoted $T_{(3,g)}$.

Next, we move to the $p = 2$ case. We compute three different solutions all of which have $T_{(3,g)}$ as shown in Figure 8.46. The root of $T_{(3,g)}$ is always directly connected to sink r_3. Sink r_2 is driven by the source s or a buffer of type 1 or by a buffer of type 2. We choose $T_{(2,g)}$ to be the best amongst the three configurations in Figure 8.46 and the two-level tree of Figure 8.45(b). The algorithm continues in the same manner for the $p = 1$ case, which is the last iteration.

Overall, the complexity of the LT-tree based algorithm is polynomial in the number of sink nodes and buffers in the library. However, the algorithm is not optimal because the two-level fan-out tree generation algorithm is not optimal.

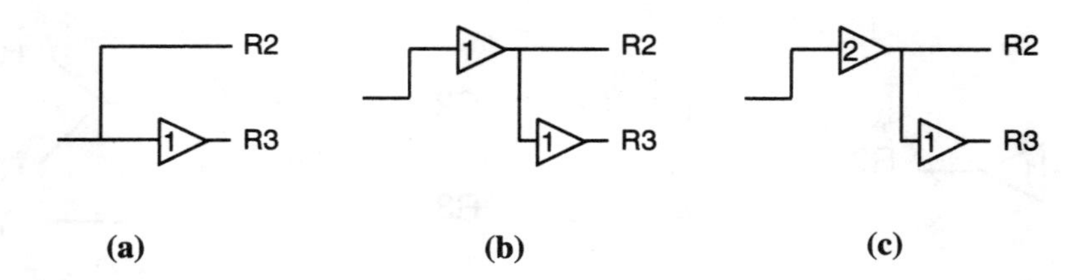

Figure 8.46: Choices in the dynamic programming step with p = 2

Problems

1. Give an example of a circuit with fixed delays where a path π beginning from an input l is statically sensitized on a vector v_2, but applying $\langle v_1,\ v_2 \rangle$ where $v_1 = v_2 - \{l\} \cup \bar{l}$ does not propagate a transition to the primary output of the circuit.

2. Give a timed calculus over $\{0, 1, 2\}$ for a three-input OR gate that can be used for true floating mode delay computation.

3. Run the timed test generation procedure on the carry-bypass circuit of Figure 8.3, attempting to justify a 0(10) at the `c2` output. Assume the gate delays and input arrival times given in the analysis following the figure. Assume a backtrace procedure which selects primary inputs that are at unknown values in the order `a0`, `b0`, `c0`, `a1`, and `b1`. Draw the decision tree for the run of the procedure. How many backtracks are required before the procedure completes?

4. The **PODEM** strategy can be embellished using backward implication in the **IMPLY** procedure. Given a required logical value at the output of the circuit, logical values can be inferred for intermediate lines in the circuit. For example, if the output of an AND gate is required to be a 1, we can infer that the inputs all have to be 1s. Similarly, if the output of an AND gate is a 0, a single input to the AND gate is unknown, and the remaining inputs are already set to 1s, we can infer that the unknown input has to be a 0. Thus, beginning from the logical value required at the primary output of the circuit and the given par-

tial input setting, values can be deduced for intermediate lines. If the deduced values conflict with the values produced by the current partial input setting, then we know that the primary output cannot be set to the required logical value under the current partial input setting.

Develop a timed backward implication strategy for use in timed test generation. Given an AND gate with delay d, a required output logical value L in $\{0, 1\}$ and a required delay δ, consider all possible logical settings in $\{0, 1, 2\}$ for the AND gate inputs. Consider also the cases where each input i has an upper bound on its delay value u_i with $u_i < \delta - d$ or $u_i \geq \delta - d$. For each case, state whether logical values and/or required delays can be inferred for the inputs, or whether a logical or time conflict has been detected.

5. Consider the situation where backward implication has inferred a logical value of 0 for the output of a three-input AND gate, whose inputs are set by forward implication to 1, 2, and 2, respectively. (In this case, forward implication cannot set the value of the AND gate output to a known value.) If one is given the lower and upper bounds for the delays at each of the inputs, is it possible to use the required value at the gate output to tighten the bounds provided by the timed calculus of Table 8.1? By tightening the bounds, we mean that the upper bound has been made lower or the lower bound has been made higher or both.

6. Prove that the routine **AND_DECOMP** will produce a balanced AND tree with k levels if given a 2^k-input AND gate whose inputs all have equal arrival times.

7. Prove that the complexity of the two-level tree fan-out optimization algorithm is $O(dn^{1.5})$, where d is the number of buffers in the library and n is the number of sinks.

8. Prove that the number of LT-trees of type 1 is equal to $(d + 1)^{n-2}$, where d is the number of buffers in the library and n is the number of sinks.

REFERENCES

[1] R. Brayton, R. Rudell, A. Sangiovanni-Vincentelli, and A. Wang. MIS: A Multiple-Level Logic Optimization System. *IEEE Transactions on Computer-Aided Design of Integrated Circuits*, CAD-6(6):1062–1081, November 1987.

[2] K. Chaudhary and M. Pedram. A Near Optimal Algorithm for Technology Mapping Minimizing Area under Delay Constraints. In *Proceedings of the* 29th *Design Automation Conference*, pages 492–498, June 1992.

[3] H. C. Chen and D. H. Du. Path Sensitization in Critical Path Problem. In *Proceedings, Tau 90: 1990 ACM Workshop on Timing Issues in the Specification and Synthesis of Digital Systems*, August 1990.

[4] S. Devadas, K. Keutzer, and S. Malik. Delay Computation in Combinational Logic Circuits: Theory and Algorithms. In *Proceedings of the International Conference on Computer-Aided Design*, pages 176–179, November 1991.

[5] S. Devadas, K. Keutzer, S. Malik, and A. Wang. Certified Timing Verification and the Transition Delay of a Combinational Logic Circuit. In *Proceedings of the Design Automation Conference*, pages 549–555, June 1992.

[6] S. Devadas, K. Keutzer, S. Malik, and A. Wang. Computation of Floating Mode Delay in Combinational Logic Circuits: Practice and Implementation. In *Proceedings of the International Symposium on Logic Synthesis and Microprocessor Architecture*, pages 68–75, July 1992.

[7] S. Devadas, K. Keutzer, S. Malik, and A. Wang. Event Suppression: Improving the Efficiency of Timing Simulation for Synchronous Digital Circuits. In *Proceedings of the Brown/MIT Conference on Advanced Research in VLSI and Parallel Systems*, pages 195–209, March 1992.

[8] M. R. Garey and D. S. Johnson. *Computers and Intractability: A Guide to the Theory of NP-Completeness.* W. H. Freeman and Company, New York, NY, 1979.

[9] P. Goel. An Implicit Enumeration Algorithm to Generate Tests for Combinational Logic Circuits. *IEEE Transactions on Computers*, C-30(3):215–222, March 1981.

[10] D. Hodges and H. Jackson. *Analysis and Design of Integrated Circuits.* McGraw-Hill, New York, NY, 1988.

[11] V. Hrapcenko. Depth and Delay in a Network. *Soviet Math. Dokl.*, 19(4), 1978.

[12] P. McGeer and R. Brayton. Efficient Algorithms for Computing the Longest Viable Path in a Combinational Network. In *Proceedings of the* 26th *Design Automation Conference*, pages 561–567, June 1989.

[13] P. C. McGeer and R. K. Brayton. *Integrating Functional and Temporal Domains in Logic Design.* Kluwer Academic Publishers, Norwell, MA, 1991.

[14] C. H. Papadimitriou and K. Steiglitz. *Combinatorial Optimization: Algorithms and Complexity.* Prentice-Hall, New Jersey, 1982.

[15] R. Rudell. *Logic Synthesis for VLSI Design.* PhD thesis, University of California at Berkeley, April, 1989. ERL Memo 89/49.

[16] K. J. Singh. *Performance Optimization of Digital Circuits.* PhD thesis, University of California at Berkeley, CA, December 1992.

[17] H. Touati. *Performance-Oriented Technology Mapping.* PhD thesis, University of California at Berkeley, November 1990. ERL Memo M90/109.

Chapter 9

Testability of Multilevel Circuits

9.1 Introduction

In this chapter we will analyze testability requirements for multilevel circuits. Multilevel circuits are much more complex, topologically speaking, than two-level circuits. To facilitate testability analysis for multiple stuck-at faults and delay faults we will be using a two-level representation of a multilevel circuit called the *equivalent normal form* (ENF).

We first consider the single stuck-at fault model in Section 9.2. We give conditions for the testability of a single stuck-at fault in a multilevel circuit and describe the relationships between "don't-cares" and testability and circuit speed and testability.

We describe the ENF representation in Section 9.3. The ENF is not only used to analyze testability conditions, but also to analyze logic transformations — we define a notion of ENF reducibility in Section 9.4 that will prove useful in determining the set of logic transformations that retain testability under various fault models (see Section 9.5).

In Sections 9.6 through 9.9 we focus on testability conditions, synthesis procedures, and test generation procedures for multilevel circuits under the multiple stuck-at, path delay, and gate delay fault models. We briefly touch upon transistor stuck-open faults in Section 9.10. An example synthesis of a circuit that implements the Viterbi algorithm for speech decoding is described in Section 9.11. Unless otherwise stated, the theoretical results presented in this chapter will correspond to single-output multilevel circuits. Most of the results can be generalized to the multiple-output case.

9.2 Single Stuck-At Faults

We give conditions for a single stuck-at fault to be testable in a multilevel circuit in Section 9.2.1 and briefly touch upon test generation methods in Section 9.2.2. We describe the relationship between single stuck-at fault testability and "don't-cares" in Section 9.2.3 and the relationship between testability and circuit speed in Section 9.2.4.

9.2.1 Conditions for Testability

Most of the material in this section is taken from [6]. Consider a combinational circuit C which realizes the function $f(x_1, x_2, \cdots, x_N)$. If we denote by α the fault in which input x_i is stuck-at-0, the function, f_α, realized by the faulty circuit is the cofactor of f with respect to $\overline{x_i}$ and is given by:

$$f_\alpha(x_1, x_2, \cdots, x_N) = f(x_1, x_2, \cdots, x_{i-1}, 0, x_{i+1} \cdots, x_N) = f_{\overline{x_i}}$$

Similarly if x_i is stuck-at-1 the function is f_{x_i}

The set of tests which detect the fault α (x_i stuck-at-0) correspond to T given below.

$$\begin{aligned} T &= f \cdot \overline{f_\alpha} + \overline{f} \cdot f_\alpha \\ &= f \oplus f_\alpha \\ &= (\overline{x_i} \cdot f_{\overline{x_i}} + x_i \cdot f_{x_i}) \oplus f_{\overline{x_i}} \\ &= x_i \cdot f_{x_i} \cdot \overline{f_{\overline{x_i}}} + x_i \cdot \overline{f_{x_i}} \cdot f_{\overline{x_i}} \\ &= x_i \cdot (f_{x_i} \oplus f_{\overline{x_i}}) \end{aligned}$$

$f_{x_i} \oplus f_{\overline{x_i}}$ is referred to as the *Boolean difference* of f with respect to x_i and is denoted by $\frac{\partial f}{\partial x_i}$. It represents all conditions (associated with all inputs except x_i) under which the value of f is sensitive to x_i alone. $x_i \cdot \frac{\partial f}{\partial x_i}$ represents the set of all tests for the fault x_i stuck-at-0 since x_i applies the opposite signal value on the faulty input and the factor $\frac{\partial f}{\partial x_i}$ ensures that this erroneous signal affects the value of f. Similarly, the set of all tests which detect x_i stuck-at-1 is defined by the Boolean expression $\overline{x_i} \cdot \frac{\partial f}{\partial x_i}$.

Consider the circuit of Figure 9.1. The output of the circuit is defined by $f = (b+c) \cdot a + \overline{a} \cdot d$. The set of tests which detect the

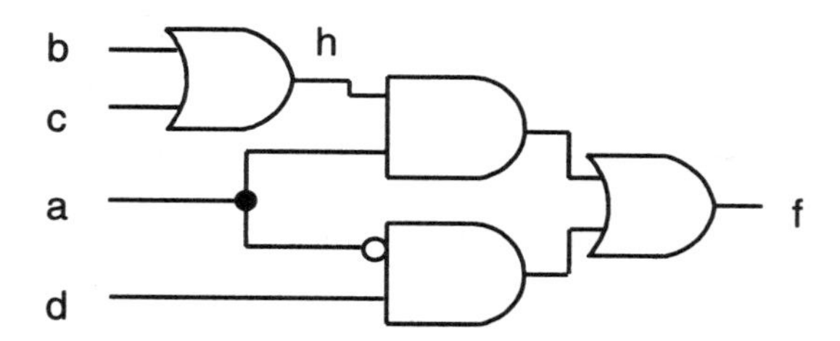

Figure 9.1: Circuit illustrating Boolean difference

fault a stuck-at-0 is defined by the Boolean expression $a \cdot \frac{\partial f}{\partial a}$ where:

$$\begin{aligned}\frac{\partial f}{\partial a} &= f_a \oplus f_{\overline{a}} \\ &= d \oplus (b + c) \\ &= \overline{b} \cdot \overline{c} \cdot d + b \cdot \overline{d} + c \cdot \overline{d}\end{aligned}$$

Thus, the set of all tests which detect this fault is defined by $T = a \cdot (\overline{b} \cdot \overline{c} \cdot d + b \cdot \overline{d} + c \cdot \overline{d})$.

The Boolean difference method can be used to derive tests for stuck-at faults that are internal to the circuit as illustrated below (taken from [6]).

Theorem 9.2.1 *Let C be a circuit which realizes a function $f(x_1, \cdots, x_N)$ and let h be an internal signal of C. Then h can be expressed as a function of the primary inputs $h(x_1, \cdots, x_N)$, and f can be expressed as a function of the primary inputs and h (by considering h as an input). Let this new function be f'. The set of all tests which detect the fault h stuck-at-0 is defined by the Boolean expression*

$$h(x_1, \cdots, x_N) \cdot \frac{\partial f'(x_1, \cdots, x_N, h)}{\partial h}$$

The set of all tests for h stuck-at-1 can also be defined similarly. The first term in the expression corresponds to justifying the opposite value on the line on which the fault resides. The Boolean difference term corresponds to the conditions for propagating the error to a primary output. Both conditions have to be satisfied in order to detect the fault.

Consider h in the circuit of Figure 9.1. $h = b + c$. The function f' expressed in terms of h and the primary inputs is $f' = a \cdot h + \overline{a} \cdot d$. By Theorem 9.2.1, the set of tests for h stuck-at-0 is $(b + c) \cdot a$.

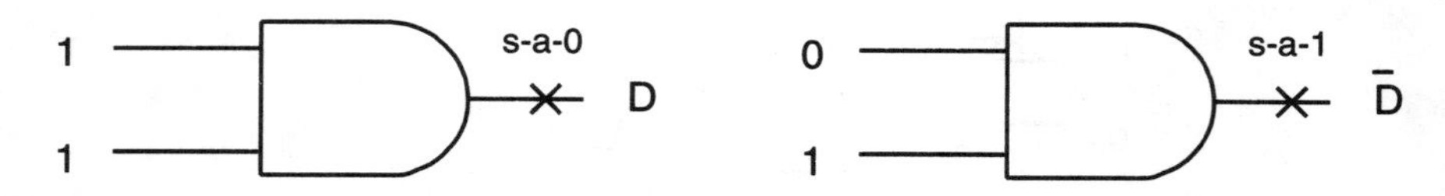

Figure 9.2: Generation of D values

9.2.2 Test Generation Methods

Test generation for combinational circuits has received a great deal of attention over the past few years. Sophisticated and efficient test pattern generators have been developed (e.g., [25, 32]), which exhibit average-case linear behavior. The basis of the satisfiability checking method given in [25] is the Boolean difference method described in the previous section. The methods used in single stuck-at fault test generation have been comprehensively described in several books (e.g., [6, 18]) and we will not describe them in detail here.

The basis of most of the current test generators in use today is the **PODEM** procedure for implicit enumeration described in Section 8.3.1. While many enhancements have been made to the basic strategy, state-of-the-art methods mimic the search strategy illustrated in Figures 8.15 and 8.16. The example of Figures 8.17 and 8.18, showing the various steps of the **PODEM** algorithm, can be viewed as an example of generating a test for the primary output stuck-at-0 fault of the circuit. Any input vector that sets the output to a 1 will be a test for this fault.

In order to generate tests for internal faults in the circuit, the standard three-valued ($\{0, 1, 2\}$) logic simulation calculus of **PODEM** has to be replaced by a D calculus [30]. (This extension is carried out much in the same way as we extended **PODEM** for floating delay computation using the timed calculus of Figure 8.1.) The D value represents the error condition that there is a 1 in the fault-free circuit and a 0 in the faulty circuit. Similarly, the $\overline{D}$ value represents the error condition that there is a 0 in the fault-free circuit and a 1 in the faulty circuit. This *generation* of D and $\overline{D}$ error values is illustrated in Figure 9.2. Error values can only be generated at the fault site, i.e., the signal which is stuck-at-0 or stuck-at-1.

Error values are *propagated* through the circuit according to the D calculus. The complete D calculus for a two-input AND gate is defined over the values $\{0, 1, 2, D, \overline{D}\}$ and is given in Table 9.1. In order for a D ($\overline{D}$) to propagate through a gate, the other inputs to the gate must be at noncontrolling values or at D ($\overline{D}$) values. If the

$i_1 \rightarrow$ $i_2 \downarrow$	0	1	2	D	$\overline{D}$
0	0	0	0	0	0
1	0	1	2	D	$\overline{D}$
2	0	2	2	2	2
D	0	D	2	D	0
$\overline{D}$	0	$\overline{D}$	2	0	$\overline{D}$

Table 9.1: D Calculus for an AND gate

inputs to a two-input AND are at D and $\overline{D}$, then the output is a 0, since the two error values cancel each other out.

By using the D calculus in **PODEM**, tests can be generated for any internal fault in a circuit. The objective is to obtain a D or $\overline{D}$ at the output of the circuit, implying that an error value has been propagated from the fault site all the way to the circuit output.

9.2.3 Don't-Cares and Testability

A Boolean network can be made prime and irredundant by minimizing each node in the network under its complete observability and satisfiability don't-care sets, where each node is a two-level Boolean function. These don't-care sets were defined in Section 7.6. Note that these don't-care sets are *internal* don't-care sets, as opposed to the *external* don't-care sets considered in Chapter 3.

We can make a precise connection between the satisfiability and observability don't-care sets and single stuck-at fault testability. A signal y_j in a Boolean network η can be tested for stuck-at-1 by finding an input test vector v such that v is contained in both $\overline{SDC_{\overline{y_j}}}$ and $\overline{ODC_j}$. The first term says that, when we evaluate the network at v the value of y_j is 0 and the other values of the intermediate variables y_i satisfy the compatibility relations $y_i = f(v, y)$. The second term says that the value of v is such that it allows the value of y_j to be observed, at least at one output. These conditions are associated with the ability to justify the fault and propagate its effect to an output as described in Section 9.2.1. If a signal y_j is not testable for stuck-at-1 (stuck-at-0), then the Boolean network is unaffected from a logical functionality standpoint if y_j is replaced with 1 (0), thereby simplifying the network. This replacement is termed *redundancy removal.*

Node minimization techniques augmented with the satisfiability and observability don't-care sets can ensure primality and irredundancy for a Boolean network and guarantee 100 percent single stuck-at fault testability [3]. While don't-care exploitation and minimization is useful for area minimization [4], the don't-care minimization procedure of [3] that makes a Boolean network prime and irredundant has found little practical use. This is because complete don't-care sets are typically very large, and therefore it can be very time consuming to generate them and use them during minimization.

Currently, the most popular method for obtaining prime and irredundant Boolean networks or fully single stuck-at fault testable multilevel circuits is to use test generation algorithms to iteratively identify and remove single stuck-at fault redundancies in combinational logic circuits. As mentioned in Section 9.2.2 the extensive work that has gone into single stuck-at fault test generation in recent years has resulted in the development of sophisticated methods for redundancy removal.

Given a multilevel circuit C and an external don't-care set D making the circuit prime and irredundant under D implies the existence of a complete single stuck-at fault test set for C outside D — the multilevel generalization of Theorem 5.3.2. Single stuck-at fault redundancy removal procedures can produce circuits that are prime and irredundant under a don't-care set D via a simple modification. A test generator is run on the circuit C and a fault f is considered. If no test vector can be found for f, f is redundant, and the wire in C corresponding to f can be replaced by an appropriate constant value. If, however, a test vector is found, we check to see if the test vector lies outside D. If so, f is testable. If not, we attempt to generate other tests for f. If we cannot find any test for f which lies outside D, then f is considered to be redundant (under D) and f is removed from C. Making a circuit prime and irredundant under an external don't-care set is an operation that is used extensively in the synthesis of sequential circuits for full single stuck-at fault testability [13, 17].

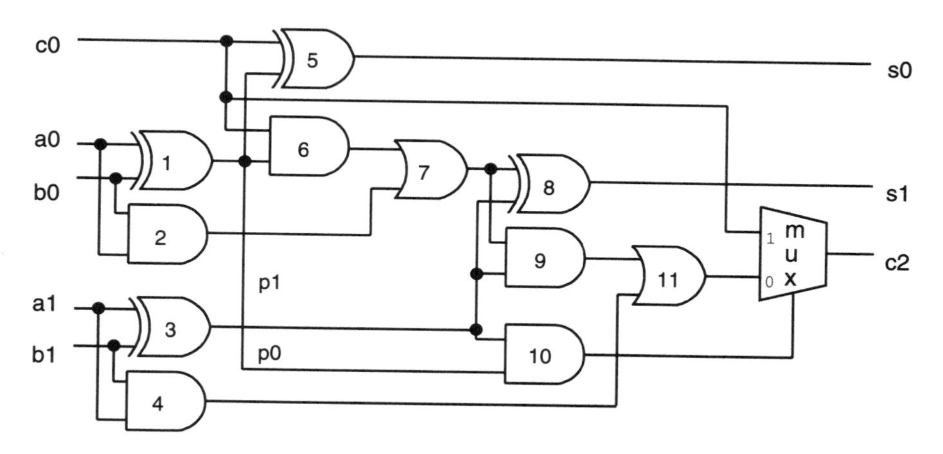

Figure 9.3: 2-bit carry-bypass adder

9.2.4 Performance and Testability

Redundancy removal improves testability and reduces the area of a circuit. However, there are circuits like the traditional implementation of a carry-bypass adder of Figure 9.3, where a stuck-at fault redundancy improves the speed of the circuit. Removing the stuck-at fault redundancy affects the speed of the circuit adversely.

The circuit of Figure 9.3 (identical to Figure 8.3) uses a conventional ripple-carry adder (the output of gate `11` is the ripple-carry output) with an extra AND gate (gate `10`), and an additional multiplexor. As described in Section 8.2.2, if the propagate signals `p0` and `p1` (the outputs of gates `1` and `3`, respectively) are high, then the carry-out of the block `c2` is equal to the carry-in of the block `c0`. Otherwise, it is equal to the output of the ripple-carry adder. The multiplexor thus allows the carry to skip the ripple-carry chain when all the propagate bits are high.

The extra AND gate and the multiplexor of the carry-bypass adder have a profound effect on the speed and testability of the circuit. First consider its impact on the speed of the circuit. Assume the primary input `c0` arrives at time $t = 5$ and all the other primary inputs arrive at time $t = 0$. Let us assign a gate delay of 1 for AND and OR gates and gate delays of 2 for the XOR gates and the multiplexor. The path that determines the worst-case delay of `c2` is the path from `a0` to `c2` through gates `1`, `6`, `7`, `9`, `11`, and the multiplexor. The output of this critical path is available after 8 gate delays. The longest path, including the late arriving input in the circuit, is the path, call

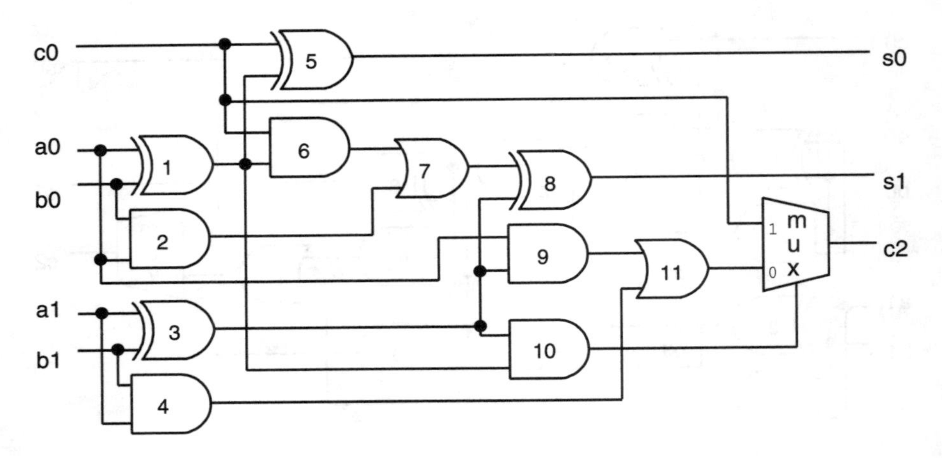

Figure 9.4: 2-bit irredundant carry-bypass adder

it P, from c0 to c2 through gates 6, 7, 9, 11, and the multiplexor (available after 11 gate delays). A transition can never propagate down this path to the output because in order for that to happen the propagate signals have to be high, in which case the transition propagates along the bypass path from c0 through the multiplexor to the output. The length of the longest path in the bypass adder is the same as the ripple-carry adder. However, the critical path is shorter.

The ripple-carry adder is fully single stuck-at fault testable, but the carry-bypass adder has a single stuck-at fault redundancy. The stuck-at-0 fault on the output of the gate 10 is not testable. If the output of the gate is replaced by the constant 0 value, then the circuit of Figure 9.3 becomes a ripple-carry circuit with *greater* critical delay. Thus, a straightforward redundancy removal method that removes stuck-at fault redundancies from a circuit in arbitrary order may lower the speed of a circuit.

A redundancy removal procedure that maintains the speed of a circuit under certain assumptions about the delay model, but can increase the area of a circuit, has been described in [22]. In Figure 9.4, the fully single stuck-at fault testable circuit obtained by applying the speed-maintaining redundancy removal algorithm to the circuit of Figure 9.3 is given. This circuit has been obtained by replacing the connection from the output of gate 7 to gate 9 by the primary input b0. The circuit is not slower than the original redundant circuit.

The operation of the procedure is illustrated in Figure 9.5(a) through Figure 9.5(c). We focus on the longest path of Figure 9.3

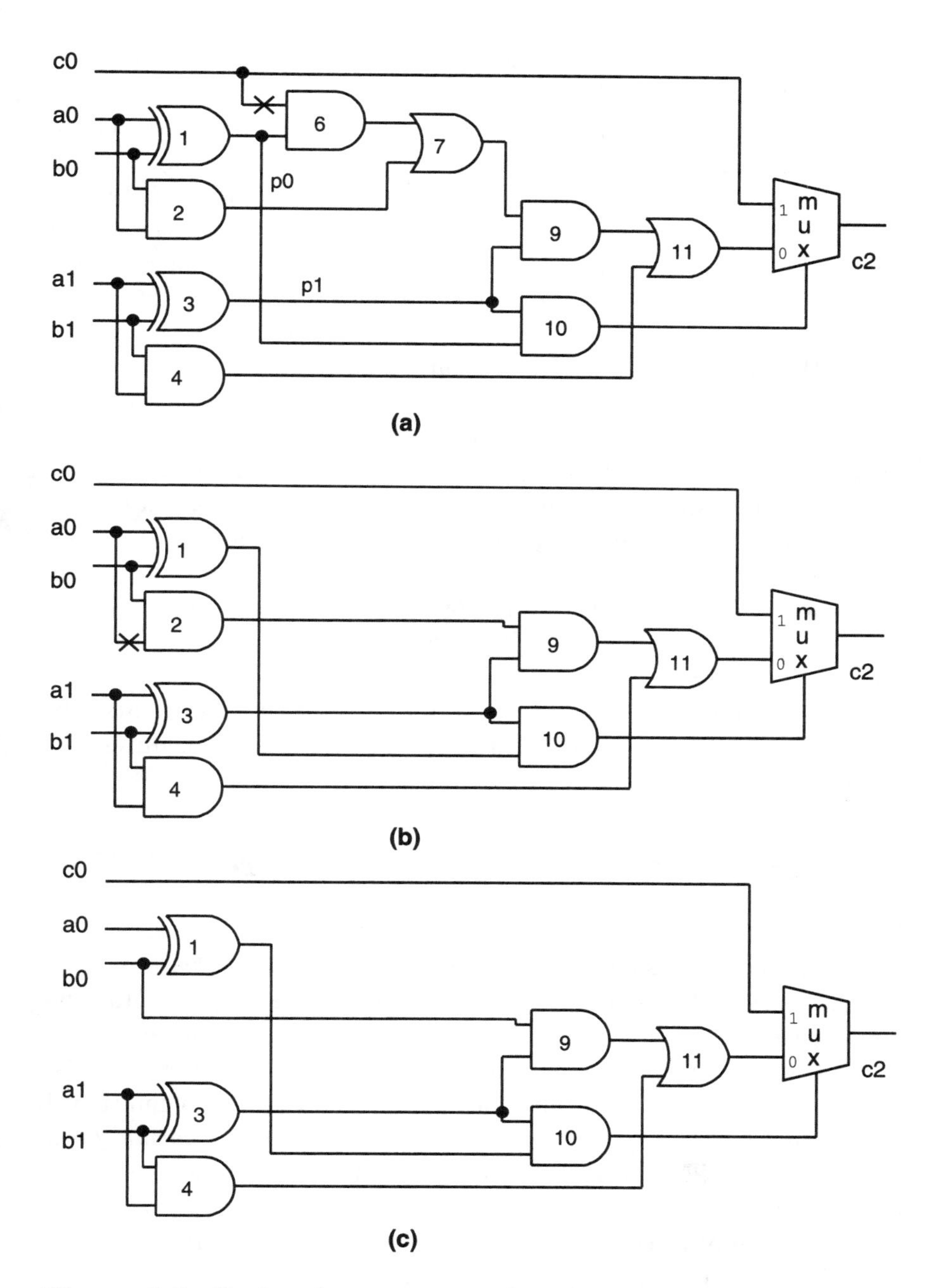

Figure 9.5: Redundancy removal maintains speed

and determine that the longest path P is not statically sensitizable. This is because in order for P to be statically sensitizable all the side-inputs to P must be at noncontrolling values. This requires p0 and p1 to be both 1, but at least one of them must be 0 for the multiplexor to sensitize P. If the longest path is not statically sensitizable, then the path is made fan-out-free. This entails logic duplication of gate 7 since it fans out to gates 8 and 9 in Figure 9.3. In the Figures 9.5(a) through 9.5(c) we do not show the whole circuit, but only the part of the circuit where the transformations are going to be applied. The subcircuit corresponding to c2 has a fan-out-free longest path and is shown in Figure 9.5(a). It can be shown that the first edge of a nonstatically sensitizable fan-out-free path will be stuck-at fault redundant, and this is the case with the marked fault in Figure 9.5(a). We replace the marked wire with a constant 0 and obtain the circuit of Figure 9.5(b). The longest path of this circuit is statically sensitizable, and if this is the case, it can be shown that we can remove the remaining redundancies in any order without affecting the speed of the circuit. The marked stuck-at-1 redundancy in Figure 9.5(b) is removed, and the irredundant circuit of Figure 9.5(c) is obtained. For a proof of the correctness of the procedure see [22].

In most cases, the longest paths of circuits are statically sensitizable and straightforward redundancy removal procedures can be applied to the circuit.

9.3 Equivalent Normal Form Representation

While the results derived for two-level circuits in Chapter 5 were of interest, similar results for multilevel circuits are of much greater utility. We therefore wish to extend the results to multilevel implementations and derive necessary and sufficient conditions for a multilevel circuit to be testable under various fault models. To aid in our analysis, our approach is to use an alternative representation of a multilevel circuit developed by Armstrong [1] called the *equivalent normal form* (ENF). The ENF of a circuit is a two-level representation in which the sensitization conditions for each path are represented by a cube with each literal in the cube annotated by information regarding its path from the primary input of the circuit to the primary output. Similar representations have been used for the analysis of hazards in asynchronous circuits [36] and in for the analysis of transients in circuits [27]. We will define ENF more formally below.

Given a logic circuit C with inputs $i_1, \ldots, i_m$, outputs $o_1, \ldots,$

o_n, and labeled gates $g_1, \ldots, g_p$, we can construct the ENF for each output o_k of C by traversing the circuit forward from the inputs and visiting each gate in topological order. Before we start the traversal, we convert the circuit into a *leaf directed acyclic graph* (leaf-DAG), i.e., each time we reach a gate that fans out to more than one destination, we duplicate that gate and the entire subcircuit behind the gate as many times as necessary until each gate has only one fan-out. Each duplicated gate retains the label of the original. When the output is reached, we have created a leaf-DAG in which only the inputs have fan-out greater than 1. Each complex gate is then translated into AND gates, OR gates, and inverters. The next step is to push all inverters back from the primary output of the subcircuit to the primary inputs. During this transformation AND gates may be converted to OR gates and vice versa. It is obvious that there is a one-to-one correspondence between each path in C and each path in the leaf-DAG circuit. Finally, the ENF expression itself is computed as follows. Each variable in an ENF expression is simply a primary input of C, such as i_j, and has a label consisting of a set of gates denoting the path this input has taken to reach the output. The ENF for a primary input i_j is simply i_j. The intermediate ENF expression corresponding to the output of an AND gate g with inputs a_α, b_β is a sum-of-products expression over tagged literals, $E_g = a_{\alpha \cup g} \cdot b_{\beta \cup g}$. The expression E_g is created by reducing the product to sum-of-products form using DeMorgan's laws and distributivity without making any Boolean reductions such as $(a + \overline{a} \equiv 1)$, $(a + a \equiv a)$, $(a \cdot a \equiv a)$, or $(a \cdot \overline{a} \equiv 0)$. Similarly, the intermediate ENF expression corresponding to the output of an OR gate h with inputs a_α, b_β is a sum-of-products expression over tagged literals, $E_h = a_{\alpha \cup h} + b_{\beta \cup h}$. As before, no Boolean reductions are made in reducing E_h to sum-of-products form.

We illustrate the derivation of the ENF of a circuit with the help of an example taken from [1]. First, the circuit of Figure 9.6 is made internal fan-out-free by unfolding it. (The bubbles at the inputs and outputs of the gates in Figure 9.6 denote inversions.) This is illustrated in Figure 9.7. The numbers inside the gates are unique identifiers for those gates. This involves duplicating gates if needed so that each copy of a gate has a single fan-out connection. The duplicated gates retain the same integer identifier as the original gate. Next, the inverters are pushed backwards toward the primary inputs using DeMorgan's laws of complementation to change the gates encountered in the process. (See Figure 9.8.)

The ENF-two-level circuit for the example circuit of Fig-

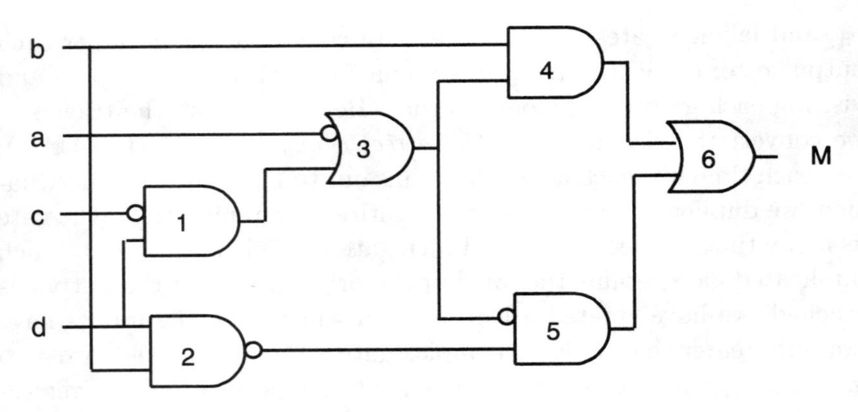

Figure 9.6: Multilevel circuit

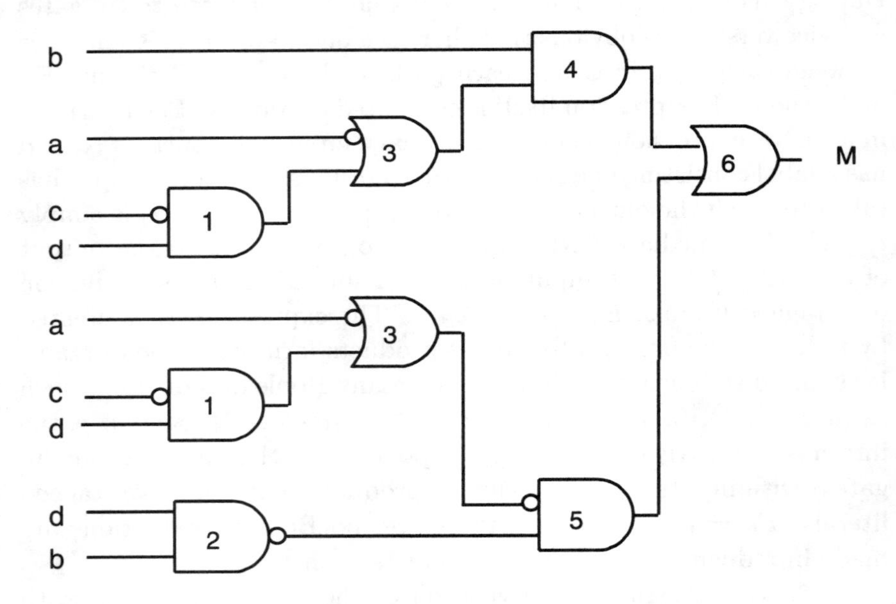

Figure 9.7: Making a circuit internal fan-out-free

ure 9.6 whose leaf-DAG is shown in Figure 9.8 is shown in Figure 9.9. This ENF has been derived using the rules for the AND and OR gates given above.

In Figure 9.10 we show a modified circuit based on Figure 9.6 which we will use as an example for illustrating path sensitization conditions. The final ENF expression for primary output M is:

$M = e_{\{5,9\}} \cdot b_{\{5,9\}} \cdot \overline{a}_{\{1,4,5,9\}} + b_{\{5,9\}} \cdot \overline{c}_{\{2,3,4,5,9\}} \cdot d_{\{3,4,5,9\}} \cdot e_{\{5,9\}}$
$+ a_{\{1,4,6,8,9\}} \cdot c_{\{2,3,4,6,8,9\}} \cdot \overline{b}_{\{7,8,9\}} + a_{\{1,4,6,8,9\}} \cdot c_{\{2,3,4,6,8,9\}} \cdot \overline{d}_{\{7,8,9\}}$

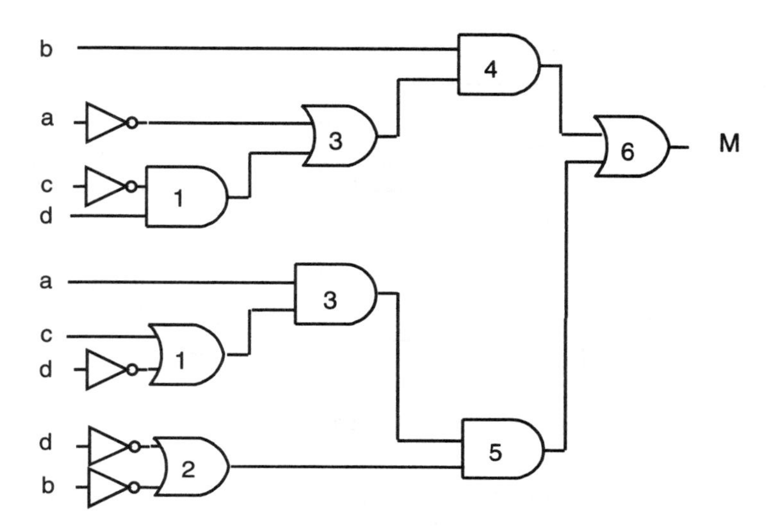

Figure 9.8: Pushing inverters to the primary inputs

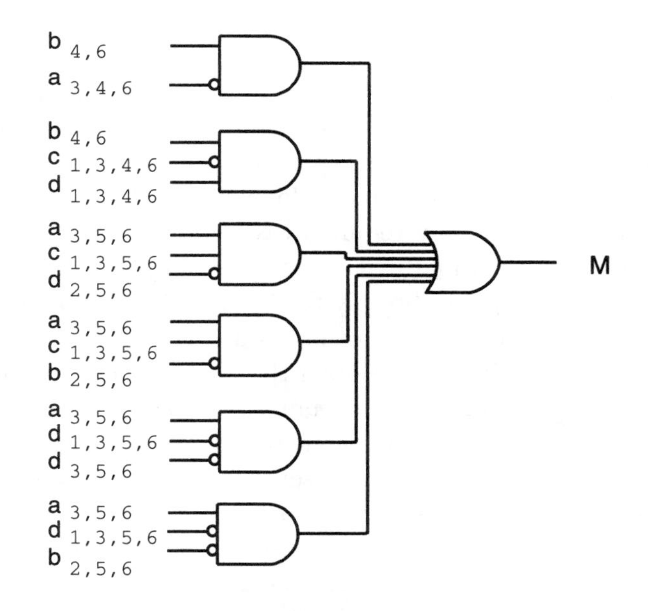

Figure 9.9: A two-level circuit representing the ENF

$+\ a_{\{1,4,6,8,9\}} \cdot \overline{d}_{\{3,4,6,8,9\}} \cdot \overline{b}_{\{7,8,9\}} + a_{\{1,4,6,8,9\}} \cdot \overline{d}_{\{3,4,6,8,9\}} \cdot \overline{d}_{\{7,8,9\}}.$

Generating the ENF of a multilevel circuit may require, in the worst case, computation time and space that grow exponentially with the number of inputs to the circuit. However, the primary motivation for using the ENF to describe the necessary and sufficient

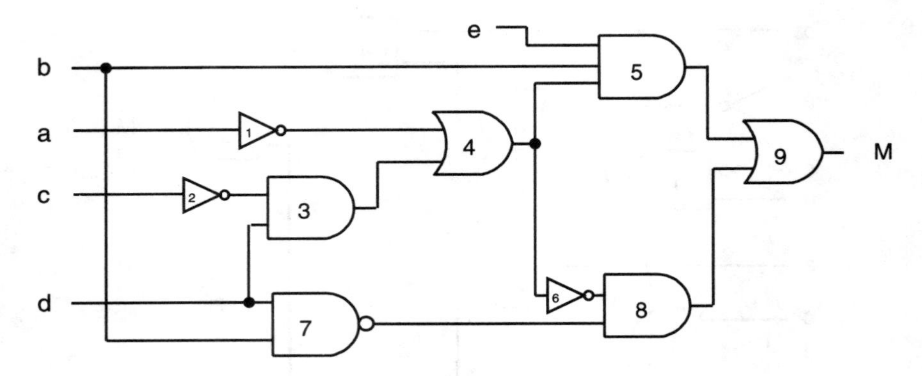

Figure 9.10: Modified multilevel circuit

conditions for testability is for clarity of exposition and for the ease of proving that synthesis procedures retain testability. ENF analysis is only used in proofs; the ENF does not have to be explicitly calculated for each circuit.

As an aside, it is interesting to note that explicit ENF computation and analysis is possible for combinational circuits of significant size. These include virtually all multilevel circuits that have been synthesized from two-level representations [14]. Furthermore, a synthesis methodology has been developed in [14] based on the theory developed here, which only requires ENF computation and analysis for small subcircuits. Finally, it is worth noting that the size of the ENF does not correspond directly to the size of a circuit but is more strongly related to the number of paths in the circuit.

Armstrong in [1] developed the ENF to describe the conditions under which a single stuck-at fault is propagated from the input of a circuit to its outputs. In particular, he showed that a test for a literal l_π in an ENF expression "sensitizes," in a way that will be made clear below, the path π in the original circuit.

Let us consider the ENF expression M given previously. The existence of literal $a_{\{1,4,6,8,9\}}$ in the cube $a_{\{1,4,6,8,9\}} \cdot c_{\{2,3,4,6,8,9\}} \cdot \overline{b}_{\{7,8,9\}}$ indicates that a 1 is needed on primary input a to *sensitize to a 1* the path from input a along $a_{\{1,4,6,8,9\}}$. (If the cube contained $\overline{a}_{\{1,4,6,8,9\}}$, then a 0 would be needed to sensitize to a 1 the path from input a.) In addition, to sensitize to a 1 the path $a_{\{1,4,6,8,9\}}$, a noncontrolling value is needed at the other input of gate 4, so we need a 1 on primary input c. This is reflected by the literal $c_{\{2,3,4,6,8,9\}}$. Finally, we need a noncontrolling value on the other input to gate 8, and this is reflected by the literal $\overline{b}_{\{7,8,9\}}$. A noncontrolling value on the other input to gate 8 could also be accomplished by setting primary input d to a 0. This is reflected by the existence of literal $\overline{d}_{\{7,8,9\}}$ in the cube $a_{\{1,4,6,8,9\}} \cdot c_{\{2,3,4,6,8,9\}} \cdot \overline{d}_{\{7,8,9\}}$.

An input assignment that sensitizes a 1 along a path need not *statically sensitize a 1* along a path. The condition for statically sensitizing a 1 is stronger than sensitizing a 1, as defined in Section 5.5. One way to sensitize a 1 along the path $b_{\{7,8,9\}}$ is reflected by the cube $a_{\{1,4,6,8,9\}} \cdot c_{\{2,3,4,6,8,9\}} \cdot \overline{b}_{\{7,8,9\}}$, i.e., by the partial assignment $\langle a = 1, b = 0, c = 1\rangle$. A vector may sensitize more than one path to a 1. This partial assignment is consistent with the partial assignment $\langle a = 1, c = 1, d = 0\rangle$. Thus, the input vector $\langle a = 1, b = 0, c = 1, d = 0, e = 0\rangle$ sensitizes to a 1 both the paths $b_{\{7,8,9\}}$ and $d_{\{7,8,9\}}$.

The paths are not *statically sensitized to a 1* because both paths present controlling values to the input of gate 7.

Furthermore, it is not the case that only one path can be *statically sensitized to a 1* by a vector. The vector $\langle a = 0, b = 1, c = 1, d = 1, e = 1\rangle$ statically sensitizes to a 1 the path $a_{\{1,4,5,9\}}$ and the path $b_{\{5,9\}}$. It must be the case, however, that for two paths π and ρ to be statically sensitized to a 1 by the same vector w, the paths must meet at a gate g that has a noncontrolled value on w.

9.4 ENF Reducibility

We say that two cubes, c and d, are *syntactically identical*, denoted $c \equiv_s d$, if and only if they are equal as sets of literals. We say that two sum-of-products expressions, C and D, are *syntactically identical*, denoted $C \equiv_s D$, if and only if they are equal as sets of cubes. An ENF expression, E, considered as an untagged sum-of-products expression is denoted $NOTAGS(E)$. We say that two ENF expressions, E and F, are *syntactically identical up to a renaming of tags*, denoted $E \geq_t F$, if and only if $NOTAGS(E) \equiv_s NOTAGS(F)$ and there is a many-to-one mapping from the tagged literals of E to the tagged literals of F. We will call this property of ENF equivalence up to a many-to-one mapping of tags *ENF reducibility*.

What we are attempting to capture in the notion of ENF reducibility between circuits C and D, i.e., $C \geq_t D$, is the notion of a syntactic equivalence between the unreduced collapsed networks of C and D together with the notion that there is a many-to-one mapping from the paths (tags) of C to the paths (tags) of D. For example, consider the following ENF expressions: $E = a_{\{1,3\}} \cdot b_{\{1,3\}} + a_{\{2,3\}} \cdot c_{\{1,3\}}$ corresponding to $a \cdot b + a \cdot c$ and $F = a_{\{2\}} \cdot b_{\{1,2\}} + a_{\{2\}} \cdot c_{\{1,2\}}$ corresponding to $a \cdot (b + c)$. $NOTAGS(E) \equiv_s NOTAGS(F)$, and there is a simple many-to-one mapping from the tags of E to those of F (remember that a function or mapping is just a set of ordered pairs): $\langle\{a,1,3\},\{a,2\}\rangle$, $\langle\{a,2,3\},\{a,2\}\rangle$, $\langle\{b,1,3\},\{b,1,2\}\rangle$, $\langle\{c,1,3\},\{c,1,2\}\rangle$. Thus, $E \geq_t F$. It is easy to see that $\geq_t$ is a transitive relation.

9.5 ENF Reducibility Preserving Transforms

We have a property, ENF reducibility, which, as we will show in the following sections, ensures retaining testability under various fault models. Our task is now reduced to finding a class of operations that retain ENF reducibility. A first step in this direction was presented in [12, 15] where the operations of cube extraction, kernel extraction, and algebraic resubstitution without complement were shown to retain hazard-free robust path delay fault testability. The arguments there implicitly used the property of ENF reducibility.

In this section we first show that the operation of algebraic resubstitution without the use of the complement and algebraic resubstitution with a constrained use of the complement retains ENF reducibility. We consider technology mapping in Section 9.5.3. Regarding the other operations used in area optimization, we will consider elimination, Boolean resubstitution, and simplification in Section 9.5.4. We will show that these operations do not preserve ENF reducibility and, in Section 9.7.3, that they may introduce paths that are not robustly path delay fault testable. Thus, it can be shown that the operations commonly used for area optimization (see Chapter 7) can be divided into two classes — those that provably maintain ENF reducibility and those that do not.

9.5.1 Algebraic Resubstitution without Complement

Theorem 9.5.1 *Let η be a Boolean network of N nodes, and for one node k in η let $F_k = G + H \cdot C$, where G, H, and C are covers and H and C have no variables in common. Let η' be a Boolean network of $N+1$ nodes obtained by factoring C from F_k. In other words, for $1 \leq i \leq N$, $i \neq k$, ${F'}_i = F_i$. Let the new node be ${F'}_{N+1} = C$ with associated output variable ${y'}_{N+1}$ and let ${F'}_k = G + H \cdot {y'}_{N+1}$. Then* *the ENF expression of η is related to the ENF expression of η' by $E \geq_t E'$.*

Proof. Since C and H do not share any variables, C is an algebraic factor and no simplification occurs when the node F'_{N+1} is eliminated in η'. Thus, $NOTAGS(E) \equiv_s NOTAGS(E')$. F_k is a cover. If H and C are cubes, then there is a one-to-one correspondence between the tags of E and E'. However, if H and C are covers, then the collapsed $H \cdot C$ in E will have a many-to-one mapping of tagged literals to E'.

Consider the case of H and C being a disjunction of literals. In the collapsed $H \cdot C$ in E each distinct $l \cdot m$ cube where $l \in H$ and $m \in C$ will have a distinct tag. The number of distinct tagged literals is twice the number of distinct cubes, i.e., $2 \cdot ||H|| \cdot ||C||$, since each cube contains two literals. In the factored $H \cdot C$ in E' we have a single tag corresponding to all the literals in H and a single tag corresponding to all the literals in C. The number of distinct tagged literals is $||H|| + ||C||$. Each tagged literal of E' whose literal l is an element of H ($l \in H$) corresponds to $||C||$ tagged literals in E and each tagged literal of E' whose literal m is an element of C ($m \in C$) corresponds to $||H||$ tagged literals in E. We thus have a many-to-one mapping of tagged literals from E to E'. The above holds true for when H and C are covers as well. □

As an example let η be $F_1 = a \cdot c + a \cdot d + b \cdot c + b \cdot d$. Let the ENF E be $a_{\{1,5\}} \cdot c_{\{1,5\}} + a_{\{2,5\}} \cdot d_{\{2,5\}} + b_{\{3,5\}} \cdot c_{\{3,5\}} + b_{\{4,5\}} \cdot d_{\{4,5\}}$. Cover extraction produces $F'_1 = (a + b) \cdot (c + d)$, where $H = a + b$ and $C = c + d$, and ENF E' is $a_{\{1,5\}} \cdot c_{\{2,5\}} + a_{\{1,5\}} \cdot d_{\{2,5\}} + b_{\{1,5\}} \cdot c_{\{2,5\}} + b_{\{1,5\}} \cdot d_{\{2,5\}}$. The ENFs are syntactically identical, and there is a many-to-one mapping from the tagged literals of E to E'. F_1 and F'_1 are shown pictorially in Figures 9.11(a) and (b), respectively. Each tagged literal in E and E' corresponds to a path in F_1 and F'_1, respectively.

Theorem 9.5.2 *Let η be a Boolean network of N nodes, and let $F_j = C$ and $F_k = C$, where C is a cover, be two nodes in η. Let η' be a Boolean network of $N - 1$ nodes with F_j resubstituted for F_k. In other words, for $1 \leq i \leq N - 1$, $i \neq k$, let $F'_i = F_i$, except let all instances of literal y_k in F'_i be replaced by instances of the literal y_j. Then the ENF expression of η, E, is syntactically identical up to a renaming of tags to the ENF expression of η', E'.*

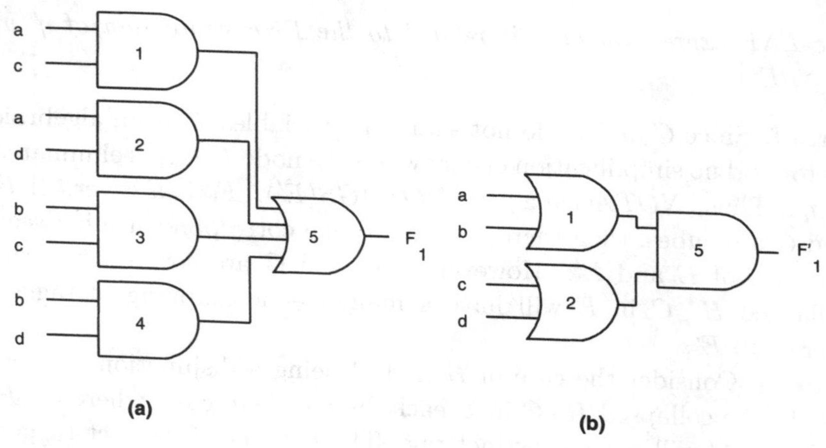

Figure 9.11: Many-to-one mapping of tagged literals and paths

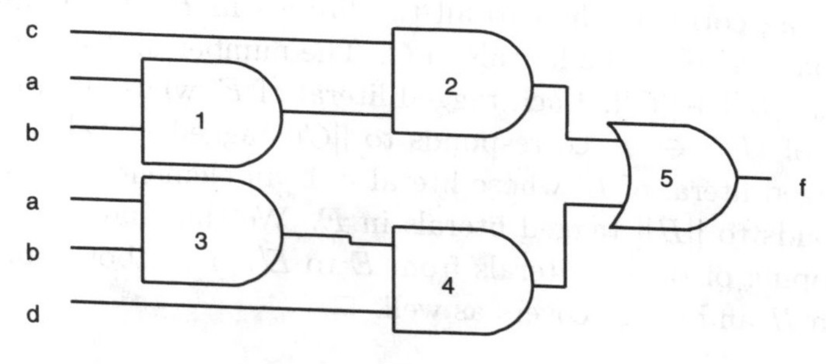

Figure 9.12: Node resubstitution

Proof. From the construction of the ENF, it is clear that the untagged sum-of-products expression $NOTAGS(E) \equiv_s NOTAGS(E')$. Furthermore, there is a one-to-one correspondence between the tagged literals or paths in E to those in E'. Subpaths to the inputs to node F_k are destroyed, but identical subpaths to the inputs of F_j pass through F_j and go through to the fan-outs of both y_k and y_j. □

Consider the ENF $a_{\{1,2,5\}} \cdot b_{\{1,2,5\}} \cdot c_{\{2,5\}} + a_{\{3,4,5\}} \cdot b_{\{3,4,5\}} \cdot d_{\{4,5\}}$ of the circuit of Figure 9.12. $C = a \cdot b$, $F_j = a_{\{1,2,5\}} \cdot b_{\{1,2,5\}}$ and $F_k = a_{\{3,4,5\}} \cdot b_{\{3,4,5\}}$. Resubstitution, i.e., discarding gate 3, produces an ENF $a_{\{1,2,5\}} \cdot b_{\{1,2,5\}} \cdot c_{\{2,5\}} + a_{\{1,4,5\}} \cdot b_{\{1,4,5\}} \cdot d_{\{4,5\}}$. The ENFs are syntactically identical, and there is a one-to-one mapping from the tagged literals of E to E'.

Theorem 9.5.3 *Let C be a single-output combinational circuit with*

ENF expression E_C. Let M be C after the resubstitution of a single uncomplemented algebraic factor. Then $E_C \geq_t E_M$.

Proof. Algebraic resubstitution without complement is equivalent to a cube or kernel extraction followed by a merging of identical nodes. By Theorems 9.5.1 and 9.5.2, each of these operations retains ENF reducibility, and since the relation $\geq_t$ is transitive, we have the required result. □

9.5.2 Algebraic Resubstitution with Complement

We now explain how to constrain algebraic resubstitution with complement in such a way that ENF reducibility is retained.

Lemma 9.5.1 *Let η be a Boolean network of N nodes, and for one node k in η let the cover of F_k be G. Let η' be a Boolean network of $N+1$ nodes obtained by pushing an inverter out from F_k. In other words, for $1 \leq i < N+1$, $i \neq k$, $F_i' = F_i$. Let the new node be $F_{N+1}' = \overline{y_k}$ and let $F_k' = \overline{G}$.*

If when the node F_k' in η' is collapsed into node F_{N+1}' the resulting cover is syntactically identical to the node F_k in η, then the ENF expression of η is syntactically identical up to a renaming of tags to the ENF expression of η'.

Proof. In the construction of the ENF, as described in Section 9.3, the node F_k' in η' will be collapsed forward into F_{N+1}'. The relevant parts of η and η' are shown in Figures 9.13(a) and (b), respectively. If the cover resulting from this collapsing is syntactically identical to the node F_k in η, then the intermediate ENFs will also be syntactically identical up to a renaming of tags. The only difference between the ENFs will be that paths through node F_k' in η' will have an additional tag for the inverter at node F_{N+1}' that paths through node F_k in η will not. As η and η' are identical in every other respect, the ENF of η is syntactically identical up to a renaming of tags to the ENF of η'. □

Theorem 9.5.4 *Let C be a single-output circuit with ENF expression E_C. Let M be C after the resubstitution of a single complemented algebraic factor as constrained by Lemma 9.5.1. Then $E_C \geq_t E_M$.*

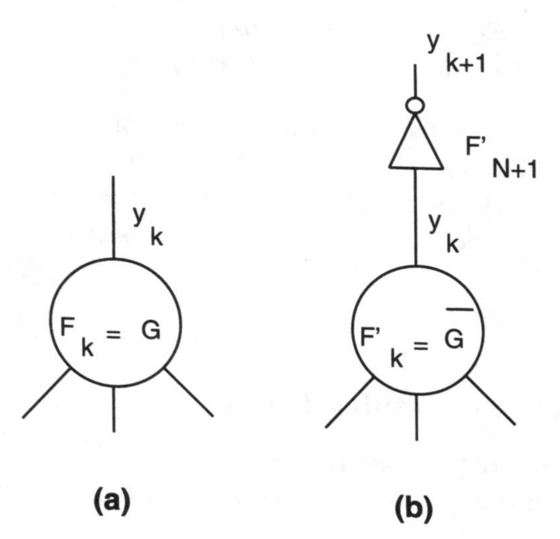

Figure 9.13: Algebraic factorization in a Boolean network

Proof. Let M_1 be C after moving an inverter up in C. From Lemma 9.5.1 we know that $E_C \geq_t E_{M_1}$. Let M be M_1 after a single instance of algebraic resubstitution without complement. By Theorem 9.5.3 $E_{M_1} \geq_t E_M$. The relation $\geq_t$ is transitive, so $E_C \geq_t E_M$. □

For example, consider the sum-of-products expression $C = a\overline{b}d\overline{e} + cd\overline{e} + a\overline{b}f + cf + \overline{a}\,\overline{c}e + \overline{a}\,\overline{c}g + b\overline{c}e + b\overline{c}g$. Let $B = (c + a\overline{b})(d\overline{e} + f) + (\overline{a}\,\overline{c} + b\overline{c})(e + g)$. Let F_k be $(\overline{a}\,\overline{c} + b\overline{c})$. Moving an inverter up gives $F'_k = (a\,\overline{b} + c)$ and $F'_{N+1} = \overline{y_k}$. The node F'_k can now be algebraically resubstituted without complement into the other instance of $(a\,\overline{b} + c)$ in B. The resulting circuit, call it A, is given in Figure 9.14. The ENF of A is $E_A = a_{\{3,4,10,11\}} \cdot \overline{b}_{\{1,3,4,10,11\}} \cdot d_{\{8,9,10,11\}} \cdot \overline{e}_{\{2,8,9,10,11\}} + c_{\{4,10,11\}} \cdot d_{\{8,9,10,11\}} \cdot \overline{e}_{\{2,8,9,10,11\}} + a_{\{3,4,10,11\}} \cdot \overline{b}_{\{1,3,4,0,11\}} \cdot f_{\{9,10,11\}} + c_{\{4,10,11\}} \cdot f_{\{9,10,11\}} + \overline{a}_{\{1,3,4,6,7,11\}} \cdot \overline{c}_{\{4,6,7,11\}} \cdot e_{\{5,7,11\}} + \overline{a}_{\{3,4,6,7,11\}} \cdot \overline{c}_{\{4,6,7,11\}} \cdot g_{\{5,7,11\}} + b_{\{1,3,4,6,7,11\}} \cdot \overline{c}_{\{4,6,7,11\}} \cdot e_{\{5,7,11\}} + b_{\{1,3,4,6,7,11\}} \cdot \overline{c}_{\{4,6,7,11\}} \cdot g_{\{5,7,11\}}$.

Note that the ENF of the circuit A is syntactically identical to the sum-of-products expression C if the tags on the literals are ignored.

The necessity of the constraint on algebraic resubstitution with complement can also be seen in this example. Starting over again with B, if we chose $F_k = (c + a\overline{b})$, then the complement of F_k can also be expressed as $F'_k = (\overline{a}\,\overline{c} + b\overline{c})$. The node F'_k can now

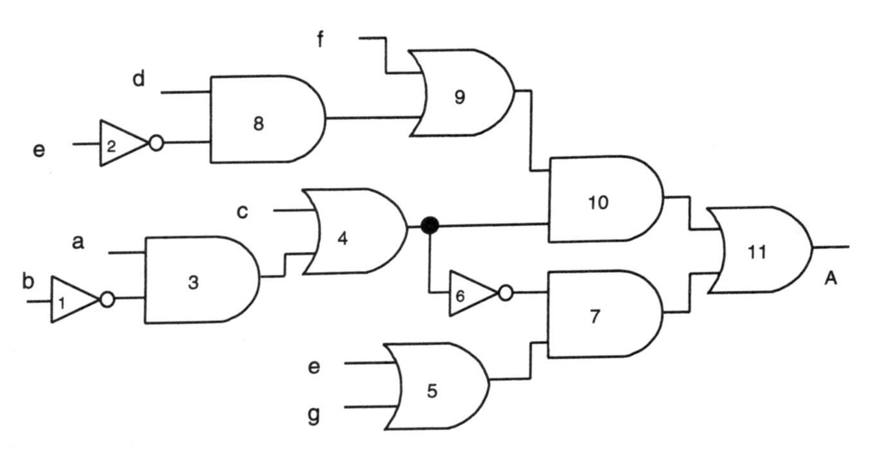

Figure 9.14: Algebraic factorization example using inverse

be algebraically resubstituted (without complement) into the other instance of $(\overline{a}\,\overline{c} + b\overline{c})$. This choice of $F_k' = (\overline{a}\,\overline{c} + b\overline{c})$ does not meet the requirements of Lemma 9.5.1 nor does it result in a circuit whose ENF is syntactically identical to C. The ENF corresponding $\overline{(\overline{a}\,\overline{c} + b\overline{c})}$ contains a $\overline{c}\,\overline{c}$ term and is not syntactically identical to $(c + a\overline{b})$.

9.5.3 Technology Mapping

So far our results treated technology-independent optimizations only. In this section we state some theorems that show that common technology mapping procedures such as those used in DAGON [21] or MIS [11] retain ENF reducibility.

In tree covering based technology mapping each fan-out-free subnetwork (tree) t of a combinational circuit C is replaced by a logically equivalent and single-fault testable tree t' in M, the mapped circuit[1]. Furthermore, the fan-out points of M are exactly the fan-out points of C. For each input i in a fan-out-free combinational circuit C there is a unique path π from that input to the output of that circuit. If M is a logically equivalent tree covering technology mapped version of C, then for each path π in C there is a *corresponding* path ρ in M.

[1]Strictly speaking, tree covering may be applied in such a way that a tree t in the input circuit is replaced by a nontree structure after covering. In this discussion we are explicitly disallowing any such coverings.

Lemma 9.5.2 *Let C be a* NAND *gate and* INVERTER *implementation of a circuit, and let M be a tree covering based technology mapped version of C. A vector sequence $\langle v_1, v_2 \rangle$ event sensitizes a path π in C if and only if $\langle v_1, v_2 \rangle$ event sensitizes the corresponding path ρ in M.*

Proof. We again make the observation that the ENF expression for C, call it E_C, is the same, up to renaming of tags, as the ENF expression for M, call it E_M. Moreover, for each path ρ in M there is a corresponding path π in C (not a many-to-one correspondence) and therefore for each path or tagged literal l_ρ in E_M there is a corresponding tagged literal l_π in E_C. Thus, there exists a vector v which tests for l_π precisely when v tests for l_ρ, and similarly vector sequence $\langle v_1, v_2 \rangle$ event sensitizes π precisely when $\langle v_1, v_2 \rangle$ event sensitizes ρ. □

9.5.4 Nonretainment of ENF Reducibility

While in the previous section we showed that variations of algebraic factorization resulted in ENF reducibility being preserved by the transformations, in this section we briefly note other transformations which do not retain ENF reducibility.

Eliminating a node by pushing it into its fan-out nodes can destroy ENF reducibility. An example is the circuit $(a + c)\overline{b} + b\overline{c}$. The ENF of this circuit, given an appropriate ordering of the gates, is $a_{\{1,2,4\}} \cdot \overline{b}_{\{2,4\}} + \overline{b}_{\{2,4\}} \cdot c_{\{1,2,4\}} + b_{\{3,4\}} \cdot \overline{c}_{\{3,4\}}$. Eliminating the node $a+c$ results in a circuit with ENF $a_{\{1,4\}} \cdot \overline{b}_{\{1,4\}} + \overline{b}_{\{2,4\}} \cdot c_{\{2,4\}} + b_{\{3,4\}} \cdot \overline{c}_{\{3,4\}}$. ENF reducibility is not preserved across elimination.

Figure 9.15 shows the two-level implementation of a function $f = a \cdot b + \overline{a} \cdot \overline{c} \cdot d + a \cdot c \cdot d + \overline{a} \cdot \overline{b} \cdot c \cdot \overline{d}$. Choose $g = c \cdot \overline{d} + \overline{c} \cdot d$ as a Boolean factor. After performing Boolean division, f can be written as $f = (c \cdot \overline{d} + \overline{c} \cdot d) \cdot (\overline{a} \cdot \overline{b}) + a \cdot b + \overline{a} \cdot \overline{c} \cdot d + a \cdot c \cdot d$. This implementation of f is shown in Figure 9.16. The ENF for this implementation is given by $f = a_{\{8,12\}} \cdot d_{\{8,12\}} \cdot c_{\{8,12\}} + \overline{a}_{\{1,11,12\}} \cdot c_{\{5,7,11,12\}} \cdot \overline{d}_{\{4,5,7,11,12\}} \cdot \overline{b}_{\{2,11,12\}} + \overline{a}_{\{1,11,12\}} \cdot d_{\{6,7,11,12\}} \cdot \overline{c}_{\{3,6,7,11,12\}} \cdot \overline{b}_{\{2,11,12\}} + \overline{a}_{\{1,10,12\}} \cdot \overline{c}_{\{3,10,12\}} \cdot d_{\{10,12\}} + a_{\{9,12\}} \cdot b_{\{9,12\}}$. This ENF is clearly not syntactically identical to the ENF of the two-level circuit.

Modifying the cover of any node in a Boolean network necessarily changes the ENF expression and therefore does not retain ENF reducibility. Furthermore, as illustrated in Chapter 5, different

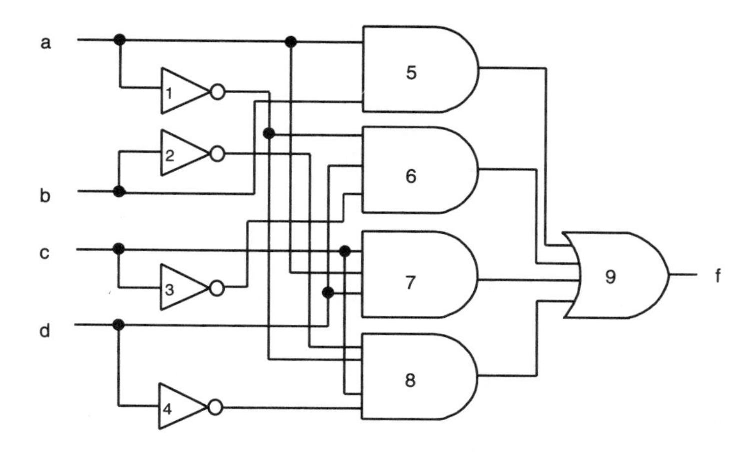

Figure 9.15: Two-level circuit for a Boolean factorization example

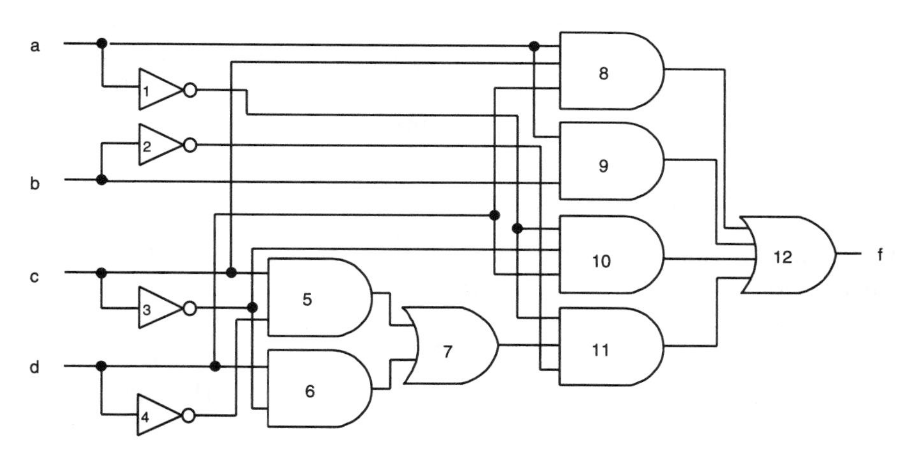

Figure 9.16: Circuit resulting from Boolean factorization

covers of the same Boolean function have differing path delay fault testabilities, so modifying the cover may also change the delay fault testability.

9.6 Multiple Stuck-At Faults

The sheer number of multiple stuck-at faults (multifaults) in a circuit, $3^k - 1$ multifaults in a circuit with k wires, makes explicit test generation for multifaults impossible. The goal in multifault testing has been to obtain sufficiency conditions for the complete multifault

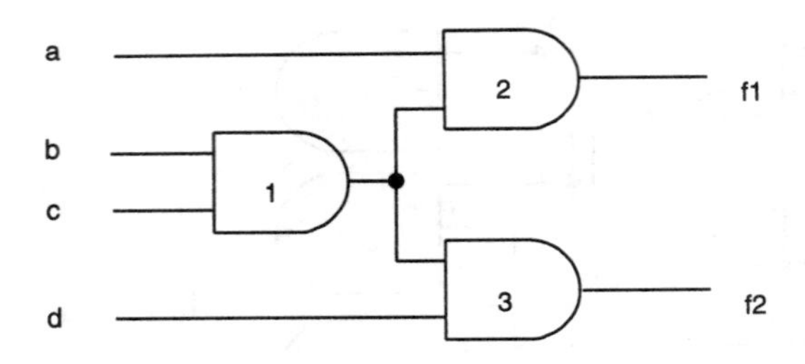

Figure 9.17: Example for link definition in ENF

testability of two-level and multilevel circuits that provide efficient test generation mechanisms. We give a sufficiency condition for a multilevel circuit to be completely multifault testable in Section 9.6.1 and show that transformations that satisfy ENF reducibility retain multifault test sets in Section 9.6.3. We describe a method of multifault test generation that does not require explicit consideration of multifaults in Section 9.6.2. Compositional methods for synthesizing multifault testable networks and the multifault testability of multiplexor-based networks are the subjects of Section 9.6.4.

9.6.1 Conditions for Testability

Given a combinational circuit C with ENF expression E_C and a tagged literal l_π, the *complete path-cube-complex associated with* l_π, denoted P_{l_π}, is the set of all cubes $q \in E_C$ such that $l_\pi \in q$.

In the tags on the literals in the ENF a *link* in the circuit is represented as an adjacent pair of gate numbers in the tag or as a primary input and the first gate in the tag for that literal. Consider the simple circuit shown in Figure 9.17, with $E_{f1} = a_{\{2\}} \cdot b_{\{1,2\}} \cdot c_{\{1,2\}}$ and $E_{f2} = b_{\{1,3\}} \cdot c_{\{1,3\}} \cdot d_{\{3\}}$. The link between gates 1 and 2 is specified by the pair of gates 1, 2 in the tags of literals b and c in E_{f1}. The link between primary input b and gate 1 is specified by the occurrence of $b_{\{1}$ in the tags of the literal b in both E_{f1} and E_{f2}.

The *link-cube-complex* of a link λ in a circuit C with ENF E_C is defined to be the set of all cubes containing tagged literals associated with a path through λ, or more formally, $L_\lambda = \{q \in E_C \mid (l_\pi \in q) \wedge (\lambda \in \pi)\}$. For instance, the link 2, 3 in the ENF $a_{\{1,3\}} \cdot b_{\{1,3\}}$ $+ c_{\{2,3\}} \cdot d_{\{2,3\}}$ has an associated link-cube-complex $c_{\{2,3\}} \cdot d_{\{2,3\}}$.

We now define primality and irredundancy tests for tagged literals in an ENF E_C. A primality test v for a tagged literal l_π in a cube $q \in P_{l_\pi}$ is such that the value of l in v is 0, for all literals

$m \neq l \in q$, $m(v) = 1$, and for all cubes $r \neq q$ there exists some literal $k \in r$ such that $k(v)$ is 0.[2] As an example, consider $E_C = a \cdot b + \overline{a} \cdot c$, where we have not shown the tags for the various literals. A primality test for $a \in a \cdot b$ is $v = \overline{a} \cdot b \cdot \overline{c}$ or $v = \overline{a} \cdot b \cdot c$.

An irredundancy test v for a complete path-cube-complex corresponds to a relatively essential vertex for the set of cubes in the complete path-cube-complex, i.e., v is contained in some cube q belonging the complete path-cube-complex (link-cube-complex) and is not covered by any of the cubes not in the complete path-cube-complex. A similar definition follows for the link-cube-complex.

We now give a basic sufficiency condition for multifault testability.

Theorem 9.6.1 *Let C be a single-output combinational circuit with ENF expression E_C. A test set S comprising of the primality tests for each tagged literal l_π in each cube of P_{l_π} and a set of irredundancy tests for each link-cube-complex L_λ in E_C will detect all the multifaults of C.*

Proof. Consider the effect of a multifault m on E_C. For each single fault f in the multifault m some edge e in C gets stuck to 1 or 0. This results in every literal in E_C that is tagged with a path that passes through e being set to 1 or 0 in each cube in which it appears. The effect of the fault m falls into three classes: Tagged literals are uniformly stuck-at-0, and as a result cubes are uniformly removed from E_C. Tagged literals are uniformly stuck-at-1, and as a result cubes are uniformly expanded (i.e., literals are removed from the cubes) in E_C. Some tagged literals are stuck-at-1 and some are stuck-at-0, and as a result cubes are simultaneously removed from E_C and expanded in E_C.

In the first case if m strictly removes cubes from E_C, then we argue that any set of removed cubes has to contain all the cubes in some link-cube-complex L_λ. Let f be a single fault in the multifault m and let f reside on a link λ in C. Any tagged literal l_π, such that $\lambda \in \pi$, is affected by f (and therefore by m). Therefore, all the cubes in L_λ are also affected by f (and therefore m). In general, these cubes may be expanded or removed due to f. However, we are considering only the case where the effect of m corresponds to a strict removal of cubes in E_C, so any set of removed cubes has to contain all the cubes in L_λ. By the hypothesis to the theorem, there exists

[2]Note that k can be l.

an irredundancy test in S, call it v, for L_λ. The vector v is a member of the ON-set that is uniquely covered by L_λ, and so the removal of other cubes in E_C cannot possibly result in covering v. Hence the faulty network gives a value of 0 on the vector v. Thus, v detects the multifault m.

In the second or third cases, if m strictly expands literals in E_C or expands literals in cubes that are not removed, then there exists a link λ in C that results in the expansion of literals in at least one cube in some tagged literal that passes through λ. By the hypothesis to the theorem each tagged literal in the complete path-cube-complex of this tagged literal has a primality test, v. This means that $E(v) = 0$; but if l_π is expanded in q in the faulty ENF, call it E_f, then $E_f(v) = 1$. Therefore, v detects m.

Thus, in all three cases we can find a $v \in S$ that detects m, and therefore S detects all multifaults of C. □

While the requirements of the above theorem may seem constrained, it should be noted that the ENF of any prime and irredundant two-level circuit, or any algebraically factored multilevel circuit will satisfy the conditions.

9.6.2 Test Generation Methods

Single stuck-at fault test generation algorithms can be easily modified to generate tests for a particular multifault in a circuit. For instance, one might wish to generate tests for all the double-faults in a circuit. However, this approach requires the explicit consideration of each double fault and rapidly becomes infeasible for faults of greater multiplicity.

A method that avoids the explicit consideration of multifaults, using the notion of unconditional testability [10], is described here by means of an example.

A combinational circuit C is assumed to have fault-free gates, and is initially assumed to have possible stuck-at-1 or stuck-at-0 faults on any primary input fan-out stem or gate output fan-out stem. The fault status of each fan-out stem is $(ff,\ s_0,\ s_1)$, where ff stands for a fault-free condition, s_0 stands for stuck-at-0, and s_1 stands for stuck-at-1.

An example illustrating the techniques of [10] is shown in Figures 9.18 and 9.19. A vector pair $\langle v_1,\ v_2 \rangle$ is fault simulated on a combinational circuit C using a forward propagation algorithm that

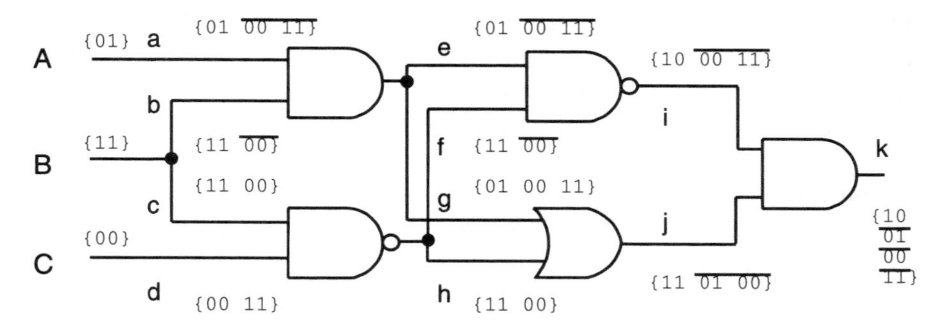

Figure 9.18: Multifault test generation pair 1

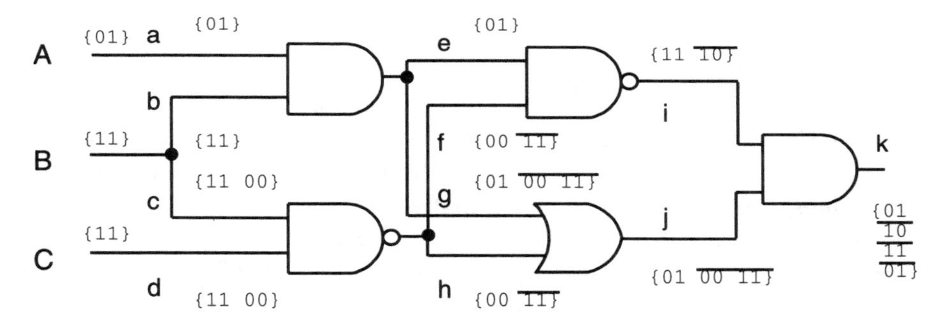

Figure 9.19: Multifault test generation pair 2

determines all possible values for each line in the circuit, assuming all possible faulty conditions. The first value in each set of line values is the fault-free value. The vector pair chosen in our example is ⟨010, 110⟩ Forward propagation results in the possible values shown in Figure 9.18. Line **a** can take the values of {01, 00, 11}, where 01 is the fault-free value of input **A**, and 00 and 11 are the stuck-at-0 and stuck-at-1 values, respectively. Similarly, line **b** can take values from {11, 00}. The output of the AND gate with inputs **a** and **b** can take on values from {01, 00, 11}. We compute this by taking the AND of each pair of values in the value sets for lines **a** and **b**. The possible values at the circuit output are {10, 01, 00, 11}.

Next, given the value obtained at the output (the fault-free value for the purposes of test generation) backward implication is performed in order to deduce which of the possible values were actually present. From these deduced values the fault status of some lines are updated. Some faults on some lines may be *unconditionally tested*, regardless of other fault conditions in the network. A new vector pair is chosen, and the process is iterated until the required fraction of stuck-at faults are unconditionally tested. Backward implication, given that the circuit output has the fault-free value of 10, is illustrated in Figure 9.18 by crossing out the values that do not occur. The three values of 01, 00, and 11 are crossed out at the circuit output. Given that the value at the output is 10, we can infer that the only possible value for line **i** is 10 and the only possible value for line **j** is 11 (for instance, 00 at line **i** would produce 00 at the circuit output). Backward implication is not performed from line **j** back to the OR gate because the deduced value of **j** could come about either because it is stuck-at-1 or because the stem's value is 11. Other deductions are performed all the way back to line **a**, which is deduced to be at 01. We deduce that lines **a**, **e**, **i**, and **k** are fault-free and that lines **b**, **f**, and **j** are not stuck-at-0.

Note that unconditional testing assumes a certain fault status for the lines in the circuit. For the first vector pair there are 3^k possible conditions corresponding to the k lines in the circuit. Since four lines were deduced to be fault-free by the first vector pair in our example and three lines were deduced to be either stuck-at-1 or fault-free, for the second vector pair there are $3^{k-4} \cdot 2^3$ possible conditions. More stuck-at faults on lines may now be unconditionally tested given that we will apply a different vector pair and that we have fewer possible conditions.

In our example, we now apply the vector pair ⟨011, 111⟩ Forward propagation is carried out after dropping the faults that were detected by the first pair. This is illustrated in Figure 9.19. Backward implication, again illustrated in Figure 9.19 by crossing out line values, deduces that lines **f**, **g**, and **j** are fault-free, and that line **h** is not stuck-at-1. At this point we have unconditionally detected both stuck-at faults on lines **a**, **e**, **f**, **g**, **i**, **j**, and **k**, the stuck-at-0 fault on **b**, and the stuck-at-1 fault on **h**.

The generation of input vector pairs is a heuristic process. Input vector pairs that differ in a single bit and which produce an output transition are generated. The bit that they differ in is varied heuristically. The implementation described in [10] was able to generate unconditional tests for approximately 90 percent of the stuck-at faults in a popular combinational benchmark suite [7].

9.6.3 Synthesis for Full Testability

In Section 9.5, we showed that various commonly used logic transformations used in area optimization of multilevel logic retained ENF reducibility. We are now in a position to utilize the theorems of Section 9.5 to arrive at synthesis and optimization methods that produce fully multifault testable designs. To this end, we will show that ENF reducibility preserving transformations retain multifault testability.

Theorem 9.6.2 *Let C be a single-output circuit with ENF expression E_C. Let M be a single-output circuit with ENF expression E_M. If $E_C \geq_t E_M$, then the complete multifault test vector set S for E_C constructed as in Theorem 9.6.1 detects all multifaults in M.*

Proof. Let π be a path in M. There is a corresponding tagged literal l_π in E_M. Corresponding to l_π in E_M there is a nonempty set of corresponding tagged literals $\{l_{\pi_1}, \ldots, l_{\pi_m}\}$ in E_C. We have primality tests in S for each l_{π_i} in each cube of E_C. This implies we have primality tests for l_π in each cube of E_M. We have an irredundancy test for each link-cube-complex L_λ in E_C. Any link-cube-complex in E_M has to be larger than or equal to some link-cube-complex in E_C. This is because a link-cube-complex L_λ in E_C is a set of path-cube-complexes, and each path-cube-complex is E_M is larger than or equal to some path-cube-complex in E_C. Therefore, S satisfies the conditions necessary for being a complete multifault test set for M as per Theorem 9.6.1. □

We can begin from a prime and irredundant single-output two-level circuit that is fully multifault testable and apply various ENF reducibility (and, therefore, multifault testability) preserving logic transformations to improve the area and speed characteristics of the circuit.

9.6.4 Compositional Techniques for Full Testability

So far the results of this section have been developed for individual modules of combinational logic that can be flattened to two-level circuits. We now consider how these results can be practically applied in an environment in which flattenable combinational logic modules are cascaded with other, perhaps nonflattenable, combinational logic modules. Experience has shown that if a combinational logic module is the result of the synthesis of a *finite state machine* (FSM) description, then it can be flattened to a two-level circuit. On the other hand, some have argued that most register-bounded combinational blocks, even in control portions of designs, are not flattenable. How can these two points of view be consistent? A code fragment describing a portion of a hardware system is shown below. This description would typically be synthesized into a combinational logic block by a system such as CONES [35].

```
switch(state){
   ...
   case state_22:
     if(packet_odd_parity)
         ...
   case state_30:
     if(add_overflow)
         ...
}
```

In this control module `packet_odd_parity` and `add_overflow` are treated as primary inputs, but in the final design they are intermediate signals. The signal `packet_odd_parity` is the output of a 32-bit parity generator module and the `add_overflow` is the final carry-out output of a 16-bit adder module. Both of these signals are the outputs of combinational logic modules and will feed the combinational logic module associated with the FSM. If the hierarchy is removed and the register-bounded combinational circuit resulting from the adder and parity generator feeding the controller is examined, it will be found that the resulting combinational logic block is not flattenable to a two-level circuit. This is because of the explosion in product terms due to the flattening of the parity generator and adder. Thus, the synthesis procedures developed so far cannot be directly applied because the single combinational logic block is not flattenable to two levels.

What is the impact of this on the testability of the resulting circuit? Let us recall that when optimizing for circuit area, the savings are clearly compositional. In other words, the savings in optimizing over k blocks is the sum of the savings for each individual block. The same is true to a lesser extent for timing optimization. As long as a timing optimization system is able to take in variable arrival times for inputs into account, two cascaded combinational logic blocks can be optimized independently. Unfortunately, in the context

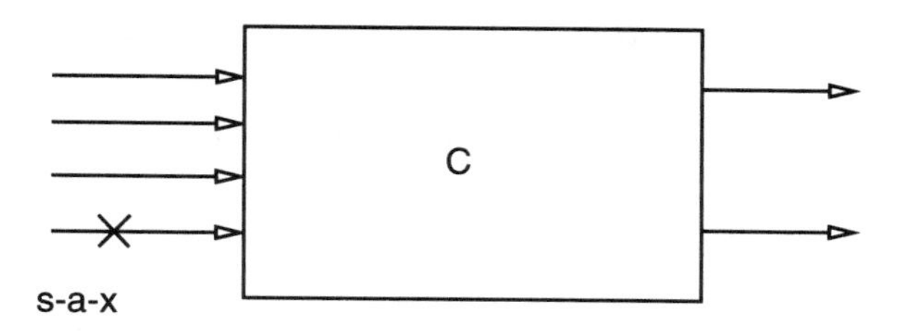

Figure 9.20: Circuit with stuck-at fault on a primary input

of circuit testing, hierarchy is even more difficult to manage and *the composition of two completely testable circuits may be poorly testable.*

One solution to this problem is to develop sets of *composition rules* that allow us to prove conditions such that when one multifault testable circuit module feeds another multifault testable circuit module, then the resulting circuit is also multifault testable. We begin with a modest undertaking in this regard, but one that may be applied to cover a significant number of the cases of cascading combinational logic modules in control-dominated designs that may not be flattened to two-level circuits.

In order to prove the composition rule presented in this section, we will first prove a related lemma. Consider the circuit of Figure 9.20. It has a stuck-at fault on one of its primary inputs. Assume that the circuit C is fully multifault testable and its ENF satisfies the conditions of Theorem 9.6.1. We are concerned here with detecting this stuck-at fault on the primary input, in the presence of other arbitrary fault conditions in C. The following lemma gives us a compact set of test vectors that would detect the primary input stuck-at fault, regardless of other fault conditions in C.

Lemma 9.6.1 *Given a circuit C with inputs $x_1, x_2, \ldots, x_N$, whose ENF E_C satisfies the conditions of Theorem 9.6.1, if an input x_i is stuck-at-1 (stuck-at-0), then the following set of tests will detect this stuck-at fault in the presence of any multifault in C.*

1. *Primality tests for all x_i ($\overline{x_i}$) literals in E_C.*
2. *Primality tests for all literals (other than x_i or $\overline{x_i}$) in each cube in E_C that does not contain x_i or $\overline{x_i}$.*
3. *A link-cube-complex irredundancy test for the link corresponding to the x_i primary input.*

Proof. We wish to detect a multifault m in C, which contains a stuck-at-1 fault on x_i. Due to the stuck-at-1 fault in C the x_i literals in E_C will be expanded and the cubes corresponding to the $\overline{x_i}$ literals will be removed.

We have primality tests for each x_i literal in each cube of E_C, namely Set (1). This implies that if a single cube expanded in x_i remains, we will detect m. Consider the case where all of the cubes expanded in x_i are removed due to m. We can have two cases. Cubes may be uniformly removed from E_C, and this implies that the link-cube-complex of the input x_i has been completely removed from E_C. This is because all the x_i literal cubes and the $\overline{x_i}$ literal cubes have been removed. In this case a link-cube-complex irredundancy test for the link corresponding to input x_i, namely Set (3), will detect m. The other case is that some cubes that do not contain x_i or $\overline{x_i}$ literals are expanded but not removed. In this case some vector in Set (2) will detect m.

The stuck-at-0 case can be argued similarly. □

Note that the set of test vectors above could be substantially smaller than the complete multifault test for E_C. Consider an extreme case of x_i or $\overline{x_i}$ being present in all cubes in E_C. Set (1) only requires primality tests for one input x_i rather than the N inputs. Set (2) is empty. Set (3) only contains one test. Further, the primality tests in Set (1) can be compacted by determining vectors that are primality tests for multiple literals in different cubes. For example, given the expression $a \cdot b + a \cdot c$ the vector $a = 0$, $b = 1$, and $c = 1$ is a primality test for both occurrences of a.

We will now give a simple composition rule (depicted in Figure 9.21) and prove its usefulness in the following theorem.

Theorem 9.6.3 *Consider the circuit of Figure 9.21. Assume that C_1 and C_2 are multifault testable combinational circuits, the input variables of C_1 and C_2 are disjoint, and the ENF of C_2, namely E_{C_2}, satisfies the properties of Theorem 9.6.1. If a circuit C is created by connecting one output o_j of C_1 to one input i_k of C_2, then C is completely multifault testable.*

Proof. We will consider three separate cases of a multifault m in C. The multifault may be localized to circuit C_1, localized to circuit C_2, or comprised of a multifault $m_1 \in C_1$ and a multifault $m_2 \in C_2$.

If the multifault m is localized to C_1, then we can propagate the effect of the multifault either to the primary outputs of C_1, in

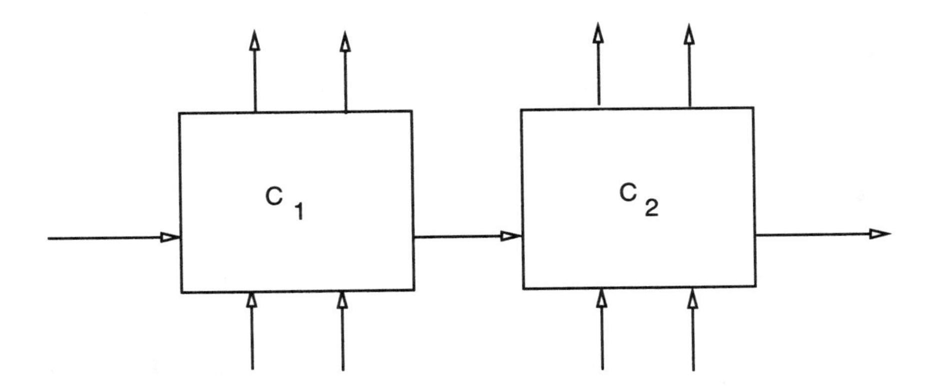

Figure 9.21: A simple composition rule

which case we detect m, or to the output o_j, using a vector v_1 on the inputs to C_1. In the case of propagation to the output o_j only we effectively have a stuck-at fault at the i_k input to C_2. Since C_2 is fully multifault testable, it is fully single stuck-at fault testable, and therefore the stuck-at fault at the input i_k can be propagated to the primary output of C_2 using some vector v_2 on the other inputs to C_2. The required values on these inputs do not conflict with the required values on the inputs to C_1 (to propagate m to o_j) since the two sets of inputs are disjoint.

If the multifault m is localized to C_2, then we simply set up the appropriate value for i_k (1 or 0) by applying a vector v_1 to the inputs of C_1 and applying an appropriate vector v_2 at the other inputs to C_2 and detecting the multifault.

We now consider the case where m is comprised of multifaults $m_1 \in C_1$ and $m_2 \in C_2$. We can propagate the effect of m_1 to the primary outputs of C_1, in which case we detect m, or to the output o_j, using a vector v_1 on the inputs to C_1. In the case of propagation to the output o_j, we effectively have a stuck-at fault at the i_k input to C_2 and a multifault $m_2 \in C_2$. The stuck-at fault on the i_k input could be a stuck-at-1 or a stuck-at-0. Since C_2 and E_{C_2} satisfy the properties required by Lemma 9.6.1, we can detect m by applying an appropriate input vector to the inputs of C_2. □

Corollary 9.6.1 *Assume that C_1 and C_2 are multifault testable combinational circuits and the ENF of C_2, namely E_{C_2}, satisfies the properties of Theorem 9.6.1. If a circuit C is created by connecting one* *output o_j of C_1 to one input i_k of C_2 in such a way that the input variables of C_1 and C_2 are disjoint, then all the multifaults in C can be detected by the following sets of test vectors:*

1. *The set $y \times M_2$, where y is a vector such that $o_j(y) = 0$.*
2. *The set $y \times M_2 \mid o_j(y) = 1$.*
3. *The set $M_1 \times M_{s-a-0,2}$.*
4. *The set $M_1 \times M_{s-a-1,2}$.*

where M_1 and M_2 are the multifault test sets for C_1 and C_2, respectively. $M_{s-a-0,2}$ ($M_{s-a-1,2}$) is the test set for C_2 with input i_k stuck-at-0 (stuck-at-1) and with other arbitrary fault conditions as given in Lemma 9.6.1.

Proof. We will use arguments similar to those in Theorem 9.6.3. A multifault m localized to C_1 that propagates to the primary outputs of C_1 is detected by $\langle y, -\rangle$ where $y \in M_1$ and the inputs to C_2 are irrelevant. A multifault m localized to C_1 that propagates to the output o_j, producing a 1 in the faulty circuit rather than a 0, is detected by $\langle y, z\rangle$ where $y \in M_1$ and $z \in M_{s-a-1,2}$. z is a single stuck-at fault test for C_2 with its input i_k stuck-at-1. If m produces a 0 in the faulty circuit rather than a 1, then we can detect m by applying $\langle y, z\rangle$ where $y \in M_1$ and $z \in M_{s-a-0,2}$.

A multifault m localized to C_2 can be detected by Set (1) or (2) above by producing an appropriate value of 0 or 1 at input i_k.

A multifault m comprised of multifaults m_1 and m_2 detectable at the primary outputs of C_1 can be tested by some vector $\langle y, -\rangle$ where $y \in M_1$ and the inputs to C_2 are irrelevant. If the effect of m_1 is propagated to output o_j, depending on whether the faulty and true values are 1 and 0, or 0 and 1, respectively, we can detect the fault using Set (3) or Set (4). □

While we could have proven quite easily that the "multiplication" of the complete multifault test sets of C_1 and C_2 would detect all the multifaults in C, we did not do so because Lemma 9.6.1 and Corollary 9.6.1 provide us with a compaction strategy that results in substantially smaller test sets. Note that if $M_{s-a-0,2}$ and $M_{s-a-1,2}$ are considerably smaller than M_2, we have a substantial savings in the multifault set for C, as opposed to $M_1 \times M_2$. Sets (3) and (4) above do not require the complete M_1, rather the part of M_1 corresponding to fault propagation to output o_j will suffice. The remaining part of M_1 has to be applied to C_1 with any combination of values at the other inputs to C_2.

We only require the ENF of C_2, namely E_{C_2} to satisfy the conditions of Theorem 9.6.1, while E_{C_1} can be an arbitrary ENF as long as C_1 is fully multifault testable. This is important since we may have to repeatedly apply the composition rule, for instance to create a fully testable ripple-carry adder. Such an adder can be created if the full adder cell used satisfies the conditions of Theorem 9.6.1, since we can then compose the adder beginning from the least significant bit to the most significant bit.

We show in Section 9.7.6 that a fully hazard-free robust path delay fault testable network is fully multifault testable as well. This implies that we can use proven composition rules for path delay fault testability [9] to synthesize fully multifault testable networks. A generalization of the above rule has been presented that allows multiple o_j's connecting C_1 to C_2 [16]. The composition topology of Figure 9.21 has been shown to preserve single stuck-at fault testability if the individual blocks are single stuck-at fault testable [8]. (The individual blocks are not required to be multifault testable.)

Finally, we give a multiplexor composition rule that has important applications in multifault-testability-driven synthesis. This is because we can identify a rich class of networks that are completely multifault testable by a repeated application of the composition rule.

The topology of the composition rule corresponds to that of Figure 9.22. Note, however, that logic can be shared between the two circuits C_1 and C_2. The theorem below is similar to a theorem given in [23] (see Theorem 4 in [23]) that proved that multifault testability was preserved under Shannon decomposition. We prove the theorem by using deductive arguments about the faults that are unconditionally tested upon the application of a vector pair to a circuit.

Theorem 9.6.4 *Given two multifault testable circuits C_1 and C_2 whose inputs are $x_1, x_2, \ldots, x_N$, if the outputs of C_1 and C_2 are connected to a multiplexor with a control input x_{N+1}, then if the composition C is fully single stuck-at fault testable, it is fully multifault testable. Furthermore, the test vectors for C correspond to the following sets of vectors:*

1. *Vectors $w_1 = \langle v, x_{N+1}\rangle$ and $w_2 = \langle v, \overline{x_{N+1}}\rangle$ where $C_1(v) = 1, C_2(v) = 0$.*
2. *Vectors $w_3 = \langle v, x_{N+1}\rangle$ and $w_4 = \langle v, \overline{x_{N+1}}\rangle$ where $C_1(v) = 0, C_2(v) = 1$.*

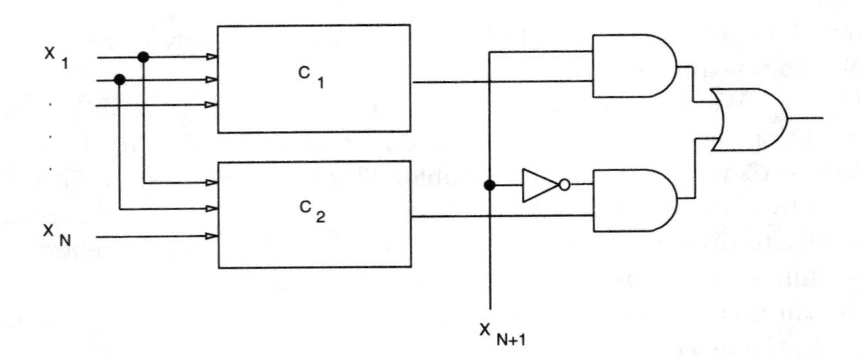

Figure 9.22: A multiplexor composition rule

3. *Vector set* $\langle M_1, x_{N+1} \rangle$, *where* M_1 *is the multifault test set for* C_1.
4. *Vector set* $\langle M_2, \overline{x_{N+1}} \rangle$, *where* M_2 *is the multifault test set for* C_2.

Proof. Since the composition C is fully single stuck-at fault testable, the single stuck-at faults within the multiplexor are also testable. In order to test the stuck-at-1 fault at the input to the AND gate connected to x_{N+1} ($\overline{x_{N+1}}$), we need a vector v over $x_1, \ldots, x_N$ that produces a 1 at the output of C_1 and a 0 at the output of C_2. Given this vector v we can toggle the x_{N+1} control input to propagate a transition to the output of C. We argue that the vector pair $\langle v, x_{N+1} \rangle$ followed by $\langle v, \overline{x_{N+1}} \rangle$ will detect any stuck-at faults on the even parity path from x_{N+1} to the output of C, in the presence of arbitrary fault conditions in C_1 and C_2. This is because $C(v, x_{N+1}) = 1$ and $C(v, \overline{x_{N+1}}) = 0$, and there is only one path with even parity from x_{N+1} to the output of C. If there is a stuck-at fault on this path, then the output of C will be a constant regardless of other fault conditions. We will therefore detect this stuck-at fault either with $\langle v, x_{N+1} \rangle$ or $\langle v, \overline{x_{N+1}} \rangle$ regardless of other fault conditions. A similar argument can be made for the stuck-at faults on the single odd parity path from x_{N+1} to the output of C. Set (1) and Set (2) above will detect the multifaults in this case.

We now consider the case where the paths from x_{N+1} are fault-free. Given a multifault $m_1 \in C_1$ and a multifault $m_2 \in C_2$, we can detect m_1 with a vector v over the inputs $x_1, \ldots, x_N$ to the output of C_1. This vector v may excite m_2 but if we set $x_{N+1} = 1$, then regardless of the true/faulty values at the output of C_2, we will

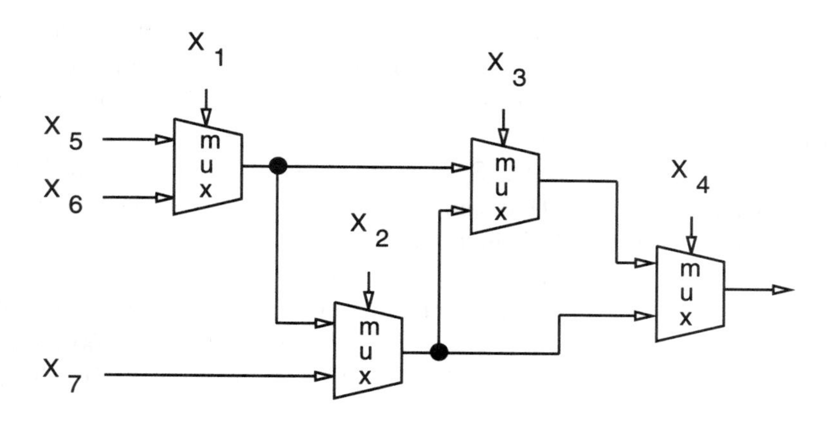

Figure 9.23: A general multiplexor-based circuit

propagate the effect of m_1 to the output of C. Similarly, setting $x_{N+1} = 0$ will propagate a faulty value at the output of C_2 to the output of C, regardless of fault conditions in C_1. Set (3) and Set (4) above will detect the multifaults in this case. □

We now use this rule repeatedly to identify a class of networks that are fully multifault testable if they are fully single stuck-at fault testable.

Consider the general multiplexor-based network of Figure 9.23. It is a general DAG, with the following property: Along any path from inputs to the circuit output we encounter a decision variable corresponding to the multiplexor input at most *once.* We can prove the following theorem about such networks (which is essentially a repeated invocation of Theorem 4 of [23]).

Theorem 9.6.5 *A multiplexor-based network, η, where no input is encountered more than once along any path from input to the circuit output, is fully multifault testable if it is fully single stuck-at fault testable.*

Proof. We can compose the multiplexor-based network η using a series of rule applications given by Theorem 9.6.4. A single multiplexor is fully multifault testable if it is fully single stuck-at fault testable. At any given point of time we have two fully multifault testable subnetworks C_1 and C_2 that are connected to a multiplexor. The inputs to C_1 and C_2, namely $x_1, \ldots, x_i$, are different from the control input to the multiplexor, namely x_{i+1}. Since η is fully single stuck-at fault testable, we can detect the stuck-at faults on the paths from x_{i+1} to the output of η. In order to do this we need to propagate the effect of the faults within the multiplexor to the multiplexor output, implying that we have a vector v such that $C_1(v) = 1$ ($C_1(v) = 0$) and $C_2(v) = 0$ ($C_2(v) = 1$). By arguments similar to those in Theorem 9.6.4 we can prove that the subnetwork rooted at the output of multiplexor is fully multifault testable.

Proceeding forward from the inputs to the outputs, we will always have two subnetworks that are completely multifault testable connected to a multiplexor whose internal single stuck-at faults are testable. Therefore, we have a fully multifault testable composition. □

The test vectors to detect all the multifaults in η can be derived by adding the multifault test vectors of the constituent subnetworks.

9.7 Hazard-Free Robust Path Delay Faults

We focus on the most restrictive testability criterion proposed thus far in the literature, namely hazard-free robust path delay faults. Complete hazard-free path delay fault testing will model and therefore detect a very large number of possible defects (single or multiple) that can occur on a fabricated circuit. We give necessary and sufficient conditions for a path in a multilevel circuit to be hazard-free robust testable in Section 9.7.1 and show how delay fault test generation can be mapped to stuck-at fault testing in Section 9.7.2. Synthesis procedures and hierarchical composition methods that maintain path delay fault testability are the subject of Sections 9.7.3 and 9.7.4, respectively. We show that hazard-free robust path delay fault testability implies multifault testability in Section 9.7.6.

9.7.1 Conditions for Testability

Given a circuit C with ENF expression E and a tagged literal l_π, the *path-cube-complex associated with* l_π is the set of all cubes q such that $l_\pi \in q$ and there is no other tagged literal $l_\rho \in q$. The reason for excluding cubes that contain more than one literal of the same variable with different tags in the path-cube-complex is that the paths

Necessity: $\langle v_1, v_2 \rangle$ is a hazard-free robust path delay fault test for π in L. This means that each side-input along path $Q \in P_{OR}$ to an OR gate along π in L is at a steady noncontrolling 0 value with no transitions on the application of v_2 after v_1. This implies that these side-inputs are insensitive to a change in l under $\langle v_1, v_2 \rangle$, i.e., asserting any constant value on the I-edge of each $Q \in P_{OR}$ leaves the value on these side-inputs unchanged under $\langle v_1, v_2 \rangle$ The same can be said for the P_{AND} case. Hence, $\langle v_1, v_2 \rangle$ remains a test for π in L_S.
Sufficiency: Only the I-edges of paths in P_{OR} and P_{AND} are set to 1 and 0, respectively, in L_S. Each side-input to OR gates along π are set to steady noncontrolling values since $\langle v_1, v_2 \rangle$ is a hazard-free robust path delay fault test for π in L_S. This implies (since there are no inverters along the way) that the constant 1 asserted on the I-edge of each path in P_{OR} does not propagate to the side-inputs of these OR gates (similarly for the P_{AND} case). Therefore, on the application of $\langle v_1, v_2 \rangle$ in L none of the transitions on these I-edges propagates to the side-inputs of OR or AND gates along π, and $\langle v_1, v_2 \rangle$ is a hazard-free robust path delay fault test for π in η. □

In effect, this theorem states that for a *hazard-free robust path delay fault* (HFRPDF) test for π, no event on any of the I-edges associated with input l (except the I-edge of π) should be allowed to propagate. Therefore, those I-edges are assigned values to block the event.

Theorem 9.7.3 *Let π be a path with input l in leaf-DAG L. v_2 is a test for the stuck-at-0 fault on the I-edge of π in the smooth network L_S for π if and only if $\langle v_1, v_2 \rangle$, where $v_2 = v_1 - \{l\} \cup \{\bar{l}\}$, is a hazard-free robust path delay fault test for the rising transition on π in L.*

Proof. Necessity: Let $\langle v_1, v_2 \rangle$ be a hazard-free robust path delay fault test for the rising transition on π in L. Then by Theorem 9.7.2 it is a hazard-free robust path delay fault test for the rising transition along π in L_S. The output of π is 1 when v_2 is applied to the smooth circuit L_S for P. Consider what happens in the presence of the stuck-at-0 fault on the I-edge of π in L_S under vector v_2. Since each side-input of π is on a noncontrolling value under v_2, the effect of the stuck-at-0 fault propagates all the way to the circuit output.
Sufficiency: Let v_2 be a test vector for the stuck-at-0 fault on the

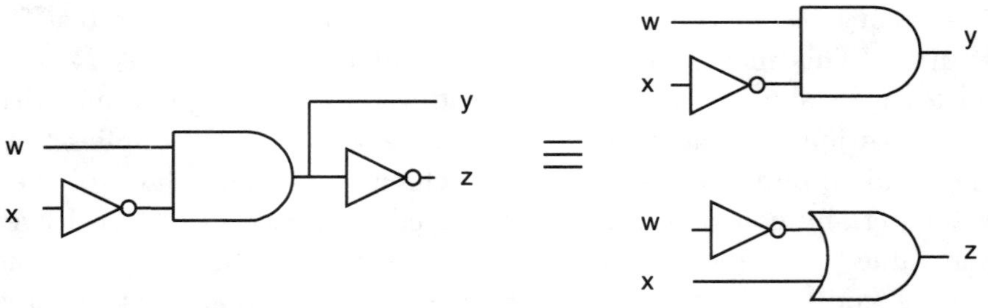

Figure 9.25: Pushing inverters to the primary inputs

I-edge of π in the smooth network L_S for π. l is the only input in L_S that changes under the vector pair $\langle v_1, \ v_2 \rangle$ Let P_S denote the paths that pass through side-inputs to π whose I-edges are not associated with l. Each side-input to π that belongs to a path in P_S must be at a noncontrolling value under v_2. This is due to the test condition for the stuck-at fault and the fact that the circuit is a leaf-DAG. By the construction of L_S no other side-inputs can exist since all paths that pass through side-inputs to π whose I-edges are associated with l have been replaced by constant values. When we apply $\langle v_1, \ v_2 \rangle$ the only change that occurs is in input l and the side-inputs to π remain at steady noncontrolling values. Therefore, $\langle v_1, \ v_2 \rangle$ is a hazard-free robust path delay fault for π in L_S and hence in L by Theorem 9.7.2. □

We have shown the equivalence of hazard-free robust path delay fault testing and single stuck-at fault testing on leaf-DAG networks. Since the size of the leaf-DAG can be exponentially larger than the size of the original circuit, we wish to develop a more efficient procedure.

Consider η' which is obtained from a multilevel circuit η by pushing all inverters in η to the primary inputs. The inverter at the output of a gate may be moved to the inputs by using DeMorgan's laws of complementation. Thus, all the inverters may be moved to the primary inputs by starting at the primary outputs and recursively applying this procedure to all gates in the circuit. As shown in Figure 9.25 a gate that is used in both inverting and noninverting phase may need to be duplicated. Since each gate is duplicated no more than once, η' is at most twice the size of η.

The only other step we need to perform is to make the path π, beginning from input l in η', for which we want to generate a delay fault test fan-out-free after l. This entails duplicating the gates on π at most once and results in a circuit η'_π which is at most four times the size of η (and is typically only a few gates larger). The number of inputs to the circuit is the same as that of η, making this approach more efficient than the test generation methods that use multiple-valued calculi. Note that the sensitization conditions of paths in η'_π and η' are the same as those for the corresponding paths in η. This can be seen by using an argument similar to that given for leaf-DAGs.

Given η'_π we can construct the smooth network η'_S for π by replacing the appropriate I-edges associated with l by 1 and 0 values. Since η'_π is inversion free after the primary inputs, each I-edge associated with i (other than the I-edge on π) either converges with π at an OR gate or an AND gate.

The above formulation of the delay fault test generation problem has been implemented and shown to produce superior results over multiple-valued calculi implementations (e.g., [26]). It allows for the direct use of sophisticated single stuck-at fault test generation and redundancy identification methods in the delay test generation problem. A disadvantage is that it requires a separate and explicit transformation of the circuit prior to generating tests for each chosen path.

9.7.3 Synthesis for Full Testability

In Section 9.5 we showed that various commonly used logic transformations used in area optimization of multilevel logic retained ENF reducibility. We can now utilize the theorems of Section 9.5 to arrive at synthesis and optimization methods that produce fully HFRPDF testable designs. We prove a theorem that relates ENF reducibility to path delay fault testability below. This theorem is the analog of Theorem 9.6.2 for path delay faults.

Theorem 9.7.4 *Let C be a single-output circuit with ENF expression E_C. Let M be a single-output circuit with ENF expression E_M. If $E_C \geq_t E_M$ and if a vector set V hazard-free robustly detects all path delay faults in C, then V hazard-free robustly detects all path delay faults in M.*

Proof. Let π be a path in M. There is a corresponding tagged literal l_π in E_M. Corresponding to l_π in E_M there is a nonempty set of corresponding tagged literals $\{l_{\pi_1}, \ldots, l_{\pi_m}\}$ in E_C. Let the path-cube-complex of l_{π_i} in E_C be L and let D be those cubes in E_C

not containing any l_{π_i}. By Theorem 9.7.1 there exists a hazard-free robust path delay fault test $\langle v_1,\ v_2\rangle$ for each π_i which has all the following properties:

1. Vertex v_2 is such that $L(v_2) = 1$ and v_2 is not covered by any cube in D.

2. Vertex $v_1 = v_2 - \{l\} \cup \{\bar{l}\}$ and v_1 is in the *OFF*-set of C.

3. For every cube d in D there exists some literal m in both v_1 and v_2 such that $d_m(v_1) = 0$ and $d_m(v_2) = 0$.

Let the path-cube-complex of l_π in E_M be L_M, and let D_M be those cubes in E_M not containing l_π. Because the path-cube-complex L_M contains all the cubes in E_M that correspond to those cubes containing l_{π_i} in E_C, $NOTAGS(L) \subseteq NOTAGS(L_M)$, so $L_M(v_2) = 1$ and v_2 is not covered by any cube in D_M. As the *OFF*-sets of E_C and E_M are exactly the same, v_1 is in the *OFF*-set of E_M. Finally, just as $NOTAGS(L) \subseteq NOTAGS(L_M)$, similarly, $NOTAGS(D) \supseteq NOTAGS(D_M)$. Thus, for every cube d in D_M there exists some literal m in both v_1 and v_2 such that $d_m(v_1) = 0$ and $d_m(v_2) = 0$. Thus, by Theorem 9.7.1 the vector pair $\langle v_1,\ v_2\rangle$ is a hazard-free robust path delay fault test for π in M. Thus, for every path π in M there exists a nonempty set of associated paths P in C such that any vector pair that tests an HFRPDF on any $\pi_i \in P$ is an HFRPDF test for π. □

While this proof shows that ENF reducibility retains HFRPDF testability this same argument may also be used to show that algebraic factorization can improve the HFRPDF testability of a circuit. For example consider $C = a\bar{b} + \bar{b}c + b\bar{c}$ and its algebraic factorization $M = (a + c)\bar{b} + b\bar{c}$. The path associated with literal $\bar{b}$ in cube $a\bar{b}$ of C, call it π_1, is not HFRPDF testable, but the path associated with $\bar{b}$ in cube $\bar{b}c$ of C, call it π_2, is HFRPDF testable. After C is factored into M, there is a many-to-one reduction from paths π_1 and π_2 to a single path π associated with literal $\bar{b}$ in the factor $(a + c)\bar{b}$, and the testability of π_2 alone is sufficient to ensure the testability of π. As a result M is completely HFRPDF testable, while C is not.

It should also be noted that ENF reducibility as defined above does not imply the preservation of *hazard-free robust gate delay fault* (HFRGDF) testability (see Section 9.9.3). The conditions for HFRGDF testability are weak and require a stronger form of ENF reducibility.

Corollary 9.7.1 *Given a completely hazard-free robust path delay fault testable network, algebraic factorization with a constrained use of the inverse as defined in Lemma 9.5.1 produces a completely hazard-free robust path delay fault testable network.*

Proof. Follows directly from Theorem 9.7.4 and Theorem 9.5.4. □

As was noted previously, if the resubstitution of Figure 9.14 is carried out in reverse, i.e., $\overline{\overline{a} \cdot \overline{c} \ + \ b \cdot \overline{c}}$ is substituted for $a \cdot \overline{b} \ + \ c$, then ENF reducibility is not preserved. If we apply the resubstitution illustrated in Figure 9.14 in reverse, then we still obtain a circuit that is completely HFRPDF testable; however, as there are more paths in the resubstituted network than in the original, the HFRPDF test vector set for the original network will not hazard-free robustly detect all path delay faults of the resulting resubstituted network. Thus, while HFRPDF testability is retained in this example, the vector set for complete coverage is not. (We will give an example where resubstitution with complement may result in a circuit that is not HFRPDF testable.) We emphasize that ENF reducibility is a sufficiency condition for retaining HFRPDF testability and not a necessary one. For instance, a XOR gate $a\overline{b}+\overline{a}b$ is HFRPDF testable. Adding an inverter to the output of the XOR gate does not affect HFRPDF testability, but the ENF of the resulting circuit $\overline{a\overline{b}+\overline{a}b}$ is more complex than the original ENF.

It can be shown that algebraic factorization without the constraint embodied by Lemma 9.5.1 may result in a circuit which is not HFRPDF testable. This phenomenon is related to the fact that if the constraint is not satisfied, the number of paths may increase after factorization. In Figure 9.26(a) we have a multilevel circuit that is completely HFRPDF testable. Algebraic factorization using the complement produces the circuit of Figure 9.26(b), where the path $\{c', 2, 6, 0, 7, 10\}$ is not hazard-free robust testable. The path $\{c, 5, 7, 10\}$ in Figure 9.26(a) has split into two paths $\{c', 2, 6, 0, 7, 10\}$ and $\{c', 3, 6, 0, 7, 10\}$ in Figure 9.26(b). The increase in the number of paths has resulted in the nonretainment of HFRPDF testability.

In experiments with algebraic factorization with the complement using the program MIS, algebraic factorization was not constrained and HFRPDF testability was retained. This was due to the fact that the number of paths typically did not increase [14].

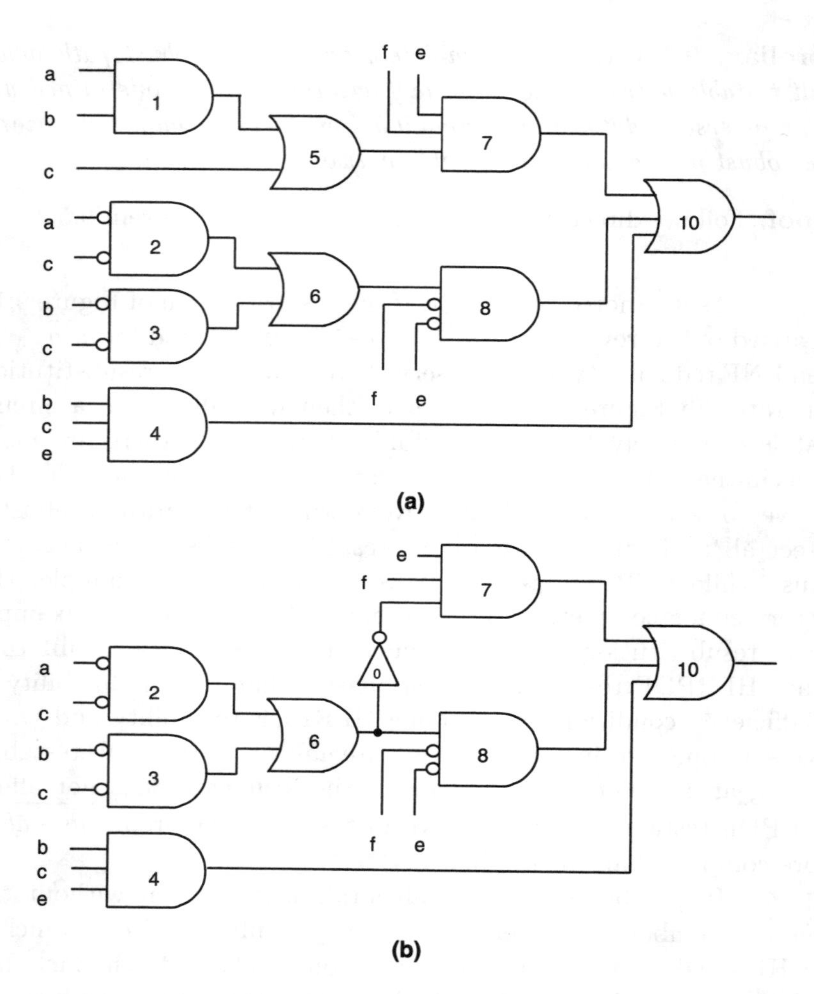

Figure 9.26: Algebraic factorization using the complement does not preserve testability

9.7.4 Compositional Techniques for Full Testability

So far the results of this section have been developed for individual modules of combinational logic that can be flattened to two-level circuits. As with the case of multifaults in Section 9.6.4 we now consider how these results can be practically applied in an environment in which flattenable combinational logic modules are cascaded with other, perhaps nonflattenable, combinational logic modules.

Our solution is to develop sets of composition rules that allow us to prove conditions such that when one HFRPDF testable circuit module feeds another HFRPDF testable circuit module, then the resulting cascade is also HFRPDF testable. We first give a composition rule that gives a sufficiency condition for the result of a composition of two HFRPDF testable circuits to be HFRPDF testable. We then apply this theorem to show that some common nonflattenable circuits are hazard-free robust path delay fault testable. We then apply the theorem in another way to show how cascading circuits consisting of both flattenable and unflattenable modules, such as those indicated by the FSM description above can also be proven to be fully hazard-free robust path delay fault testable.

Theorem 9.7.5 *If C_1 and C_2 are hazard-free robust path delay fault testable combinational circuits, the input variables of C_1 and C_2 are disjoint, and a circuit C is created by connecting one output o_j of C_1 to one input i_k of C_2, then C is also hazard-free robust path delay fault testable (see Figure 9.21).*

Proof. Let σ be a path in C. Suppose that σ was also a path in C_2. Let i_m be the input of C_2 associated with σ. Note that $i_m \neq i_k$. By the multiple-output generalization of Theorem 9.7.1 there exists a vector sequence $\langle x_1, x_2 \rangle$ that is a hazard-free robust path delay fault test for σ in C_2 such that only the value of input variable i_m changes in $\langle x_1, x_2 \rangle$ Now consider the value of input i_k in C_2 in vector x_1 (it will be the same in x_2). Without loss of generality assume that the value is 1. Let w_1 be a vector over the input variables to C_1 such that the output $o_j = 1$. Let $v_1 = \{x_1 - \{i_k\}\} \cup w_1$, and let $v_2 = \{x_2 - \{i_k\}\} \cup w_1$. The vector v_1 provides the same noncontrolling values along the side-inputs of σ as did x_1. Thus, the transition caused by the change in input i_m in vector v_2 propagates along σ just as in x_2. Thus, $\langle v_1, v_2 \rangle$ is a hazard-free robust path delay fault test for σ.

Suppose that σ is a path in C_1. In other words, σ is a path from an input of C_1 to an output o_p of C_1 such that $o_p \neq o_j$, and thus σ is not affected by the cascading of C_1 and C_2. Thus, the original vector sequence $\langle v_1, v_2 \rangle$ for σ in C_1 may be applied to give a hazard-free robust path delay fault vector sequence $\langle v_1, v_2 \rangle$ for σ in C.

Suppose that σ is not a path in C_1 or a path in C_2. Then the path σ must consist of a path, call it π beginning at some input i_n of C_1 and passing out output o_j of C_1, entering a path, call it ρ, beginning at what was formerly input i_k of C_2 and passing out of some output o_p of C_2. Our approach will be to build a vector sequence $\langle v_1, v_2 \rangle$ for testing σ by using the sequence $\langle x_1, x_2 \rangle$ used to test for π and the sequence $\langle w_1, w_2 \rangle$ used to test for ρ. By the multiple output generalization of Theorem 9.7.1 there exists a vector sequence $\langle w_1, w_2 \rangle$ that is a hazard-free robust path delay fault test for ρ in C_2 such that only the value of input variable i_k changes in $\langle w_1, w_2 \rangle$ Suppose that the value of i_k in w_1 is 0 and the value of i_k in w_2 is 1. By Theorem 9.7.1, there exists a vector sequence $\langle x_1, x_2 \rangle$ that is a hazard-free robust path delay fault test for π in C_1 such that only the value of input variable i_n changes in $\langle x_1, x_2 \rangle$ Let $v_1 = x_1 \cup \{w_1 - \{i_k\}\}$, and let $v_2 = x_2 \cup \{w_1 - \{i_k\}\}$. The vector w_1 supplied noncontrolling values along side-inputs of ρ in C_2; thus, the vector $\{w_1 - \{i_k\}\}$ does the same. The vector x_1 supplied noncontrolling values along side-inputs of π in C_1, resulting in a 0 on the output o_j of C_1. Thus, the vector $v_1 = x_1 \cup \{w_1 - \{i_k\}\}$ propagates a 0 along σ. The vector x_2 supplied noncontrolling values along side-inputs of π in C_1, resulting in a 1 on the output of C_1; thus, the vector $v_2 = x_2 \cup \{w_1 - \{i_k\}\}$ propagates a $0 \rightarrow 1$ event along σ to the output of C.

In the alternate case that the value of i_k in x_1 is 1 and the value of i_k in x_2 is 0, then the vector pair $\langle v_1, v_2 \rangle$ tests for a $1 \rightarrow 0$ event along σ to the output of C and the vector pair $\langle v_2, v_1 \rangle$ may be applied to test for a $0 \rightarrow 1$ event. □

Note that in Theorem 9.7.5 we assume complete HFRPDF testability of both C_1 and C_2. This is very important. If a single path starting with input i_k of C_2 is not hazard-free robust path delay fault testable, then *all* the paths in the subcircuit C_1 may be untestable in C. This is an instance where the Shannon decomposition techniques described in Section 9.7.5 should be used to augment algebraic factorization to ensure a completely HFRPDF testable circuit.

We now show how Theorem 9.7.5 may be applied to prove the HFRPDF testability of regular structures.

We wish to show that parity trees are HFRPDF testable. We begin by showing that a single two-input XOR gate is HFRPDF testable. Let a and b be the primary inputs to an XOR gate built from primitive gates as in the dotted box in Figure 9.27. The ENF expression for the XOR gate is $x = \overline{a}_{\{1,3,5\}} \cdot b_{\{3,5\}} + a_{\{4,5\}} \cdot \overline{b}_{\{2,4,5\}}$. We need not be concerned over the gate selection as any topologically equivalent XOR structure would have the same ENF. It is easy to see that this expression satisfies the conditions of Theorem 9.7.1, and thus a single XOR gate is HFRPDF testable. For example, con-

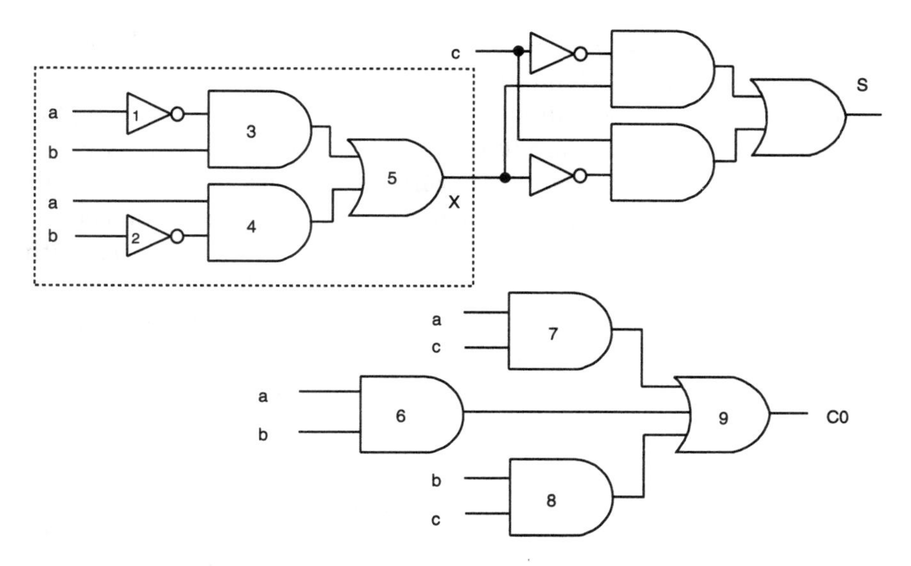

Figure 9.27: Full adder

sider the path associated with the tagged literal $\overline{a}_{\{1,3,5\}}$. The relatively essential vertex associated with the cubes containing $\overline{a}_{\{1,3,5\}}$ is $v_2 = (a = 0, b = 1)$. A member of the *OFF*-set of the XOR function distance-1 from v_2 is $v_1 = (a = 1, b = 1)$. Thus vector pair $\langle (v_1 = (a = 1, b = 1)), (v_2 = (a = 0, b = 1)) \rangle$ is a hazard-free robust delay fault test for the path from primary input a through gates $\{1, 3, 5\}$ to the output.

We may now use Theorem 9.7.5 to show that a parity tree of arbitrary size is HFRPDF testable.

Theorem 9.7.6 *If P is a parity tree built from* XOR *gates that are logically and topologically equivalent to those in the dotted box of Figure 9.27, then P is hazard-free robust path delay fault testable.*

Proof. (By induction on the number of bits): Let $S(k)$ be the statement that if P is a parity-tree of k bits built from XOR gates that are logically and topologically equivalent to those in the dotted box of Figure 9.27, then P is hazard-free robust path delay fault testable.

We choose the induction basis k to be 1. By the arguments given above a 2-bit XOR gate is hazard-free robust path delay fault testable.

Suppose $S(k)$ for all $k < n$. Consider $S(n)$. A 3-bit parity tree is shown at the top of Figure 9.27. Note that it does not matter which of the $n-1$ inputs we connect. By the induction hypothesis a parity tree of $n-1$ bits is hazard-free robust path delay fault testable. Let us call such an $n-1$ bit parity tree C_2 with gate inputs $i_1, \ldots, i_{n-1}$ and primary input variables $v_1, \ldots, v_{n-1}$. Let us call a single XOR gate C_1. C_1 is also hazard-free robust path delay fault testable. Let us connect C_1 to the input i_{n-1} in C_2 and connect the input variable v_{n-1} to one input of C_2 and a new variable v_n with the other input of C_2. As the inputs of C_1 and C_2 are disjoint and each of the circuits are hazard-free robust path delay fault testable, then by Theorem 9.7.5 the resulting circuit is also hazard-free robust path delay fault testable. (If it is required, we can build the parity tree in a balanced way.) □

We proceed to show that a full adder is HFRPDF testable. Figure 9.27 shows the labeled gate network for a full adder. We previously showed that a parity tree was HFRPDF testable, and so we argue that the sum (S) output of a full adder is also HFRPDF testable. The ENF expression for the output carry signal, CO, is $CO = a_{\{6,9\}} \cdot b_{\{6,9\}} + a_{\{7,9\}} \cdot c_{\{7,9\}} + b_{\{8,9\}} \cdot c_{\{8,9\}}$. As before any topologically equivalent gate network would have an equivalent ENF. It may be shown by inspection that this expression satisfies the conditions of Theorem 9.7.1, and thus a single full adder is HFRPDF testable. For example, consider the path associated with the tagged literal $a_{\{6,9\}}$. The relatively essential vertex associated with the cube containing $a_{\{6,9\}}$ is $v_2 = (a = 1, b = 1, c = 0)$. A member of the *OFF*-set of the output carry function distance-1 from v_2 is $(v_1 = (a = 0, b = 1, c = 0))$. Thus, vector pair $\langle (v_1 = (a = 0, b = 1, c = 0)), (v_2 = (a = 1, b = 1, c = 0)) \rangle$ is a hazard-free robust delay fault test for the path from primary input a through gates $\{6, 9\}$ to the output.

Having shown a full adder is HFRPDF testable, we continue to show that a ripple-carry adder is HFRPDF testable. An informal inductive argument is as follows.

Theorem 9.7.7 *If A is a ripple-carry adder built from full adders that are logically and topologically equivalent to that in Figure 9.27, then A is hazard-free robust path delay fault testable.*

Proof. (By induction on the number of bits): Let $P(k)$ be the statement that if A is a ripple-carry adder of k bits built from full adders that are logically and topologically equivalent to that in Figure 9.27 then A is hazard-free robust path delay fault testable.

We choose the induction basis k to be 1. A full adder is fully HFRPDF testable, and therefore a 1-bit ripple-carry adder is fully HFRPDF testable.

Suppose $P(k)$ for $k < n$. Consider $P(n)$. By the induction hypothesis the first $n - 1$ stages of the ripple-carry adder are hazard-free robust path delay fault testable. Call this subcircuit C_1. Consider the full adder associated with the n-th stage in a ripple-carry adder. Call this subcircuit C_2. The inputs to C_2 are two primary inputs a_n and b_n, and CO_{n-1}, the carry-out from C_1. Both C_1 and C_2 are hazard-free robust path delay fault testable, their primary inputs are disjoint, and one output of C_1 feeds one input of C_2. Therefore by Theorem 9.7.5 the n-stage ripple-carry adder is also hazard-free robust path delay fault testable. □

Theorem 9.7.5 was used to show that certain regular structures are HFRPDF testable. We now employ Theorem 9.7.5 to show how the FSM described in Section 9.6.4, which takes as inputs the output of a parity generator and the carry-out of an adder can be shown to be HFRPDF testable.

The first step is to synthesize an HFRPDF testable FSM circuit according to the techniques described in Section 9.7.3 by treating the signals `packet_odd_parity` and `add_overflow` as primary inputs. The parity generator module has already been shown to be HFRPDF testable, so Theorem 9.7.5 may be invoked once to show that the parity generator and the FSM circuit (with the `add_overflow` signal treated as a primary input) are HFRPDF testable. The ripple-carry adder module has been shown to be HFRPDF testable, so Theorem 9.7.5 may be invoked a second time to show that the parity generator, the FSM description, and the ripple-carry adder module all form an HFRPDF testable combinational logic circuit.

In Theorem 9.7.5 we gave a constructive proof that is essentially an algorithm for deriving the test vectors for each path in the resulting circuit C from the path delay fault test vectors for each path in C_1 and C_2. Thus, we have a procedure for building a complete set of vectors as well.

Compositional rules that preserve HFRPDF testability or HFRGDF testability have been developed for a variety of datapath modules, such as carry look-ahead, carry select, and carry bypass adders, ripple and parallel comparators, parallel multipliers, and arithmetic logic units [9]. The synthesis of a speech recognition chip, a small microprocessor, and a data encryption chip for full hazard-free robust path delay fault testability has been described in [14]. The area overheads for full testability over unconstrained designs were approximately 15 percent. We will describe the synthesis of the speech recognition chip for full HFRPDF testability in Section 9.11.

9.7.5 Shannon Decomposition for Testability

The composition methods presented in the previous section required the individual circuits in the composition to be completely HFRPDF testable. As illustrated by the function in Figure 5.18 it is not always possible to find completely HFRPDF testable two-level realizations. While algebraic factorization can improve the HFRPDF testability of a circuit, so far we have not presented any method that can guarantee a completely HFRPDF testable circuit from a starting point that is not completely HFRPDF testable. Such a method has been presented in [24]. The theorem below is similar to Theorem 6.1 of [24].

Theorem 9.7.8 *Given two hazard-free robust path delay fault testable circuits C_1 and C_2 whose inputs are x_1, x_2, $\ldots$, x_N, if the outputs of C_1 and C_2 are connected to a multiplexor with a control input x_{N+1}, then if the composition C is fully single stuck-at fault testable, C is fully hazard-free robust path delay fault testable.*

Proof. There is only one path from the output of C_1 to the output of C. In order to test a path in C that passes through the output of C_1, we simply set the control input x_{N+1} to 1 and apply the appropriate test vector pair for the path in C_1. Similarly for paths in C_2. In order to test paths from x_{N+1} we apply a transition on the input x_{N+1} and set the output of C_1 to 1 (0) and the output of C_2 to 0 (1). If we cannot set C_1 and C_2 to these values, it means that the stuck-at faults at the inputs to the AND gates in the multiplexor are redundant and we have a contradiction to the hypothesis of the theorem. □

The above theorem can be used in a synthesis strategy that begins from a two-level circuit that is not completely HFRPDF testable and produces a multilevel realization that is completely HFRPDF testable. This realization can be used as a starting point for testability-preserving logic optimization. The method used in [24] to produce a completely HFRPDF testable realization of a function f is described below.

1. Minimize f to produce a prime and irredundant two-level logic circuit.

2. Check if the path delay faults in the two-level function f are testable. If yes, stop here.

3. Identify the product terms associated with the paths that are not robustly testable. Select a binate input variable x_i that is an input to the maximum number of the identified product terms.

4. Cofactor f with respect to x_i. Apply the procedure to f_{x_i} and $f_{\overline{x_i}}$.

5. f is implemented as $x_i \cdot f_{x_i} + \overline{x_i} \cdot f_{\overline{x_i}}$.

A function that is unate in all its input variables is guaranteed to be completely HFRPDF testable, and hence we will always be able to select a binate variable in Step 3. An example of the application of the above procedure to the function of Figure 5.18 is given below. We have:

$$f = a \cdot \overline{b} + \overline{a} \cdot b + c \cdot \overline{d} + \overline{c} \cdot d + \overline{a} \cdot \overline{c}$$

We select the variable a since the paths through gate 5 are not testable. Cofactoring gives:

$$f_a = \overline{b} + c \cdot \overline{d} + \overline{c} \cdot d$$

$$f_{\overline{a}} = b + c \cdot \overline{d} + \overline{c} \cdot d + \overline{c}$$

f_a is prime and irredundant and fully HFRPDF testable. $f_{\overline{a}}$ is minimized to:

$$f_{\overline{a}} = b + \overline{d} + \overline{c}$$

which is completely HFRPDF testable. The fact that f is binate in a implies that the circuit for f formed by multiplexing f_a and $f_{\overline{a}}$ is completely HFRPDF testable by Theorem 9.7.8, since the single stuck-at faults within the multiplexor will be testable. For instance, if the stuck-at-1 fault in the AND gate that f_a is connected to is redundant, it means that f is expressible as $f_a + a \cdot f_{\overline{a}}$ and therefore f is unate in a.

This Shannon decomposition procedure serves as a preprocessing step for algebraic factorization when we are targeting 100 percent HFRPDF testability. However, it can also be used to identify a rich class of multilevel networks that are fully HFRPDF testable. We use the principle of induction and Theorem 9.7.8 to prove the result below pertaining to multiplexor-based networks.

Theorem 9.7.9 *A multiplexor-based network, η, where no input is encountered more than once along any path from inputs of the circuit to the output, is fully hazard-free robust path delay fault testable if it is fully single stuck-at fault testable.*

Proof. (By induction): We will walk from the primary inputs to the output of the circuit. At the first level we have multiplexors fed by primary inputs only. The function implemented by the output of the multiplexor is fully hazard-free robust path delay fault testable, because a multiplexor is fully hazard-free robust path delay fault testable if its control input is different from its data inputs. Assume that the functions at the ith level are fully hazard-free robust path delay fault testable. A multiplexor at the $i+1^{th}$ level is fed by hazard-free robust testable functions f_0 and f_1 whose levels are $\leq i$. The control input x_i of the multiplexor is different from the inputs to f_0 and f_1 by the definition of η. Furthermore, the function implemented by the multiplexor, namely f, is binate in x_i. This implies that Theorem 9.7.8 is applicable. We can test any path in f_0 since f_0 is hazard-free robust testable. By setting $x_i = 0$, we can hazard-free robustly propagate the transition to the output of f through the single path from f_0 to f. Similarly for paths in f_1. To test the paths from x_i we need a transition on x_i with 1 (0) and 0 (1) values for f_0 and f_1. We can achieve these values because if we could not, it means that the stuck-at faults within the multiplexor are redundant. (Alternatively we could argue that f is binate in x_i.) Thus, f is completely hazard-free robust path delay fault testable.

This completes the induction step and the proof. □

Generalizations of the above theorem can be found in [2].

9.7.6 Relationship to Multifault Testability

It is easy to see that complete hazard-free robust path delay fault testability in a circuit implies complete hazard-free robust gate delay fault testability, transistor stuck-open fault testability, and general robust path delay fault testability.

In this section we show that complete hazard-free robust path delay fault testability of a multilevel circuit implies that the circuit is completely multifault testable. Note that the reverse is not true. For example, the circuit of Figure 5.18 is completely multifault testable but not completely HFRPDF testable.

Ordinarily we use the term *fault equivalence* to describe equivalence between two faults in the same circuit. Here we wish to extend the use of the term to apply to two faults in different circuits. More precisely, given two circuits C and C' that both compute the same Boolean function, we say that a multifault m in C is *fault equivalent* to multifault m' in C' if and only if the Boolean function computed by C with m asserted is equivalent to the function computed by C' with m' asserted. We say that a network C is *multifault equivalent* to a network C' if for each multifault m in C, there is an equivalent multifault m' in C'.

Lemma 9.7.2 *Let C be a circuit and let m be a multifault in C. Let L be a leaf-DAG circuit constructed from C. Then there exists a multifault m_L in L such that the site of each single fault, $f_L \in m_L$, is the primary input connection to an* AND *or* OR *gate in L, possibly after an inverter, and C with m is multifault equivalent to L with m_L.*

Proof. For each path π in L, if the corresponding path in C has a stuck-at-0 fault as the fault closest to the primary output, we introduce a fault on the primary input connection corresponding to π in L, call it l_π (after possibly a primary input inverter). If the path from the primary input to the fault site has an even parity of inversions, then it is a stuck-at-0 fault, else it is a stuck-at-1 fault. Similarly, if the path in C has a stuck-at-1 fault closest to the output, we introduce a stuck-at-1 fault on l_π if the path from the primary input to the fault site has an even parity of inversions and a stuck-at-0 fault otherwise. The multifault m_L comprising of the set of these single faults captures the behavior of m in C. Thus, C with m is multifault equivalent to L with m_L. □

Note that m_L is nonunique, and the above procedure merely gives an m_L in L is that is equivalent to m in C. For instance, given a 3-input AND gate, a stuck-at-0 fault on the output and a stuck-at-1 fault on the first input are multifault equivalent to two stuck-at-0 faults on any two inputs.

We now proceed to show that hazard-free robust path delay fault testability implies multifault testability. Our strategy will be to first show that HFRPDF testability implies multifault testability in a leaf-DAG and then show that this implies that HFRPDF testability implies multifault testability in an arbitrary circuit. First, we give an additional definition.

We define the *set of dominant fault effects* at a gate g in a leaf-DAG circuit L and for a given multifault, m_L, as follows: The set of dominant fault effects on a primary input connection l_π of L is the stuck-at-fault on that input, l_π-stuck-at-0 or l_π-stuck-at-1, if this fault is in m_L and is the null set otherwise. If the set of dominant fault effects on the inputs to an AND gate g contain all stuck-at-1s then the set of dominant fault effects on the output of g is the union of all the sets of dominant fault effects on the inputs. Similarly, if the set of dominant fault effects on the inputs to an AND gate g contain all stuck-at-0s then the set of dominant fault effects on the output of g is the union of all the sets of dominant fault effects on the inputs. If the set of dominant fault effects on the inputs to g contain both stuck-at-1 and stuck-at-0 faults, then the set of dominant fault effects on the output of g contains only the stuck-at-0 fault effects at the inputs.

The set of dominant fault effects at an OR gate can be defined in a similar manner.

So far the dominant fault effect has been treated as a simple topological property, and it is easy to show, for example, that:

Lemma 9.7.3 *Let L be a leaf-DAG circuit, and let m_L be a multifault in L. Then for each gate in g there is a unique set of dominant fault effects, $D(g, L, m_L)$. Further, this set is nonempty as long as there is some fault $f_L \in m_L$ that is in the transitive fan-in of g.*

Proof. The uniqueness of D follows from the construction of the set and can be easily shown by induction on the depth the circuit. The procedure for the construction of $D(g, L, m_L)$ guarantees that as long as there is some input g' of g such that $D(g', L, m_L)$ is nonempty, $D(g, L, m_L)$ is nonempty. Since we are given that there is some fault $f_L \in m_L$ in the transitive fan-in of g, it follows from induction on the depth of the circuit that $D(g, L, m_L)$ is nonempty. □

In order to reason about whether a fault is detectable or not, we have to deal with the functional properties of a circuit. We say that a fault f_1 *fault masks* a fault f_2 if and only if the activation of f_1 prohibits the detection of f_2. We use the term fault masks to differentiate between masking due to fault effects and masking due to the functional property of the circuit. For example, a stuck-at-fault may be redundant strictly due to functional masking, strictly due to fault masking, or due to a combination of both effects.

Lemma 9.7.4 *Let L be a leaf-DAG circuit, and let m_L be a multifault in L. Then for each gate g, any member of $D(g, L, m_L)$ is not fault masked by any other fault in m_L.*

Proof. This follows from the procedure for the construction of $D(g, L, m_L)$. □

Note that this lemma does not imply that the fault is detectable, only that it is not fault masked. The following result immediately follows from this.

Lemma 9.7.5 *Let L be a leaf-DAG circuit and let m_L be a multifault in L. Then for each primary output gate g, if there exists some member, $f \in D(g, L, m_L)$, such that f is testable, then m_L is testable.*

Proof. By Lemma 9.7.4 we know that f is not fault masked by any member of m_L. The fact that it is testable implies that it is not functionally masked. Hence, each test for f will serve as a test for m_L. □

From the above lemma, we see that if we can test at least one component $f \in D(g, L, m_L)$, then we can test m_L. As the following lemma indicates, if a leaf-DAG is HFRPDF testable, then we can always test each such f.

Lemma 9.7.6 *Let L be a leaf-DAG circuit, and let f be a single stuck-at fault in L. If S is a complete HFRPDF test set for L, then there exists a vector pair in S that detects f.*

Proof. Let π be any path through the fault site of f, and let $V = \langle v_1, v_2 \rangle$ be the HFRPDF test for π on a $0 \rightarrow 1$ transition at the output. The vector pair V propagates a transition through the site of f. Since f is stuck-at-1 or stuck-at-0 on either v_1 or v_2, the circuit will evaluate to the incorrect value. □

Now we are in a position to combine Lemmas 9.7.5 and 9.7.6 to arrive at a result that relates m_L in a leaf-DAG circuit L to a complete HFRPDF test for L.

Theorem 9.7.10 *Let L be a leaf-DAG circuit, and let m_L be a multifault in L. If S is a complete HFRPDF test set for L, then there exists a vector pair in S that detects m_L.*

Proof. Let g be the output gate of the circuit. By Lemma 9.7.3 $D(g, L, m_L)$ is nonempty. Let f be any member of $D(g, L, m_L)$. By Lemma 9.7.4 f is not masked by any fault in m_L. Hence, with regard to detection, $f \in m_L$ acts as a single-fault in the circuit. Since L is HFRPDF testable, by Lemma 9.7.6 there exists a vector pair in S that detects f and hence m_L. □

Now we can prove our final result.

Theorem 9.7.11 *Let C be a circuit and let m be a multifault in C. If S is a complete HFRPDF test set for C, then there exists a vector pair in S that detects m.*

Proof. Let us construct the corresponding leaf-DAG circuit for C as described in Section 9.3. Call that leaf-DAG L, and call the multifault corresponding to m in L, m_L.

By the arguments of Section 9.3 the sensitization conditions for paths in L are precisely those of C. Because each path in C is HFRPDF testable, each path is single-event sensitizable by some vector pair in S. Because each path in C is single-event sensitizable by some vector pair in S, each path in L is single-event sensitizable by some vector pair in S. Thus, L is also HFRPDF testable, and S is a complete HFRPDF test set for L.

L is a completely HFRPDF testable circuit. So by Theorem 9.7.10 there exists a vector pair in S, call that pair V, that detects m_L. By the construction of m_L in Lemma 9.7.2, C is multifault equivalent to L. If a vector detects m_L, then the same vector detects m. So V detects m. □

9.8 General Robust Path Delay Faults

In this section we focus on the a less restrictive path delay fault testability criterion, namely general robust path delay faults. General robust testing allows for hazards as described in Section 5.6. We give necessary and sufficient conditions in Section 9.8.1 for a path in a multilevel circuit to be *general robust path delay fault* (GRPDF) testable and show how delay fault test generation can be mapped to stuck-at fault testing in Section 9.8.2. We show that ENF reducibility retains GRPDF testability in Section 9.8.3.

9.8.1 Conditions for Testability

We give a theorem describing necessary and sufficient conditions for a multilevel single-output circuit to be general robust path delay fault testable.

Theorem 9.8.1 *Let C be a multilevel single-output circuit with ENF E. Let π be a path in C, and let the tagged literal associated with π be l_π. Let the complete path-cube-complex of l_π be L, and let D be those cubes not containing l_π. There exists a general robust delay fault test for a $0 \rightarrow 1$ transition on π in C if and only if all the following conditions are met:*

1. *There exists a vertex v_2 such that $L(v_2) = 1$ and v_2 is not covered by any cube in D.*
2. *There exists a vertex $v_1 = v_2 - \{l\} \cup \{\bar{l}\}$ such that v_1 is in the* OFF*-set of C.*
3. *For every cube d in D there exists some literal m in both v_1 and v_2 such that $d_m(v_1) = 0$ and $d_m(v_2) = 0$.*

Proof. The only difference between this theorem and Theorem 9.7.1 is that we have defined L to be the complete path-cube-complex of l_π allowing cubes that contain both l_ρ and l_π in L. This is allowed because we are dealing with a $0 \rightarrow 1$ transition, and the slower of the two paths will set the output of C. The remainder of the proof is similar to that of Theorem 9.7.1. □

Next we consider the $1 \rightarrow 0$ transition. Note that unlike HFRPDF testability, GRPDF testability requires separate conditions for $0 \rightarrow 1$ and $1 \rightarrow 0$ transitions.

Theorem 9.8.2 *Let C be a multilevel single-output circuit with ENF E. Let π be a path in C, and let the tagged literal associated with π be l_π. Let the path-cube-complex of l_π be L. There exists a general* *robust delay fault test for a* $1 \rightarrow 0$ *transition on* π *in* C *if and only if both of the following conditions are met:*

1. *There exists a vertex* v_1 *such that* $L(v_1) = 1$.
2. *There exists a vertex* $v_2 = v_1 - \{l\} \cup \{\bar{l}\}$ *such that* v_2 *is in the* OFF*-set of* C.

Proof. Let D be those cubes not containing l_π. The difference between this theorem and Theorem 9.7.1 is that we do not need cubes in D to be turned off. Note that L does not contain cubes which contain both l_ρ and l_π. Since the output of C is set by the last cube that goes to a 0, the cube $q \in L$ such that $q(v_1) = 1$ will be the last cube to go to a 0 if there is a delay fault on π. Furthermore, the only literal in the cube q to go to 0 will be l_π, since no other l_ρ literals exist in q. The remainder of the proof is similar to that of Theorem 9.7.1. □

Variants of the above conditions were given in [28] in terms of a circuit representation that is different from the ENF.

9.8.2 Test Generation

It is possible to extend the method of Section 9.7.2 to GRPDF test generation [31]. A series of theorems can be proven that shows the equivalence of general robust path delay fault and stuck-at fault test generation.

Definition 9.8.1 *Let* π *be a path starting with primary input* l *in a leaf-DAG* L. *Let* P_{OR} (P_{AND}) *be the set of paths which pass through side-inputs to all the* OR *(*AND*) gates along* π, *such that the* I*-edge of each* $Q \in P_{OR}$ $(Q \in P_{AND})$ *is associated with input* l. *The rising-smooth (falling-smooth) circuit for path* π *is obtained by replacing the* I*-edge of each* $Q \in P_{OR}$ *by* 1 *(each* $Q \in P_{AND}$ *by* 0*).*

We have two networks corresponding to the $0 \rightarrow 1$ and $1 \rightarrow 0$ transitions on π.

In Figure 9.28 we have the initial circuit at the top, followed by the leaf-DAG, and the rising-smooth circuit for the bold path is at the bottom. The I-edge of π is the connection from the inverter to the AND gate `5`. The I-edges of paths starting at `f` through `11`, `12`, and `2` are replaced by 1.

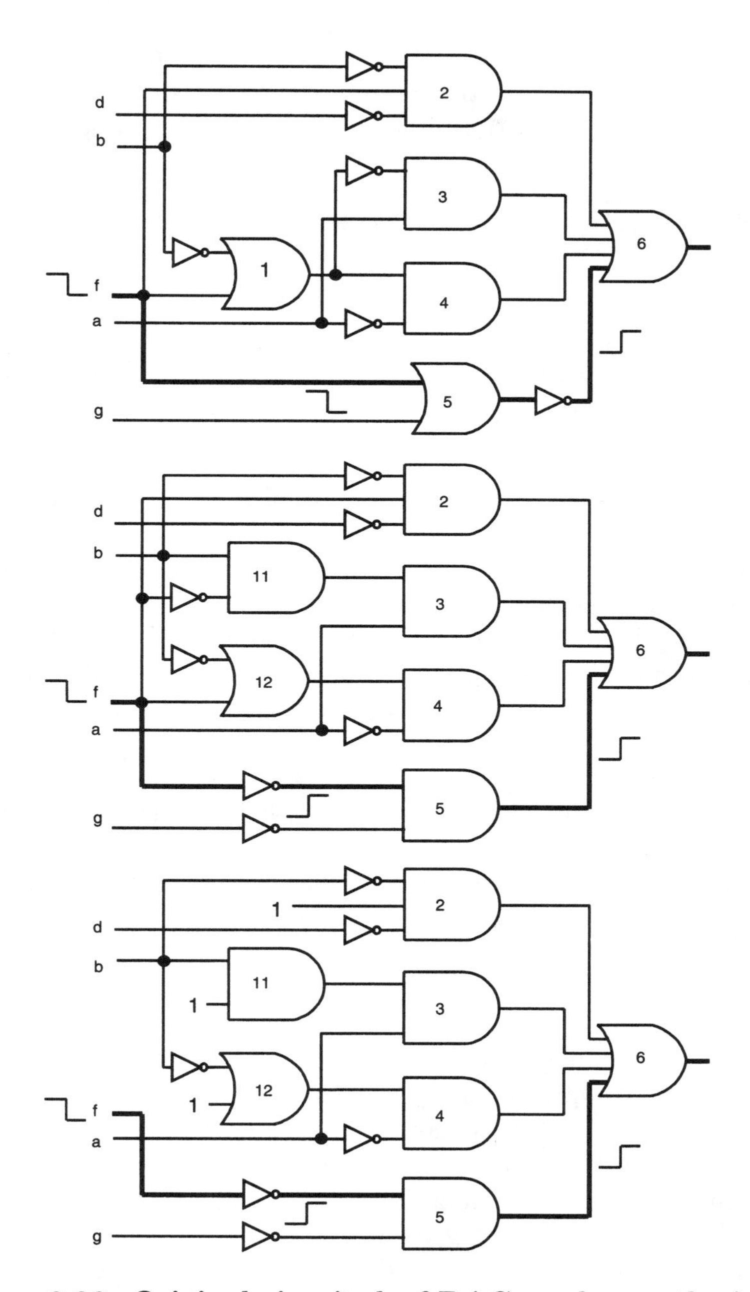

Figure 9.28: Original circuit, leaf-DAG, and smooth circuit

The falling-smooth circuit for the bold path in the leaf-DAG of Figure 9.28 is the leaf-DAG itself. We have $P_{AND} = \phi$ since other paths from input `f` all converge at OR gate 6.

Theorem 9.8.3 *Let π be a path in a leaf-DAG L. $\langle v_1,\ v_2\rangle$ is a GRPDF test for the rising transition on π in the rising-smooth circuit L_R if and only if $\langle v_1,\ v_2\rangle$ is a GRPDF test for the rising transition on π in L.*

Proof. Similar to the proof of Theorem 9.7.2. □

Theorem 9.8.4 *Let π be a path with input l in leaf-DAG L. v_2 is a test for the stuck-at-0 fault on the I-edge of π in the rising-smooth network L_R for π if and only if $\langle v_1,\ v_2\rangle$ where $v_1 = v_2 - \{l\} \cup \{\bar{l}\}$, is a GRPDF test for the rising transition on π in L.*

Proof. **Necessity**: Similar to the proof of the necessity case in Theorem 9.7.3.
Sufficiency: Let v_2 be a test vector for the stuck-at-0 fault on the I-edge of π in the rising-smooth network L_R for π. l is the only input in L_R that changes under the vector pair $\langle v_1,\ v_2\rangle$ Let P_S denote the paths that pass through side-inputs to π whose I-edges are not associated with l. Each side-input to π that belongs to a path in P_S must be at a noncontrolling value under v_2; this is due to the test condition for the stuck-at fault. Of the remaining side-inputs let I_{AND} denote the side-inputs that meet π at an AND gate and let I_{OR} denote the side-inputs that meet π at an OR gate. By the construction of L_R, I_{OR} is empty. Since a stuck-at-0 fault is being tested, each side-input to π in I_{AND} is at value 1 in L_R under v_2 (in the absence of the fault).

Now apply the vector v_1 to the circuit. The side-inputs to π in I_{AND} may be at either 0 or 1. All other side-inputs to π are at noncontrolling values. At some time greater than the delay of the network apply vector v_2. The side-inputs to π in I_{AND} may change to 1 or they may remain at 1, possibly with hazards. All the other side-inputs to π remain at hazard-free noncontrolling values, since the transition on l under the vector pair cannot propagate to these side-inputs in the rising-smooth nework L_R. The transition that propagates along π is a $0 \rightarrow 1$ transition, which propagates a delay fault at each AND gate along π irrespective of the presence of hazards on the side-inputs, provided the side-inputs have a final value of 1. Since the condition corresponding to Definition 5.6.2 holds, $\langle v_1,\ v_2\rangle$ is a general robust path delay fault test for the rising transition along π in L_R and hence in L by Theorem 9.8.3. □

Given a multilevel circuit η and a path π for which we are to generate a GRPDF test, a procedure of moving all inverters in η to the primary inputs and making π fan-out-free will produce a circuit η'_π on which the transformation corresponding to Definition 9.8.1 can be carried out. The resulting rising-smooth circuit will be at most four times the size of η and will have the same number of inputs as η. Thus, the use of multiple-valued calculi is not necessary to generate GRPDF tests for multilevel circuits.

9.8.3 Synthesis for Full Testability

We prove a theorem that relates ENF reducibility to general robust path delay fault testability below.

Theorem 9.8.5 *Let C be a single-output circuit with ENF expression E_C. Let M be a single-output circuit with ENF expression E_M. If $E_C \geq_t E_M$ and if a vector set V is a general robust test set for all path delay faults in C, then V is a GRPDF test for all path delay faults in M.*

Proof. Let π be a path in M. There is a corresponding tagged literal l_π in E_M. Corresponding to l_π in E_M there is a nonempty set of corresponding tagged literals $\{l_{\pi_1}, \ldots, l_{\pi_m}\}$ in E_C.

Consider the case of the $0 \to 1$ transition. Let the complete path-cube-complex of l_{π_i} in E_C be L, and let D be those cubes in E_C not containing any l_{π_i}. By Theorem 9.8.1 there exists a general robust path delay fault test $\langle v_1,\ v_2 \rangle$ for each π_i on a $0 \to 1$ transition which has all the following properties:

1. Vertex v_2 is such that $L(v_2) = 1$ and v_2 is not covered by any cube in D.

2. Vertex $v_1 = v_2 - \{l\} \cup \{\bar{l}\}$ and v_1 is in the *OFF*-set of C.

3. For every cube d in D there exists some literal m in both v_1 and v_2 such that $d_m(v_1) = 0$ and $d_m(v_2) = 0$.

Let the complete path-cube-complex of l_π in E_M be L_M, and let D_M be those cubes in E_M not containing l_π. Because L_M contains all the cubes in E_M that correspond to those cubes containing l_{π_i} in E_C, $NOTAGS(L) \subseteq NOTAGS(L_M)$, so $L_M(v_2) = 1$ and v_2 is not covered by any cube in D_M. As the *OFF*-sets of E_C and E_M are exactly the same, v_1 is in the *OFF*-set of E_M. Finally, just as $NOTAGS(L) \subseteq NOTAGS(L_M)$, similarly $NOTAGS(D) \supseteq NOTAGS(D_M)$. Thus, for every cube d in D_M there exists some literal m in both v_1 and v_2 such that $d_m(v_1) = 0$ and $d_m(v_2) = 0$. Thus, by Theorem 9.8.1 the vector pair $\langle v_1,\ v_2 \rangle$ is a general robust path delay fault test for a $0 \rightarrow 1$ transition on π in M.

Next consider the case of the $1 \rightarrow 0$ transition. Let the path-cube-complex of l_{π_i} in E_C be L. By Theorem 9.8.2 there exists a general robust delay fault test for a $1 \rightarrow 0$ transition on each π_i in C if and only if the following two conditions are satisfied:

1. There exists a vertex v_1 such that $L(v_1) = 1$.

2. There exists a vertex $v_2 = v_1 - \{l\} \cup \{\bar{l}\}$ such that v_2 is in the *OFF*-set of C.

Let the path-cube-complex of l_π in E_M be L_M. There may be cubes containing $l_{\pi_i} \cdot l_{\pi_j}$ literals which are not in L that are transformed into cubes containing $l_\pi \cdot l_\pi$ literals in E_M, and these cubes can be included in L_M. Because L_M contains all the cubes in E_M that correspond to those cubes containing l_{π_i} in E_C, $NOTAGS(L) \subseteq NOTAGS(L_M)$, so $L_M(v_1) = 1$. As the *OFF*-sets of E_C and E_M are exactly the same, v_2 is in the *OFF*-set of E_M. Thus, by Theorem 9.8.2 the vector pair $\langle v_1,\ v_2 \rangle$ is a general robust path delay fault test for a $1 \rightarrow 0$ transition on π in M.

Thus, for every path π in M there exists a nonempty set of associated paths P in C such that any vector pair that tests a GRPDF on any $\pi_i \in P$ is a GRPDF test for π. □

9.9 Hazard-Free Robust Gate Delay Faults

We will consider *hazard-free gate delay fault* (HFRGDF) testing in this section.

9.9.1 Conditions for Testability

We wish to extend the results of Section 5.9.1 to multilevel implementations and derive necessary and sufficient conditions for a multilevel circuit to be HFRGDF testable. The conditions for hazard-free robustly detecting a gate delay fault in a multilevel circuit are simpler to achieve than for hazard-free robustly detecting a path delay fault. As was pointed out in Lemma 5.9.1, to detect a gate delay fault it is sufficient to detect a path delay fault for *any* path that goes through the gate. Thus, unlike HFRPDF testing, it is not necessary to test all the paths in the circuit. There is one more significant weakening of the conditions necessary for detection. For a path to be hazard-free robustly tested it must be singly event sensitized. However, to hazard-free robustly detect a gate delay fault it is not necessary that any single path through a gate be singly event sensitizable. This fault will be detected even if a number of events simultaneously propagate through the gate, because if a gate delay fault exists at the gate, each of these events will be delayed.

We will use the ENF representation in our analysis.

The *gate-cube-complex* of a gate g, call it L_g, in a circuit C with ENF E is defined as the set of all cubes containing tagged literals associated with a path through g. More formally, $L_g = \{q \in E \mid (l_\pi \in q) \wedge (g \in \pi)\}$. In the cube q there could be tagged literals l_π and l_ρ such that either $g \in \pi$ or $g \in \rho$, or g in both π and ρ.

In the definition of the path-cube-complex associated with a tagged literal, given in Section 9.7.1, cubes such as $a_{\{1,4,6,8,9\}} \cdot \overline{d}_{\{3,4,6,8,9\}} \cdot \overline{d}_{\{7,8,9\}}$ were disqualified from appearing in the complex. This was required by the fact that when $\langle v_1, v_2 \rangle$ is a hazard-free robust test for a path delay fault on a $0 \rightarrow 1$ transition, $\langle v_2,\ v_1 \rangle$ is also a hazard-free robust test for the path on a $1 \rightarrow 0$ transition. In the case of cubes of the form above if we are checking for a $0 \rightarrow 1$ transition along path $d_{\{3,4,6,8,9\}}$, then we have two $0 \rightarrow 1$ transitions at the inputs of the AND gate 8, and the slower of these is passed through. However, if we are dealing with the $1 \rightarrow 0$ transition, then the faster of these transitions is passed through, and the delay fault test is invalidated.

However, in the case of a gate delay fault, if we have converging transitions at the inputs to the faulty gate, a test $\langle v_1, v_2 \rangle$ and a fault corresponding to the $0 \rightarrow 1$ transition, then the transitions at the gate are all delayed. Similarly when we use a test $\langle w_2,\ w_1 \rangle$, we are dealing with the fault corresponding to the $1 \rightarrow 0$ primary output transition, and again we ensure that the transitions at the gate are all delayed. We require that the transitions *all* pass through the faulty gate.

The requirement that HFRGDF tests do not allow hazards, races, and glitches gives the following condition. The lemma below holds for both the $0 \rightarrow 1$ and $1 \rightarrow 0$ transitions.

Lemma 9.9.1 *Let C be a multilevel single-output circuit with ENF E. Let g be a gate in C, and let the gate-cube-complex of g be L_g. Let $D = E - L_g$. If $\langle v_1, v_2 \rangle$ is an HFRGDF test for g in C, then for every cube d in D there exists some literal m in both v_1 and v_2 such that $d_m(v_1) = 0$ and $d_m(v_2) = 0$.*

Proof. This can be proved by an argument similar to the one used in the proof of Lemma 9.7.1. □

The *single-literal gate-cube-complex* associated with a gate g and a *specific* literal l is defined to be the set of all cubes in L_g such that at least one occurrence of l exists in each cube and such that *all* occurrences of literal l contain gate g in their tag. Thus, $L_g(l) = \{q \in L_g \mid (l \in q) \wedge (g \in \pi \; \forall \; l_\pi \in q)\}$. The set of cubes defined by the single-literal gate-cube-complex is used in the following theorem to give the necessary and sufficient conditions for a multilevel single-output function to be HFRGDF testable on a $0 \rightarrow 1$ primary output transition. In the theorem we will consider a rising primary output transition — depending on the number of inversions on the path from the gate output to the primary output, the gate itself will make a rising or falling transition.

The introduction of a concept developed for the analysis of hazards in asynchronous circuits will aid in proving the following theorem. Given a vertex v_1 and a vertex v_2 we define the *transition cube* for this vector pair, call it x, as follows. Let S be the set of variables that change values from v_1 to v_2. Let $x = \{l \mid (l \in v_2) \wedge (l, \bar{l} \notin S)\}$, i.e., let x be the cube resulting from the elimination of all literals in S from v_2. The cube x covers all the vertices in the N-cube that may be reached under arbitrary delays as the inputs change from v_1 to v_2.

Theorem 9.9.1 *Let C be a multilevel single-output combinational circuit with ENF E, and let g be a gate in C. There exists an HFRGDF test for gate g in C on a $0 \rightarrow 1$ primary output transition if and only if there exists some literal l such that all the following conditions are satisfied:*

1. *There exists a cube q such that $q \in L_g(l)$, where $L_g(l)$ is the single-literal gate-cube-complex associated with g and l.*

2. *There exists a vertex v_2 such that $q(v_2) = 1$ and v_2 is not covered by any cube in $D = E - L_g(l)$.*

3. *There exists a literal $l_\pi \in q$ such that $g \in \pi$ and $v_1 = v_2 - \{l\} \cup \{\bar{l}\}$ is in the* OFF*-set of C.*

4. *For every cube d in D there exists some literal m in both v_1 and v_2 such that $d_m(v_1) = 0$ and $d_m(v_2) = 0$.*

Proof. Necessity: First note that the condition that v_2 must be a relatively essential vertex of the single-literal gate-cube-complex of g, i.e., Condition 2 above, can be proved as in previous theorems.

Suppose $\langle w_1, w_2 \rangle$ is an HFRGDF test for g on a $0 \to 1$ primary output transition in C. The distance of this vector pair, i.e., the number of inputs that are different across w_1 and w_2, may be large, but we will proceed to construct a distance-1 vector pair as in the theorem. Note that in order to avoid glitches at the output, for every possible path in the Boolean N-space from w_1 to w_2, we require that the vertices encountered in the path change from being *OFF*-set vertices to *ON*-set vertices exactly once. Let x be the transition cube for the vector pair $\langle w_1, w_2 \rangle$ As noted above, under the appropriate delays on inputs, any vertex covered by x could be reached by applying the pair $\langle w_1, w_2 \rangle$ This implies that the intersection of x with each cube d in $D = E - L_g(l)$ is empty. If not, then under appropriate delays some vertex covered by another cube in D could be reached and a glitch would occur that invalidates the test for g. Thus, the cube x covers exactly members of the *ON*-set covered only by cubes in $L_g(l)$ and members of the *OFF*-set.

We wish to find two adjacent vertices in the subspace determined by x, such that one is in the *ON*-set of C and the other is in the *OFF*-set. The circuit values appearing on the inputs of C as the vector pair $\langle w_1, w_2 \rangle$ is applied may be visualized as a sequence of vertices forming a path in the Boolean N-space from vertex w_1 and vertex w_2. Furthermore, this path is restricted to be in the subspace determined by the cube x. As vertex w_1 is in the *OFF*-set and vertex w_2 is in the *ON*-set, there exist adjacent vertices v_1 and v_2 at some point on the path such that v_1 is in the *OFF*-set and v_2 is in the *ON*-set. This may easily be proved by induction on the length of the path from w_1 to w_2.

The argument above shows that it is necessary that there exists a vector pair $\langle v_1, v_2 \rangle$ which is distance-1 in some literal l and that v_2 is a relatively essential vertex of $L_g(l)$. We now wish to show that for each cube q such that $q(v_2) = 1$, there must exist one literal l in q that is tagged with a path π that passes through g. In other words, for each cube $q \in L_g(l)$ such that $q(v_2) = 1$, there is a literal l_π in q such that l_π is tagged with a path through g, and it is the transition of l_π from 0 to 1 on $\langle v_1,\ v_2\rangle$ that forces the cube q to make the transition from 0 to 1.

Suppose that there exists a cube q with no such l_π. We know that the literal l must be in q, otherwise q would not make the transition from 0 to 1. By supposition the literal $l \in q$ is not tagged with π such that $g \in \pi$ but instead with some σ such that g is not in σ. In this case we could apply the vector pair $\langle w_1,\ w_2\rangle$, and under appropriate delays we could travel from w_1 to v_1 and from v_1 (through a primary input change in the variable associated with literal l) to v_2. Now, $q(v_1) = 0$, $q(v_2) = 1$, and literal l_σ causes the transition. But g is not in σ, so when the vector pair $\langle w_1,\ w_2\rangle$ is applied, under the appropriate input delays, the path that is sensitized to a 1 does not propagate an event through gate g. But this would contradict the assertion that $\langle w_1,\ w_2\rangle$ is an HFRGDF test for g. So there must exist such an $l_\pi \in q$ as in the theorem.

Condition 4 of the theorem, i.e., for every cube d in D there exists some literal m in both v_1 and v_2 such that $d_m(v_1) = 0$ and $d_m(v_2) = 0$ was shown in Lemma 9.9.1.

Sufficiency: After applying v_1 the output of C settles to a 0. When the vector v_2 is applied, the only paths that are sensitized to a 1 are those that pass through g. The restriction that for every cube d in D, there exists some literal m in both v_1 and v_2 such that $d_m(v_1) = 0$ and $d_m(v_2) = 0$ ensures that neither *ON*-set cubes in E nor cubes that do not cover *ON*-set members can cause a glitch at the output masking the event. Thus, if all transitions at g are delayed, then the transition to a 1 at the output is delayed, and the pair $\langle v_1,\ v_2\rangle$ is an HFRGDF test for g. □

For example, consider an HFRGDF test for gate 8 in Figure 9.29, consisting of $w_1 = \langle 1, 1, 0, 1, 0\rangle$ and $w_2 = \langle 1, 0, 0, 0, 0\rangle$ Then $S = \{b, d\}$ and $x = a \cdot \overline{c} \cdot \overline{e}$. Under the appropriate delays any vertex in the subspace covered by $a \cdot \overline{c} \cdot \overline{e}$ is reachable. This is a distance-2 test, but we wish to find a distance-1 test as in the theorem. We can choose any path from w_1 to w_2 that we like; one route is $\langle 1, 1, 0, 1, 0\rangle$,$\langle 1, 1, 0, 0, 0\rangle$, and $\langle 1, 0, 0, 0, 0\rangle$. The first vertex in this sequence $\langle 1, 1, 0, 1, 0\rangle$ is in the *OFF*-set. The next vertex $\langle 1, 1, 0, 0, 0\rangle$ is in the *ON*-set. Thus,

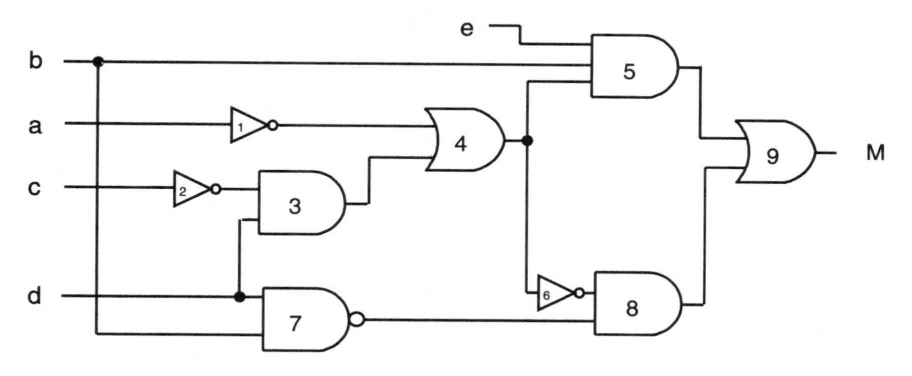

Figure 9.29: Multilevel circuit to illustrate HFRGDF testability requirement

$v_1 = \langle 1,1,0,1,0 \rangle$ and $v_2 = \langle 1,1,0,0,0 \rangle$ is a distance-1 test with the variable d changing from v_1 to v_2. The ENF of the circuit is $M = e_{\{5,9\}} \cdot b_{\{5,9\}} \cdot \overline{a}_{\{1,4,5,9\}} + b_{\{5,9\}} \cdot \overline{c}_{\{2,3,4,5,9\}} \cdot d_{\{3,4,5,9\}} \cdot e_{\{5,9\}} + a_{\{1,4,6,8,9\}} \cdot c_{\{2,3,4,6,8,9\}} \cdot \overline{b}_{\{7,8,9\}} + a_{\{1,4,6,8,9\}} \cdot c_{\{2,3,4,6,8,9\}} \cdot \overline{d}_{\{7,8,9\}} + a_{\{1,4,6,8,9\}} \cdot \overline{d}_{\{3,4,6,8,9\}} \cdot \overline{b}_{\{7,8,9\}} + a_{\{1,4,6,8,9\}} \cdot \overline{d}_{\{3,4,6,8,9\}} \cdot \overline{d}_{\{7,8,9\}}$. One cube q (in fact the only one) such that $q(v_2) = 1$ is $a_{\{1,4,6,8,9\}} \cdot \overline{d}_{\{3,4,6,8,9\}} \cdot \overline{d}_{\{7,8,9\}}$. Both tagged literals $\overline{d}_{\{3,4,6,8,9\}}$ and $\overline{d}_{\{7,8,9\}}$ have paths through gate 8, and both are responsible for cube $a_{\{1,4,6,8,9\}} \cdot \overline{d}_{\{3,4,6,8,9\}} \cdot \overline{d}_{\{7,8,9\}}$ making the transition from $0 \rightarrow 1$. It is interesting to note that $\langle \langle 1,1,0,1,0 \rangle, \langle 1,1,0,0,0 \rangle \rangle$ is not a hazard-free robust path delay fault test because paths $d_{\{3,4,6,8,9\}}$ and $d_{\{7,8,9\}}$ are both sensitized to a 1. But as they go through gate 8, they will both be delayed if gate 8 has a delay fault.

As mentioned previously, to avoid glitches at the output, we require that any HFRGDF test pair $\langle w_1, w_2 \rangle$ be such that for every possible path in the Boolean N-space from w_1 to w_2, the vertices encountered in the path change from being *OFF*-set vertices to *ON*-set vertices exactly once. It can be argued that glitches that propagate through the gate being tested do not invalidate a test even if they appear at the output. Relaxing the requirement above regarding the relationship between $\langle w_1, w_2 \rangle$ and analyzing the nature of the delay fault tests under this relaxed relationship bears further investigation. This corresponds to the general robust gate delay fault model. An example is the test pair $w_1 = \langle 1,1,0,1,0 \rangle$ and $w_2 = \langle 1,0,0,0,0 \rangle$ for the circuit of Figure 9.10.

The conditions for HFRGDF testability imply that the vector pair $\langle v_2, v_1 \rangle$ will be an HFRGDF test for g for the $1 \to 0$ transition.

9.9.2 Test Generation Methods

The HFRPDF test generation methods described in Sections 9.7.2 and 9.8.2 do not generalize in a straightforward way to HFRGDF test generation. This is because we allow transitions along multiple paths all of which pass through the gate g being tested in HFRGDF testing.

A test generation method for gate delay faults that is a simple modification of the path delay fault test generation method is to pick a particular path π that passes through the gate g beginning from input l and attempt to generate an HFRPDF test for π. While the test vector obtained in this fashion is guaranteed to be a robust test for g, such an approach is needlessly restrictive. An alternate, less restrictive approach is outlined below.

We choose a particular input l that is in the transitive fan-in of gate g. We will attempt to generate a $\langle v_1, v_2 \rangle$ test for gate g that is distance-1 in l. If this fails, then we will pick another input that is in the transitive fan-in of g, and so on.

Determine the set of paths that begin from input l in the leaf-DAG that all pass through any copy of g. Call this set Π. Pick a particular copy of g; this has a unique subpath σ to the circuit output. Call the set Π^σ the set of paths that pass through this copy of g. Now, the smooth circuit can be obtained by replacing the I-edge of any path that begins from l and converges with σ (each path in Π^σ) at an OR gate by 1 and replacing the I-edge of any path that begins from l and converges with σ (each path in Π^σ) at an AND gate by 0. (Note that if we had not restricted Π to Π^σ, then a path beginning from l may converge with a path in Π at an OR gate and converge with another path at an AND gate.) If we cannot generate a test for the chosen σ we can choose another copy of g with a different extension to the circuit output.

9.9.3 Algebraic Factorization for Full Testability

We now wish to construct a procedure which synthesizes an HFRGDF testable multilevel circuit from an HFRGDF testable two-level circuit. Because path delay fault testability was such a strong property, it was possible to retain it using unconstrained algebraic factorization.

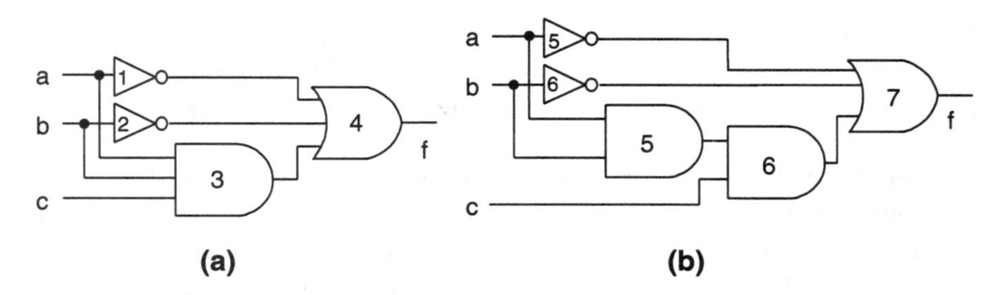

Figure 9.30: Algebraic factorization can lower HFRGDF testability

However, a constrained algebraic factorization procedure is required to retain HFRGDF testability in a multilevel circuit.

We will illustrate using a simple example how algebraic factorization decreases HFRGDF testability. Consider the two-level circuit shown in Figure 9.30(a). All four gates are hazard-free robust testable for gate delay faults. However, when the algebraic factor ab is extracted from the cube abc, the circuit is no longer HFRGDF testable. The algebraically factored circuit is shown in Figure 9.30(b). In this circuit, gate 5 is untestable. In the original circuit the only path through gate 3 which was testable was that originating at literal c. Thus, there is no way to test the new gate corresponding to the factor ab when it is extracted from the cube abc in the original cover. A constrained factorization would be required in order to preserve testability of the gate delay faults in the circuit.

It has been observed that more often than not algebraic factorization *increases* HFRGDF testability, as illustrated by the following example. Consider a function whose sum-of-products expression is $f = a\overline{b} + \overline{a}b + c\overline{d} + \overline{c}d + \overline{a}d$. The AND gate associated with the cube $\overline{a}d$ is not hazard-free robust testable. However, if the circuit is implemented as the algebraically factored expression $f = a\overline{b} + c\overline{d} + \overline{c}d + \overline{a}(b + d)$, it will be completely hazard-free robust testable for gate delay faults.

An algebraic factorization procedure that retains HFRGDF testability of the initial two-level circuit must essentially ensure that in each cube of each factor (cube or kernel) there is at least one literal with the properties of literal l in Theorem 9.7.1. Of course, if all literals in the initial two-level circuit have this property, as in the case of an HFRPDF testable two-level circuit, then no constraints are required. The theorem below gives a succinct sufficiency condition for the retainment of HFRGDF testability after algebraic factorization, beginning from a two-level circuit. This condition ensures that at least one path through each gate is hazard-free robust testable.

Theorem 9.9.2 *Given a two-level network C, assume that at least one path through each* AND *gate is hazard-free robust testable. Let E_C be the ENF of circuit C. Let A be a multilevel circuit resulting from an algebraic factorization of C, having an ENF E_A. If the tagged literals in E_C which satisfy the conditions of Theorem 9.7.1 correspond to a set of tagged literals in E_A (paths in A) that cover all the gates in A, then A is completely HFRGDF testable.*

Proof. By Theorems 9.7.4 and 9.5.4, there is a one-to-one correspondence between the literals of E_A and E_C. Thus, if a tagged literal in E_C satisfies the conditions of Theorem 9.7.1, then the corresponding tagged literal in E_A does also. Since the set of paths in A correspond to tagged literals in E_A that satisfy the conditions of Theorem 9.7.1 and these paths cover all the gates in A, hazard-free robust testing of these paths will hazard-free robustly test all the gate delay faults in A. □

It is possible, as illustrated in Figure 9.30, that the hazard-free robust testable paths in the two-level circuit do not produce corresponding paths which cover all the gates after algebraic factorization. Constraints have to be imposed on cube and kernel extraction in order to ensure that this condition is satisfied. Making this check after each cube or kernel extraction step during multilevel logic optimization will ensure that HFRGDF testability is maintained.

If we choose to test all link delay faults in a circuit, rather than gate delay faults, then as we will show in Section 9.10, unconstrained algebraic factorization will retain testability.

Lemma 9.5.2 may be used to show the delay fault testability of technology mapped networks:

Theorem 9.9.3 *Let C be a* NAND *gate and* INVERTER *implementation of a circuit, and let M be a tree covering technology mapped version of C. Then, if C was completely hazard-free robust or general robust gate or path delay fault testable, then so is M.*

Proof. Given a one-to-one correspondence between the paths of C and M, hazard-free robust or general robust path delay fault testability is maintained. In the case of C being HFRGDF (but not HFRPDF testable), in order to test a gate in C, multiple paths may have to be sensitized. The very same paths can be simultaneously sensitized in M to test the corresponding mapped gate in M. □

Tree covering based technology mapping, as described in Chapter 7, maps a NAND and INVERTER representation of a circuit into a library of gates (such as a standard-cell library) by decomposing a circuit into trees (fan-out-free circuits) and then mapping each tree into a logically equivalent tree built from library elements. Thus, by the theorem the tree covering based approach to technology mapping as used in [5, 11, 21] retains delay fault testability.

9.10 Hazard-Free Robust Stuck-Open Faults

Recall that a transistor stuck-open fault in a primitive gate within a combinational circuit can be robustly detected by a robust test for the link delay fault corresponding to the transistor in the gate (see Section 5.6).

Test generation and synthesis for testability procedures for gate delay faults can be adapted to hazard-free robust transistor stuck-open faults by using a finer grain *link delay fault* model, where a link is synonymous with an edge (see Section 5.5). Alternatively, one can introduce a buffer at the inputs to every gate in a combinational circuit and test the circuit under the gate delay fault model.

9.10.1 Conditions for Testability

The link-cube-complex L_λ associated with a link λ was defined in Section 9.6.1. We will define the *single-literal link-cube-complex* associated with a link λ and a *specific* literal l is to be the set of all cubes in L_λ such that at least one occurrence of l exists in each cube and such that *all* occurrences of literal l contain link λ in their tag. Thus, $L_\lambda(l) = \{q \in L_\lambda \mid (l \in q) \wedge (\lambda \in \pi \; \forall \; l_\pi \in q)\}$. The single-literal link-cube-complex with respect to literal b or literal c for the link 1,2 in Figure 9.17 is $b_{\{1,2\}} \cdot c_{\{1,2\}}$.

The collection of cubes defined by the single-literal link-cube-complex is used in the following theorem to give the necessary and sufficient conditions for a multilevel single-output function to be hazard-free robustly stuck-open fault testable.

Theorem 9.10.1 *Let C be a multilevel single-output circuit with ENF E. Let λ be a link in C. There exists a hazard-free robust stuck-open fault test for link λ in C if and only if there exists some literal l such that all the following conditions are satisfied:*

1. *There exists a cube q such that $q \in L_\lambda(l)$. where $L_\lambda(l)$ is the single-literal link-cube-complex associated with λ and l.*
2. *There exists a vertex v_2 such that $q(v_2) = 1$ and v_2 is not covered by any cube in $D = E - L_\lambda(l)$.*
3. *There exists a vertex v_1 such that $v_1 = v_2 - \{l\} \cup \{\bar{l}\}$ is in the OFF-set of C.*
4. *For every cube d in D there exists some literal m in both v_1 and v_2 such that $d_m(v_1) = 0$ and $d_m(v_2) = 0$.*

Proof. Similar to the proof of Theorem 9.9.1. □

9.10.2 Synthesis for Full Testability

If a two-level circuit is completely robustly testable for all link delay faults (or transistor stuck-open faults), then unconstrained algebraic factorization will retain complete testability. This is in contrast to the gate delay fault case (see Figure 9.30).

Theorem 9.10.2 *Given ENF reducibility across a multiple-output two-level and a multilevel network, if the two-level network is fully hazard-free robust transistor stuck-open fault testable, then so is the multilevel network. Furthermore, the test vectors that hazard-free robustly detect all stuck-open faults in the two-level network, hazard-free robustly detect all faults in the multilevel network.*

Proof. The conditions of Theorem 9.10.1 have to be satisfied for every link in the two-level network. This means that for every cube in the ENF of the two-level network and for each literal in the cube, Conditions 2, 3, and 4 of Theorem 9.10.1 have to be satisfied for some output. Given a many-to-one correspondence between the tags of the ENFs of the two-level and multilevel networks, the conditions for testing links are more relaxed, or at worst, the same as for the links in the two-level network. That is, any single-literal link-cube-complex for any link λ_m in the multilevel circuit will contain the same number or more cubes than the single-literal link-cube-complex of some link λ_t in the two-level network. This implies that any test for λ_t will test λ_m. □

The theorem above could have been proven in another way. By using the equivalence of hazard-free robust path delay fault testability and stuck-open fault testability in two-level circuits, Theorem 9.7.4 can be used to show that the resulting multilevel circuit is completely path delay fault testable. Finally, using the obvious result that complete path delay fault testability implies complete transistor stuck-open fault testability, the conclusion of the theorem can be derived.

Theorem 9.10.2 leads naturally to an efficient synthesis-for-testability procedure for multilevel stuck-open fault testable circuits, namely algebraic factorization with a constrained use of the inverse. The above theorem requires a two-level circuit that is completely testable as a starting point. A conjecture that has not been proved or disproved yet is that ENF reducibility across multilevel circuits maintains link delay fault testability.

A comprehensive treatment of testing for transistor stuck-open faults with a test methodology similar to stuck-at faults can be found in [19]. An interesting point worthy of note is that transistor stuck-open fault testability as defined in [19] is *not* retained by algebraic factorization.

9.11 The Viterbi Processor

9.11.1 Introduction

It is widely accepted that hidden Markov models are currently the most efficient technique to model speech for use in automatic speech recognition [20]. The states of a hidden Markov model speech process correspond to generic speech sounds (e.g., a phoneme or part of a phoneme). The task that has to be performed during speech recognition is to find, for a given HMM, the state sequence that most likely could have produced the speech that has to be recognized. A very efficient search algorithm for the most likely state sequence is the *Viterbi* algorithm.

A hardware design capable of performing the Viterbi equation for up to 50,000 states in real time has been presented [34]. In

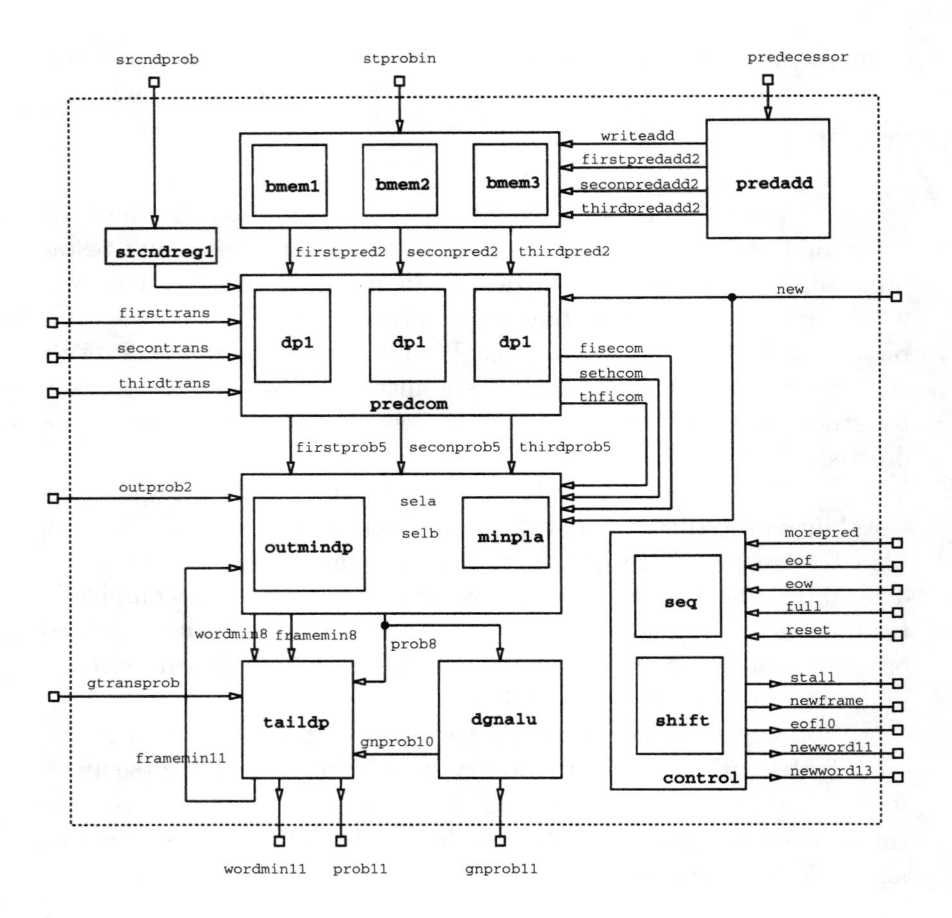

Figure 9.31: The Viterbi processor

Figure 9.31, a block diagram of the Viterbi processor is shown. The hierarchical composition and interconnection of the various blocks are of primary interest. The processor has up to 11 levels of pipelining. The basic blocks are eight datapaths, three bidirectional memories, and a control unit with data stationary control architecture. The chip was fabricated using a MOSIS 2μm *complementary metal oxide semiconductor* (CMOS) technology. Including 204 pads, it has a die size of 11.6×9.8 square mm with 25,000 transistors.

Here, we are concerned primarily with the synthesis of the combinational logic in the processor (including the arithmetic units in the datapaths). Logic between the registers is synthesized to be fully HFRPDF testable.

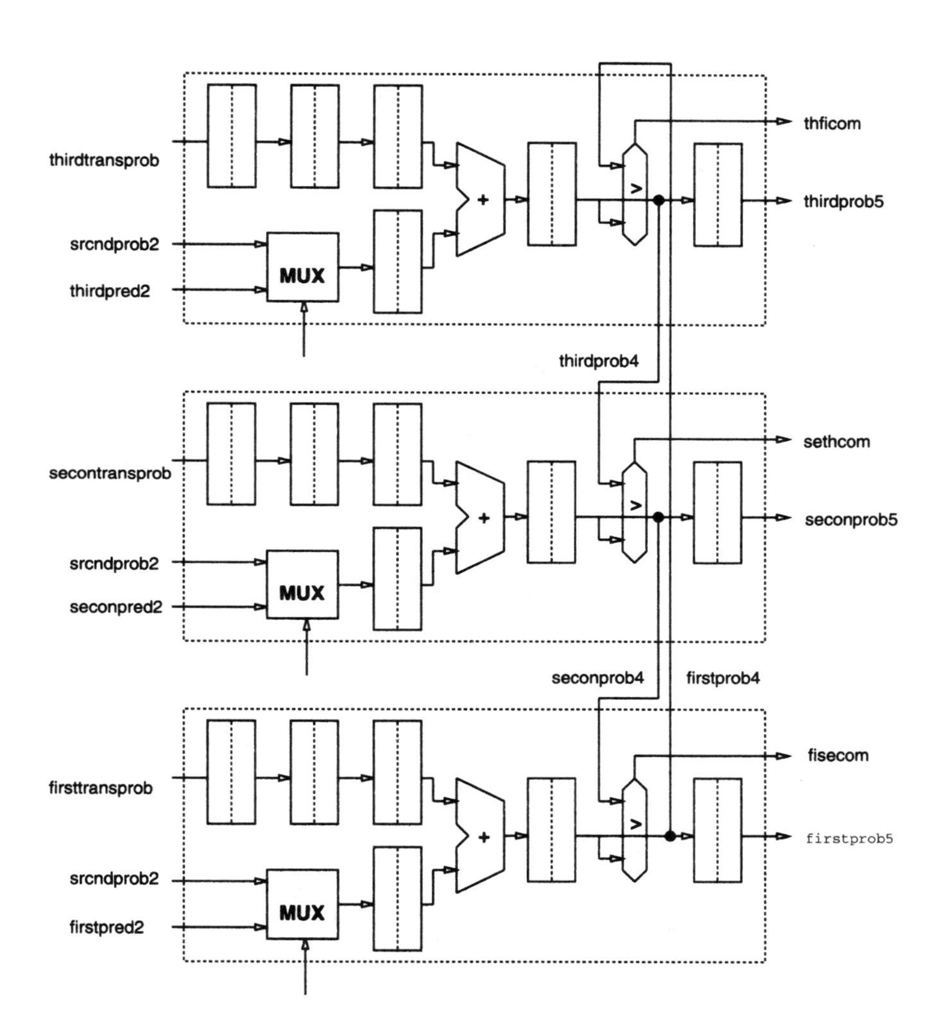

Figure 9.32: The predcom datapath

9.11.2 The Datapaths

The eight datapaths in the Viterbi processor have varying complexities. In Figure 9.32, the topology of the datapath **predcom** is shown, which in itself is composed of three smaller datapaths. The > blocks correspond to comparators, and the + blocks to arithmetic logic units. As can be seen, the interconnection topology conforms to the composition rule, and hence synthesizing a fully testable comparator suffices to ensure full HFRPDF testability for **predcom**.

In Figure 9.33 the topology of the datapath **outmindp** is shown. The datapath has a comparator that feeds a multiplexor.

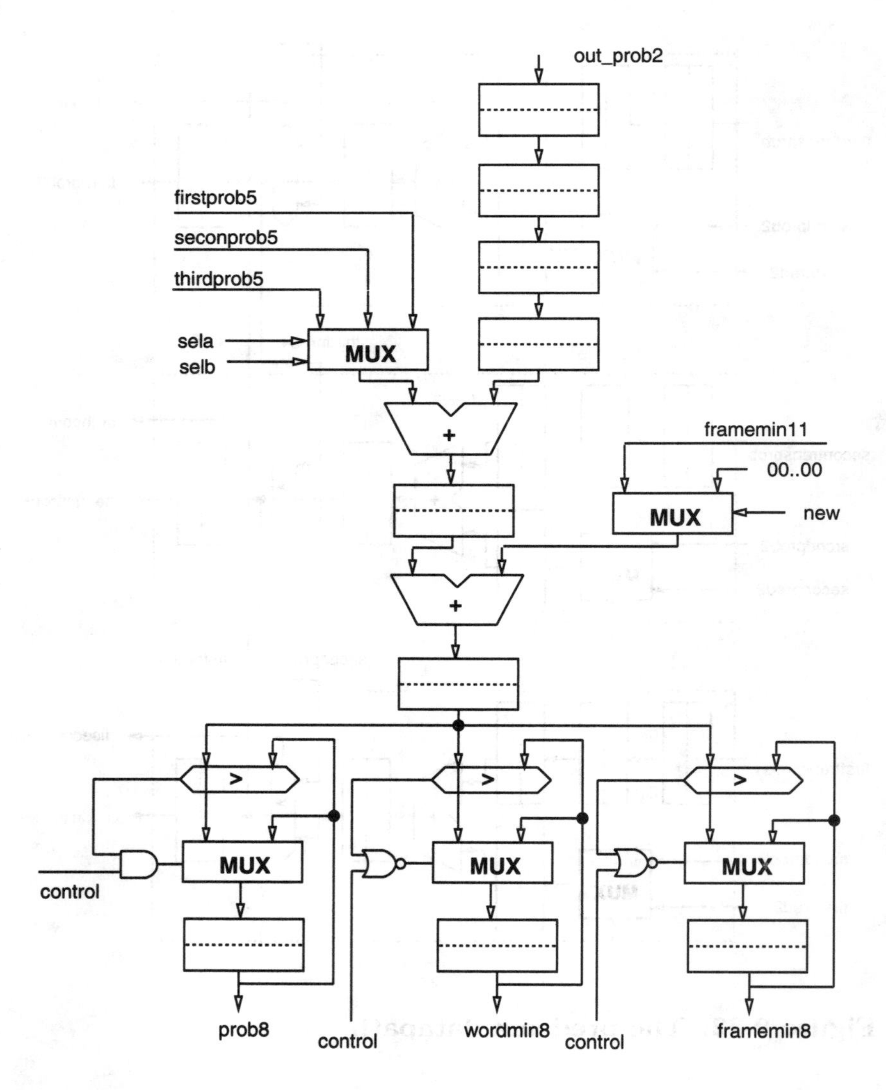

Figure 9.33: The outmindp datapath

For example, a comparator receives as inputs the output of a + block and **prob8**, and its output controls a multiplexor whose data inputs are the same as those of the comparator. This topology implements the maximum operation and is not hazard-free robust path delay fault testable. We can, however, synthesize a fully testable maximum function by composing modified subtractor cells (see Section 9.11.5).

The other datapaths are very similar and can also be synthesized to be fully testable.

9.11.3 The Controller

The controller in the Viterbi processor is a 16-state finite state machine with pipeline registers that produce delayed control outputs. In fact, all the control output signals but one, namely, `stall`, are outputs of registers. The signal `stall` shown in Figure 9.31 is both a primary output of the chip and feeds the `minpla` block which is a small piece of combinational logic. The outputs of the `minpla` block are latched with registers.

9.11.4 Putting Them Together

As mentioned previously, other than the signal `stall`, the control outputs are all latched. This implies an effective decoupling between the datapaths and control for testability purposes. There is, however, quite a lot of interaction between the datapaths. For example, datapath **predcom** drives datapath **outmindp** in Figure 9.31. Looking at Figures 9.32 and 9.33 we note that the driving signals from datapath **predcom** are all outputs of registers. This implies that we can compose the datapaths disregarding the other datapaths. The other interactions between the remaining datapaths are also handled quite easily.

9.11.5 Synthesis Results

There are mild testability constraints on the synthesis of the different datapaths and controllers which to a large extent were followed in the manual design [34] as well.

All the datapaths have to be synthesized separately, as in the manual design. The adders are carry look-ahead adders with 4 bits of look-ahead. Since each adder is 12 bits wide, three such modules are required. A 4-bit carry look-ahead adder can be synthesized to be fully HFRPDF testable with minimal area overhead in the combinational portion. The finite state machine controller was specified as a state transition graph (with 16 states). The control logic was flattened and algebraically factored to produce a fully HFRPDF testable circuit. Algebraic factorization in this case involved no area penalty. Some comparator cells (e.g., [29]) are not fully testable when hierarchically composed. Alternate designs that are fully HFRPDF testable are presented in [9].

The comparison $A > B$ can be performed simply by checking if $A - B > 0$. (Similarly for $A < B$.) This is implemented using a

UNCONSTRAINED			TESTABLE			
No. lits.	No. paths	Fault coverage	No. lits.	No. paths	Fault coverage	CPU time
1941	45×10^6	$< 25\%$	2215	46×10^6	100%	2.2h

Table 9.2: Testability of the Viterbi processor

subtractor and inspecting the final borrow bit. A subtractor, like an adder (see Section 9.7.4), can be composed in a bit-sliced design for full HFRPDF testability. A maximum function can be based on this design as well. The subtractor cell in the comparator does not have to implement the sum output – we are only interested in the final borrow. However, for the maximum function, we have an output for each cell i representing the ith bit of $MAX(A,\ B)$.

Thus the entire chip can be synthesized to be fully path delay fault testable.

The adder and comparator circuits used in the manual design [34] are not fully path delay fault testable. This implies that the manual design has very low hazard-free robust path delay fault testability ($<$ 20 percent). Testability can be increased to 100 percent with approximately 10 percent area overhead in the combinational logic. It is interesting to note that over 46 million paths exist in the synthesized processor (Table 9.2).

Problems

1. Prove Theorem 9.2.1.

2. Consider the topology of Figure 9.21. If $x_1,\ x_2,\ \cdots,\ x_N$ are the inputs to C_1, $f(x)$ is the output of C_1, and the other inputs to C_2 are $y_1,\ y_2,\ \cdots,\ y_K$, then the function computed at the output of C_2 can be given by $g(f(x),\ y)$, where x and y have no variables in common. Prove that

$$\frac{\partial g}{\partial x_i} = \frac{\partial g}{\partial f} \cdot \frac{\partial f}{\partial x_i}$$

3. Prove that in an irredundant realization of a combinational function $f(x_1, x_2, \cdots, x_N)$ no single stuck-at fault can change the output to correspond to the function $\overline{f(x_1, x_2, \cdots, x_N)}$.

4. Prove, using ENF analysis, that if a fan-out-free circuit is fully single stuck-at fault testable, it is fully multifault testable.

5. Prove, using ENF analysis, that if a fan-out-free circuit is fully single stuck-at fault testable, it is fully hazard-free robust path delay fault testable.

6. Prove that a fan-out-free circuit can be completely tested for multifaults by a complete hazard-free robust gate delay fault test set.

7. Prove that a primality test for a tagged literal l_π in any cube q in the ENF E of a multilevel circuit C statically sensitizes the corresponding path $\pi \in C$ to a 0, and vice versa.

8. Prove that an irredundancy test for a cube q in the ENF E of a multilevel circuit C statically sensitizes all paths in C corresponding to tagged literals $l_\pi \in q$.

9. Prove that a relatively essential vertex of the complete path-cube-complex of a tagged literal l_π in the ENF E of a multilevel circuit C statically sensitizes π in C to a 1.

10. Prove that if a fan-out-free path in a circuit C is not statically sensitizable, then the path has a stuck-at fault redundancy on it.

11. Give an example of a fully single stuck-at fault testable Boolean network where merging two nodes results in a network with a single stuck-at fault redundancy.

Hint: The merging should turn a redundant double fault in the circuit into a redundant single fault.

12. Prove that complete general robust path delay fault testability of a multilevel circuit implies complete single stuck-at fault testability.

13. Give an example of the composition topology of Figure 9.21 where if C_1 and C_2 are completely hazard-free robust gate delay fault testable, then the composition is not fully hazard-free robust gate delay fault testable.

14. Prove for the composition topology of Figure 9.21 that if C_1 and C_2 are completely single stuck-at fault testable, the composition is fully single stuck-at fault testable.

15. Prove or disprove for the composition topology of Figure 9.21 that if C_1 and C_2 are completely hazard-free robust transistor stuck-open fault testable, the composition is fully robust stuck-open fault testable.

16. Give necessary and sufficient conditions for a gate in a multilevel circuit to be general robust testable for $0 \rightarrow 1$ primary output transition, based on the ENF of the circuit. Do the same for the $1 \rightarrow 0$ transition.

17. * Prove or disprove the following statement. Given two multilevel circuits C_1 and C_2, with ENFs E_1 and E_2, respectively, if $E_1 \geq_t E_2$, then if C_1 is completely hazard-free robust link delay fault testable, then so is C_2.

18. * Prove or disprove for the composition topology of Figure 9.21 that if C_1 and C_2 are completely general robust path delay fault testable, the composition is fully general robust path delay fault testable.

19. * Prove or disprove that general robust path delay fault testability of a multilevel circuit implies complete multifault testability.

20. Prove, using induction, that the set D of Lemma 9.7.3 is unique.

21. Prove Lemma 9.7.4.

22. Prove Lemma 9.9.1.

REFERENCES

[1] D. B. Armstrong. On Finding a Nearly Minimal Set of Fault Detection Tests for Combinational Logic Nets. *IEEE Transactions on Computers*, EC-15(2):66–73, February 1966.

[2] P. Ashar, S. Devadas, and K. Keutzer. Testability Properties of Multilevel Logic Networks Derived from Binary Decision Diagrams. In *Proceedings of the Santa Cruz Conference on Advanced Research in VLSI*, pages 35–54, March 1991.

[3] K. Bartlett, R. K. Brayton, G. D. Hachtel, R. M. Jacoby, C. R. Morrison, R. L. Rudell, A. Sangiovanni-Vincentelli, and A. R. Wang. Multilevel Logic Minimization Using Implicit Don't-Cares. *IEEE Transactions on Computer-Aided Design of Integrated Circuits*, 7(6):723–740, June 1988.

[4] D. Bostick, G. D. Hachtel, R. Jacoby, M. R. Lightner, P. Moceyunas, C. R. Morrison, and D. Ravenscroft. The Boulder Optimal Logic Design System. In *Proceedings of the International Conference on Computer-Aided Design*, pages 62–65, November 1987.

[5] R. Brayton, R. Rudell, A. Sangiovanni-Vincentelli, and A. Wang. MIS: A Multiple-Level Logic Optimization System. *IEEE Transactions on Computer-Aided Design of Integrated Circuits*, CAD-6(6):1062–1081, November 1987.

[6] M. A. Breuer and A. D. Friedman. *Diagnosis and Reliable Design of Digital Systems.* Computer Science Press, Woodland Hills, CA, 1976.

[7] F. Brglez and H. Fujiwara. A Neutral Netlist of 10 Combinational Benchmark Circuits and a Target Translator in FORTRAN. In *International Symposium on Circuits and Systems (Special Session on ATPG and Fault Simulation)*, Kyoto, Japan, June 1985.

[8] M. J. Bryan. Synthesis Procedures to Preserve Testability of Multilevel Combinational Logic Circuits. Master's thesis, Massachusetts Institute of Technology, June 1990.

[9] M. J. Bryan, S. Devadas, and K. Keutzer. Analysis and Design of Regular Structures for Robust Dynamic Fault Testability. *VLSI Design: An International Journal of Custom-Chip Design, Simulation and Testing*, 1(1):45–60, August 1993.

[10] H. Cox and J. Rajski. A Method of Fault Analysis for Test Generation on Fault Diagnosis. *IEEE Transactions on Computer-Aided Design of Integrated Circuits*, 7(7):813–833, July 1988.

[11] E. Detjens, G. Gannot, R. Rudell, A. Sangiovanni-Vincentelli, and A. Wang. Technology Mapping in MIS. In *Proceedings of the International Conference on Computer-Aided Design*, pages 116–119, November 1987.

[12] S. Devadas and K. Keutzer. Necessary and Sufficient Conditions for Robust Delay-Fault Testability of Logic Circuits. In *Proceedings of the Sixth MIT Conference on Advanced Research on VLSI*, pages 221–238, April 1990.

[13] S. Devadas and K. Keutzer. A Unified Approach to the Synthesis of Fully Testable Sequential Machines. *IEEE Transactions on Computer-Aided Design of Integrated Circuits*, 10(1):39–51, January 1991.

[14] S. Devadas and K. Keutzer. Synthesis of Robust Delay-Fault Testable Circuits: Practice. *IEEE Transactions on Computer-Aided Design of Integrated Circuits*, 11(3):277–300, March 1992.

[15] S. Devadas and K. Keutzer. Synthesis of Robust Delay-Fault Testable Circuits: Theory. *IEEE Transactions on Computer-Aided Design of Integrated Circuits*, 11(1):87–101, January 1992.

[16] S. Devadas, K. Keutzer, and S. Malik. A Synthesis-Based Approach to Test Generation and Compaction for Multifaults.

Journal of Electronic Testing: Theory and Applications, 4(1):91–104, February 1993.

[17] S. Devadas, H-K. T. Ma, A. R. Newton, and A. Sangiovanni-Vincentelli. Irredundant Sequential Machines via Optimal Logic Synthesis. *IEEE Transactions on Computer-Aided Design of Integrated Circuits*, 9(1):8–18, January 1990.

[18] H. Fujiwara. *Logic Testing and Design for Testability*. MIT Press, Cambridge, MA, 1985.

[19] N. K. Jha and S. Kundu. *Testing and Reliable Design of CMOS Circuits*. Kluwer Academic Publishers, Norwell, MA, 1990.

[20] B. H. Juang and L. R. Rabiner. An Introduction to Hidden Markov Models. *IEEE ASSP Magazine*, pages 14–16, January 1986.

[21] K. Keutzer. DAGON: Technology Mapping and Local Optimization. In *Proceedings of the* 24th *Design Automation Conference*, pages 341–347, June 1987.

[22] K. Keutzer, S. Malik, and A. Saldanha. Is Redundancy Necessary to Reduce Delay? *IEEE Transactions on Computer-Aided Design of Integrated Circuits*, 10(4):427–435, April 1991.

[23] S. Kundu, S. M. Reddy, and N. K. Jha. On the Design of Robust Multiple Fault Testable CMOS Combinational Logic Circuits. In *Proceedings of the International Conference on Computer-Aided Design*, pages 240–243, November 1988.

[24] S. Kundu, S. M. Reddy, and N. K. Jha. Design of Robustly Testable Combinational Logic Circuits. *IEEE Transactions on Computer-Aided Design of Integrated Circuits*, 10(8):1036–1048, August 1991.

[25] T. Larrabee. Efficient Generation of Test Patterns Using Boolean Difference. In *Proceedings of the International Test Conference*, pages 795–801, August 1989.

[26] C. J. Lin and S. M. Reddy. On Delay Fault Testing in Logic Circuits. *IEEE Transactions on Computer-Aided Design of Integrated Circuits*, CAD-6(5):694–703, September 1987.

[27] E. J. McCluskey. Transients in Combinational Logic Circuits. In R. H. Willson and W. C. Mann, editors, *Redundancy Techniques for Computing Systems*, pages 9-46, Spartan Books, New York, NY, 1962.

[28] A. Pramanick and S. Reddy. On the Design of Path Delay Fault Testable Combinational Circuits. In *Proceedings of the* 20th *Fault Tolerant Computing Symposium*, pages 374–381, June 1990.

[29] C. H. Roth. *Fundamentals of Logic Design.* West Publishing Company, New York, NY, 1979.

[30] J. P. Roth. Diagnosis of Automata Failures: A Calculus and a Method. IBM *Journal of Research and Development*, 10(4):278–291, July 1966.

[31] A. Saldanha, R. Brayton, and A. Sangiovanni-Vincentelli. Equivalence of Robust Delay-Fault and Single Stuck-Fault Test Generation. In *Proceedings of the* 29th *Design Automation Conference*, pages 173–176, June 1992.

[32] M. Schulz, E. Trischler, and T. Sarfert. SOCRATES : A Highly Efficient Automatic Test Pattern Generation System. *IEEE Transactions on Computer-Aided Design of Integrated Circuits*, 7(1):126–137, January 1988.

[33] G. L. Smith. A Model for Delay Faults Based on Paths. In *Proceedings of the International Test Conference*, pages 342–349, September 1985.

[34] A. Stoelzle. A VLSI Wordprocessing Subsystem for a Real Time Large Vocabulary Continuous Speech Recognition System. Master's thesis, University of California at Berkeley, September 1989.

[35] C. E. Stroud, R. R. Munoz, and D. A. Pierce. CONES: A System for Automated Synthesis of VLSI and Programmable Logic from Behavioral Models. In *Proceedings of the International Conference on Computer-Aided Design*, pages 428–431, November 1986.

[36] S. H. Unger. *Asynchronous Sequential Switching Circuits.* John Wiley and Sons, New York, NY, 1969.

Chapter 10

Ongoing Work and Future Directions

The focus of this book has been to describe logic synthesis and optimization algorithms that target area, speed, and testability. In doing so we have given detailed descriptions of general optimization algorithms that target area optimality and described modifications to the basic methods to target circuit speed. We have given conditions for testability under various fault models and indicated how popular synthesis algorithms that target area may have to be modified in order to target testability under a chosen fault model.

Circuit representations and data structures cut across all facets of computer-aided design, and representations of combinational and sequential circuits affect virtually all problems in logic synthesis, testing and verification. We summarize the combinational logic representations presented in this book in Section 10.1 and indicate the directions of current research in this area.

We summarize current research in the development of combinational logic optimization methods that target area, speed, testability, and power dissipation in Section 10.2.

While we have not described methods for sequential logic synthesis in the book, we briefly summarize relevant work in this area in Section 10.3. A good deal of ongoing research in the logic synthesis area is focused on sequential circuit optimization for area, speed, and testability.

As evinced by the material in previous chapters, a large body of theory has evolved in the area of logic synthesis and testing in recent years, and the developed theoretical frameworks have resulted in efficient algorithms that solve synthesis for testability problems under various fault models. It is relevant to examine what use the developed theory and algorithms have been put to in the design of real-life very-large scale integrated systems. Algorithms have to be continually improved in order to keep pace with the increasing scale of logic designs that are carried out in industry. Therefore, in Section 10.4 we summarize the current status of the use of logic synthesis in academic, industrial, and commercial tools and indicate trends that are likely to occur in the future.

10.1 Combinational Circuit Representations

Five different representations of logic circuits have been used in this book. The sum-of-products representation is useful both for circuit implementation as a programmable logic array (see Section 3.3) and as an underlying representation for Boolean manipulation. While Boolean operations can be carried out on cover representations quite efficiently if the covers are of reasonable size, many functions have sum-of-products representations that are unmanageably large (see Section 3.7). However, the sum-of-products representation is very easy to analyze, as evinced by the simplicity of the testability theorems of Chapter 5.

In order to analyze multilevel circuits easily, we defined the leaf-DAG, and the *equivalent normal form* (ENF) of a multilevel circuit (see Section 9.3) which is an annotated two-level representation. The ENF is only used for analysis purposes in this book. The theorems for two-level testability conditions can be generalized to the multilevel case if the ENF representation is used. Furthermore, the logic transformations used in multilevel logic optimizers can be analyzed using the ENF representation.

Binary decision diagrams (BDDs), and in particular, reduced, ordered BDDs (see Section 6.3) are a more efficient representation for a large class of combinational functions as compared to two-level representations. They are also quite easy to manipulate and can be used as the underlying representation for test generation methods. However, current two-level logic minimization algorithms still use two-level representations, and multilevel logic optimization methods primarily use the algebraically factored form. BDDs correspond to a circuit implementation that is a multiplexor-based network, and several interesting testability properties can be proved for such networks.

Lastly, the general multilevel circuit is indispensable as an area-efficient and speed-efficient realization of a combinational function. Unfortunately, it is difficult to analyze and manipulate. A special case of a multilevel circuit is an algebraically factored circuit, and algebraically factored forms such as those described in Chapter 7 are more amenable to manipulation and optimization. A decade of effort has resulted in average-case efficient methods for single stuck-at fault test generators that operate directly on multilevel circuits. We gave an overview of the PODEM algorithm for test generation (more details can be found in [4, 16]) and showed how test generation under other fault models can be mapped to this well-researched problem. The leaf-DAG and ENF representations were used to perform the mapping (see Section 9.7.2).

The search for more efficient Boolean representations is ongoing. Generalizations of BDDs that allow variables to appear multiple times on a path from a terminal vertex to the root vertex have appeared in the recent literature (e.g., [1, 24]). Removing the ordering restriction on reduced BDDs is another promising direction of research [23].

10.2 Combinational Logic Optimization

10.2.1 Area and Performance

There are several ongoing research efforts in using alternative Boolean representations for logic optimization problems. For example, recently it has been shown that BDDs can be used to efficiently represent any collection of implicants of a two-level logic function. A method to generate all the prime implicants and the essential prime implicants of a logic function that does not require the explicit enumeration of each implicant has been given in [10]. Prime generation is thus possible even for functions with millions of primes. Essential prime implicants can be implicitly detected, and only the covering problem corresponding to the relatively essential primes (see Chapter 4) has to be explicitly solved.

The observability and satisfiability "don't-cares" of Section 7.6 can be very large in a two-level representation for even moderately sized multilevel circuits. An algorithm has been developed to compute local don't-cares for each node in the network (in terms of the immediate input variables of each node) [33]. These local don't-cares are derived from the total don't-care set using BDD-based image computation techniques [9]. The local don't-cares can then be used to minimize each node. Procedures for calculating the full observability don't-care set for each node have been developed that can be used for large circuits [32].

There are no constructive methods existing today that can efficiently identify Boolean factors of a logic expression which are nonalgebraic, and which result in superior factorizations. Techniques which do result in Boolean factorizations operate by taking an initial circuit structure and resubstituting Boolean factors using strong division. A first step in the direction of a constructive method was taken in [13] where the use of multiple-valued minimization to identify Boolean factors was advocated. However, the method relies on a heuristic selection of inputs and operates on sum-of-products representations of logic functions. With the advent of implicit prime generation methods it may be possible to arrive at methods for efficient Boolean factorization.

The main limitation of current logic synthesis systems is that a description style has to be enforced on the input *register-transfer level* (RTL) description in order to produce quality results. Furthermore, there is no formal definition of such a description style. Rather, as described in Chapter 2, examples that are effective starting points constitute the definition of the style. If a description is badly written, in many cases, the final optimized circuit is unacceptable from an area or speed standpoint.

One approach to solving this problem is to increase the power of the optimization step so it can recover from badly written input. Another solution is to precisely define the characteristics of a "good" RTL description that is suitable as a starting point for logic synthesis and manipulate the given RTL description into this form. Both of these directions are being followed in the development of the next generation of synthesis systems.

10.2.2 Testability

We described synthesis for testability methods in Chapters 5 and 9. A variety of techniques that make combinational logic highly testable exist today, and it is possible to constrain logic optimization to achieve testability under a variety of fault models, such as the multiple stuck-at and delay fault models. The practical use of synthesis for testability techniques is growing, as evinced by the recent surge in the incorporation of testability under the single stuck-at fault model in industrial and commercial logic synthesis tools.

Synthesis for random pattern testability under the single stuck-at fault model is a useful and attractive area of research. Random and pseudo-random patterns have been used for the testing of VLSI circuits for many years. For a review of such techniques see [3]. The random patterns are generated, in most cases, using a *linear feedback shift-register* (LFSR). The response of the circuit under test is compressed into a signature, once again in a shift-register, and compared with the expected response of the circuit under fault-free conditions. This expected response is obtained by fault simulation of the random patterns. The simplicity of the test generation and comparison units makes their integration into the circuit under test very easy, which makes this approach attractive for *built-in self test* (BIST) applications. However, for this approach to be effective, the circuit under test must be inherently random pattern testable. Unfortunately, not all circuits are random pattern testable.

Techniques to improve the random pattern testability of circuits have been presented (e.g., [3, 5, 14, 20, 22]). Most of these methods rely on the addition of control and observation points and are relatively ad hoc. A systematic method for adding control and observation points was presented a few years ago [5] and was further improved recently [22]. The improvements are mainly in the way redundant faults in a circuit are handled.

It has been shown that the use of correlated random patterns can achieve 100 percent fault coverage of circuits synthesized using standard logic optimization techniques, and the number of random patterns required is significantly smaller than when equally probable random patterns are used [30]. This work is, however, restricted to circuits synthesized from a two-level representation as a starting point. The generalization to multilevel circuits is a subject of current research.

10.2.3 Power Dissipation

With increased density on integrated circuits there is increased power dissipation per unit area of silicon. Because of this the problem of packaging VLSI circuits has become increasingly difficult. Designing for low power has become important for all types of VLSI circuits and is essential for applications such as personal communications systems and portable computers.

In *complementary metal oxide semiconductor* (CMOS) circuits power dissipation is mainly due to switching activity. When a transistor switches, current flows from a capacitor to the supply or to ground causing energy dissipation. It is important to be able to estimate the power dissipation of a circuit at the register-transfer or gate levels without having to fabricate the circuit and physically measure its power dissipation. Methods have been developed recently to estimate the average switching activity of combinational and sequential circuits [17, 29].

Designing for low power requires a multiple-tiered approach, where power dissipation constraints are propagated down from the architectural through the logic, transistor, and layout levels [8]. From a logic synthesis perspective it is necessary for the optimization program to be able to minimize the power dissipation by appropriately restructuring the circuit so as to minimize an estimate of the switching activity. A first effort in this direction has already been made [34]; however, versatile and efficient methods have yet to be developed.

10.3 Sequential Logic Synthesis

10.3.1 Area and Performance

There are two different types of sequential modules in VLSI circuits. The first type corresponds to finite state machine controllers that are typically described as *state transition graphs* (STGs) such as the one of Figure 10.1. (It is possible to write a VHDL description of the STG as well.) *Encoding* the STG description, i.e., assigning binary codes to each of the symbolic states, produces a combinational logic specification, which along with the appropriate number of flip-flops can implement the desired behavior.

The problem of obtaining a gate-level circuit corresponding to an STG specification entails many optimization steps such as state minimization, finite state machine decomposition, and state encoding. Obtaining efficient algorithms for these optimization problems is a subject of intensive research. For a comprehensive survey of state-of-the-art methods we refer the reader to [2].

The second type of sequential circuits are datapath modules. One can directly write HDL specifications for such sequential circuits as interconnections of gates and flip-flops. For instance, a single-bit counter was described in VHDL as an interconnection of gates and flip-flops in Figure 2.9. It is possible to view sequential circuits such as counters and datapaths at the STG level, and such a description

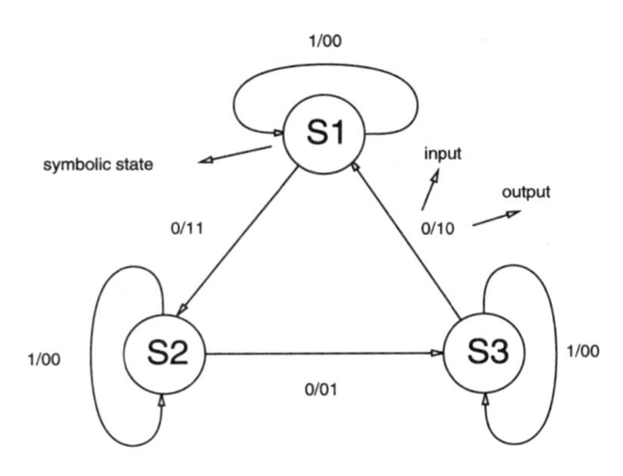

Figure 10.1: An example of an STG

can be extracted from the gate-level circuit. However, the STG description can be very cumbersome for such circuits. Moreover, there is no freedom in encoding or reencoding the states of such circuits.

One optimization that can be performed at the gate level that modifies the functionality of the combinational logic but does not change the terminal behavior of the sequential circuit is *retiming*. Retiming a circuit corresponds to changing the positions of flip-flops in a sequential circuit with the constraint that the number of flip-flops in each path from primary inputs to the primary outputs of the circuit remains constant. However, the speed of the circuit may be improved because the maximum delay between any pair of flip-flops is reduced. Algorithms for finding an optimal retiming of a circuit for maximum speed were first presented in [27], and since then these algorithms have been generalized in many ways. Methods to improve the speed of sequential circuits by integrating combinational logic optimization methods with retiming techniques are currently a subject of extensive investigation.

10.3.2 Testability

There is a large body of work in the area of synthesis of sequential circuits for testability under the single stuck-at fault model (for a survey see [18]). Don't-care conditions, whose exploitation is necessary for full sequential testability, have been defined [12, 11]. In order to compute these don't-care conditions, the entire state space of the given sequential circuit has to traversed.

As mentioned previously, the manipulation and design of

control modules are possible using an explicit STG specification like the one of Figure 10.1. However, such a representation cannot be generated even for moderately sized datapath modules, because the number of states of the module could be very large. An alternate representation for such sequential circuits that is significantly more compact is the characteristic function representation [7, 9]. In the characteristic function representation sets of states are represented using Boolean functions. The Boolean functions can in turn be manipulated using BDDs. Given a large datapath circuit, the entire set of states of the circuit can be computed using the symbolic traversal methods of [7, 9] without having to explicitly enumerate each state in the circuit. The computation of the don't-care conditions necessary for full testability currently relies on these symbolic traversal techniques for efficiency, and circuits with about 200 memory elements can be made fully testable using current techniques.

Another representation that has been used for sequential synthesis for testability and that has succeeded in handling some classes of large circuits (that cannot be handled even by the symbolic approach) is the RTL representation of a sequential circuit. As we have seen in Chapter 2 an RTL representation is typically very compact; however, it is not easy to manipulate. Methods to manipulate these descriptions so as to generate the invalid state and equivalent state don't-care set required for full sequential testability have been presented in detail elsewhere [18]. These methods have been used successfully for machines with up to 1000 memory elements and hold the greatest promise with the increasing use of RTL synthesis systems. However, these methods are currently heavily dependent on the particular syntax of the RTL description and the set of predefined modules used.

To improve the quality of fabricated integrated circuits, it has become clear that some form of at-speed or delay testing is required [6]. The simplest method of testing an integrated circuit is to apply a stream of input vectors to its primary inputs, clock the circuit at speed, and monitor its output response. The goal of sequential logic synthesis for testability approaches is to design circuits that achieve a high degree of stuck-at and delay fault coverage, given the above test methodology. Most of the current research in synthesis for sequential testability is focused on this problem, and much more work needs to be done in order to achieve this goal. In the meantime, standard or enhanced scan design can be used to achieve a high degree of multiple stuck-at and delay fault coverage for se-

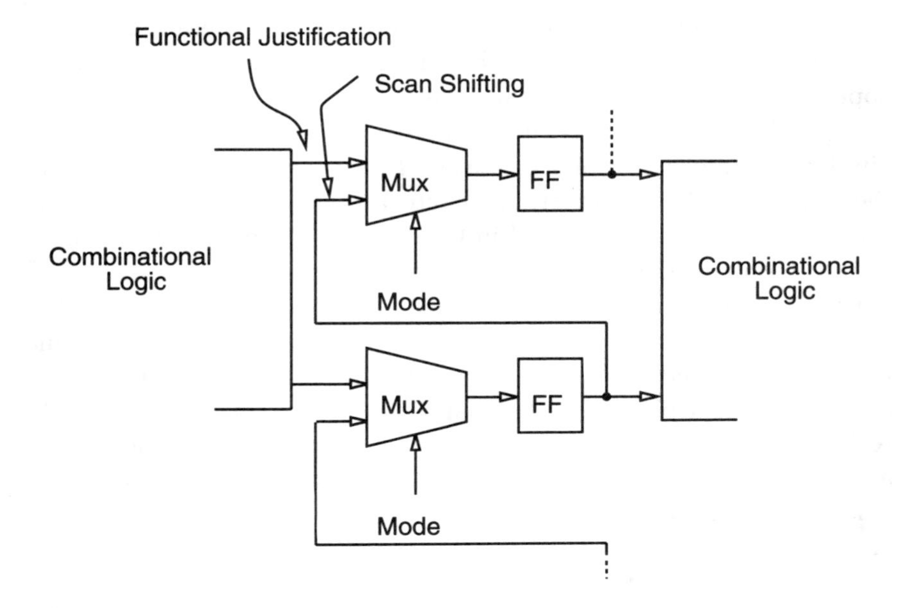

Figure 10.2: Scan design structure

quential circuits. We will briefly describe these design methodologies which convert the sequential test problems into their combinational counterparts.

Standard scan design converts the problem of single and multiple stuck-at fault test generation for sequential circuits into a purely combinational one. Thus, the techniques of Chapter 9 can be directly applied to sequential circuits.

The standard scan concept is illustrated in Figure 10.2. The flip-flops in the circuit are connected in a *scan* chain. An additional multiplexor is required before every flip-flop. If the `Mode` switch is high, then the flip-flop receives its input from the predecessor in the scan chain. If the `Mode` switch is low, then the circuit behaves like the original, except with an additional multiplexor delay. This additional multiplexor delay is a penalty that has to be incurred, which may not be acceptable in some design scenarios.

The delay test of sequential circuits is harder than single stuck-at fault testing because we have to ensure that a vector pair $< v_1, v_2 >$ can be applied to the circuit. If the circuit is being run at speed, and if i_1 and s_1 (i_2 and s_2) correspond to the primary input and state fields, respectively, for v_1 (v_2), then we require that the machine when placed in s_1 should move to s_2 upon the application of i_1. Since the $< v_1, v_2 >$ test is derived by operating on the combinational logic portion of the circuit, it may not be possible to apply the test to the sequential circuit.

Enhanced scan structures have been proposed where each flip-flop can store two bits of state. These scan structures convert the sequential delay test problem into a purely combinational one; thus, the techniques presented in Chapter 9 can be directly used.

Enhanced scan structures are more complex than the standard scan structure described immediately above. An extra latch to each existing scan double-latch can be added to store v_1, while v_2 is being shifted in (see [28]). An enhanced latch that allows the application of arbitrary vector pairs has been proposed for nMOS technologies in [19]. Unfortunately, this may result in a nonrobust application, but if the combinational logic is hazard-free robust path delay fault testable, then the sources of test invalidation are considerably restricted.

Note that these scan structures are more expensive than traditional scan structures. For some circuits, quality and reliability considerations will be predominant, and therefore the additional cost of the specialized scan structure will be worthwhile.

10.4 Logic Synthesis Systems

10.4.1 Area and Performance

Several logic synthesis systems have been developed in universities and industry, and are being routinely used for the automated design of *application-specific integrated circuits* (ASICs). Most systems (e.g., SYNOPSYS, CADENCE and MIS) begin from an RTL description of a circuit written in a language like VHDL or VERILOG and use methods similar to those described in Chapters 7 and 8 to optimize for area and speed. As described earlier, the main limitation of current systems is that an informal description style on the input RTL description has to be currently enforced in order to produce quality results.

The use of synthesis in the design of general-purpose microprocessors is less extensive. High circuit speed is the predominant focus in the design of such VLSI circuits, and synthesis methods are not always able to achieve results comparable to manual design. One of the reasons for this is that, as technologies move to the submicron levels, it has become increasingly difficult to quickly and accurately characterize the speed of the circuit. Another reason is that under a cell library approach, it is harder to take advantage of transistor-level optimizations, e.g., using pass gates or precharged gates to improve speed.

Other limitations of current systems are dealing with partially asynchronous operation, analog components and transistor-level digital components that do not have completely satisfactory logic-level abstractions.

10.4.2 Testability

Several logic synthesis for testability systems have been developed in academia and industry. Prototype implementations of all the synthesis for test methods for the various fault models described in this book have been carried out at universities (e.g., M.I.T. and U. C. Berkeley). Industrial logic synthesis for testability systems currently have a narrower testability focus, namely the single stuck-at fault model.

Commercially available logic synthesis systems (e.g., SYNOPSYS and AT&T) offer the feature of synthesizing fully testable combinational logic under the single stuck-at fault model. Don't-care optimization and redundancy removal methods are used to ensure 100 percent testability. For sequential circuits standard scan flip-flops may be inserted automatically under a complete standard scan design methodology. Thus, 100 percent fault coverage can be obtained, albeit with an associated area and/or speed penalty (depending on the technology). Partial standard scan design methodologies have been recently implemented in commercial software (e.g., SYNOPSYS) in an effort to reduce the area and speed overhead associated with standard scan flip-flops. Only those flip-flops that are not on the critical path of the circuit are replaced with standard scan flip-flops.

The incorporation of testability into a synthesis system has an additional advantage in that higher-level specifications such as the RTL description of a circuit can be used to aid logic synthesis for testability. Several companies have produced sequential test generators that exploit the use of RTL descriptions and therefore perform significantly better than purely logic-level test generators (e.g., [15, 25, 26, 31]). The theoretical results presented in [18] have been used to develop synthesis for testability methods that begin from RTL descriptions.

Targeting the delay fault model is the next major step being taken in the development of industrial synthesis for testability systems. One approach is to design the combinational logic of the circuit to be highly delay fault testable and implement the circuit under a complete or partial scan design methodology.

10.5 Future Design Methodologies

There are many open research problems in logic synthesis and its related areas; however, there is no doubt that current techniques have reached sufficient maturity to provide an acceptable design methodology for integrated circuit design.

At the same time semiconductor processing continues to improve, and today's designers are already uncertain of their ability to design circuits of a million gates relying only on currently available design methods. There are a number of approaches that hold some promise for providing another significant boost to designer productivity and we will briefly review each of these.

10.5.1 Design Reuse

Reusing portions of circuitry that have been previously designed is the easiest way to improve designer productivity. Cell libraries are one of the simplest examples of design reuse. One of the major obstacles to design reuse in earlier eras was the effort required to carefully characterize the electrical properties of a predesigned module. Using logic optimization and technology mapping a designer can be insulated from the details of the electrical design of a circuit and concentrate on whether the module has the functionality required. Logic optimization steps will also tailor the module to the specific application.

10.5.2 Domain Specific Synthesis

For typical integrated circuit designs it is not clear that simply migrating from a VHDL model described at the RTL to a VHDL model described at the behavioral level will be sufficient to provide a significant boost in productivity. What does offer a better chance of improving productivity is designing using representations very closely tailored to the domain.

One example of such a representation is the *statechart* for describing control dominated design. Statecharts offer a hierarchical structure that avoids specifying cycle-by-cycle behavior in more detail than is necessary. The company ILOGIX also offers software for translating statecharts to VHDL which can subsequently be synthesized by a logic synthesis system. A few industrial designers have already demonstrated significant productivity gains from such a design methodology.

Another example of such a representation is the use of *signal flow graphs* for describing signal processing systems. One popular representation of these flow graphs is in the language SILAGE [21]. Signal flow graphs are a natural way of describing the flow of data through a signal processing system. Translation of signal flow graphs into descriptions implementable by synthesis systems has been pursued academically by a group at U. C. Berkeley [35] and is commercially distributed by the company COMDISCO. This approach has also found advocates in industry.

10.5.3 Migration to Software

As time-to-market becomes an increasingly important factor, we see in industrial statistics an increasing amount of system functionality showing up in software relative to hardware. When software running on a processor provides acceptable performance, there is no doubt that software design is more productive than hardware design. Furthermore, faster microprocessors are making processor-oriented solutions cover more and more problems that previously required custom hardware solutions. Nevertheless, these processors also require significant design efforts, and logic synthesis is playing an increasingly large role in their design.

10.5.4 Future Role of Logic Synthesis

In the beginning of the computer revolution support for programming computers evolved very quickly: from machine language to assembler; from assembler to macroassembler, and from macroassembler to a tremendous variety of languages. Although the C programming language was not as sophisticated as many of its predecessors, the emergence of a standard led to a much slower development of commercial programming languages and subsequent developments were more incremental, e.g., C++. We anticipate a similar development in design methodologies for integrated circuits. To realize future productivity demands all the techniques described in this section will be required. The use of these techniques does not undermine the importance of logic synthesis but rather underscores it. Each of the productivity improving techniques described in this section on logic synthesis rely on logic synthesis for final implementation of the circuit, and we anticipate that synthesis from a standard HDL will be the dominant means of designing hardware for some time.

REFERENCES

[1] P. Ashar, S. Devadas, and A. Ghosh. Boolean Satisfiability and Equivalence Checking Using General Binary Decision Diagrams. In *Proceedings of the International Conference on Computer Design: VLSI in Computers and Processors*, pages 259–264, October 1991.

[2] P. Ashar, S. Devadas, and A. R. Newton. *Sequential Logic Synthesis.* Kluwer Academic Publishers, Norwell, MA, 1991.

[3] P. H. Bardell, W. H. McAnney, and J. Savir. *Built-In Test for VLSI: Pseudo-Random Techniques.* John Wiley and Sons, New York, NY, 1987.

[4] M. A. Breuer and A. D. Friedman. *Diagnosis and Reliable Design of Digital Systems.* Computer Science Press, Woodland Hills, CA, 1976.

[5] A. J. Briers and K. A. E. Totton. Random Pattern Testability by Fast Fault Simulation. In *Proceedings of the International Test Conference*, pages 274–281, September 1986.

[6] O. Bula, J. Moser, J. Trinko, M. Weismann, and F. Woytowich. Gross Delay Defect Evaluation for a CMOS Logic Design System Product. In IBM *Technical Memorandum*, May 1989.

[7] J. R. Burch, E. M. Clarke, K. L. McMillan, and D. Dill. Sequential Circuit Verification Using Symbolic Model Checking. In *Proceedings of the* 27th *Design Automation Conference*, pages 46–51, June 1990.

[8] A. Chandrakasan, T. Sheng, and R. W. Brodersen. Low Power CMOS Digital Design. *IEEE Journal of Solid State Circuits*, 27(4):473–484, April 1992.

[9] O. Coudert, C. Berthet, and J. C. Madre. Verification of Sequential Machines Using Boolean Functional Vectors. In *IMEC-IFIP International Workshop on Applied Formal Methods for Correct VLSI Design*, pages 111–128, November 1989.

[10] O. Coudert and J-C. Madre. Implict and Incremental Computation of Primes and Essential Primes of Boolean Functions. In *Proceedings of the* 29th *Design Automation Conference*, pages 36–39, June 1992.

[11] S. Devadas, H-K. T. Ma, and A. R. Newton. Redundancies and Don't Cares in Sequential Logic Synthesis. *Journal of Electronic Testing: Theory and Applications*, 1(1):15–30, February 1990.

[12] S. Devadas, H-K. T. Ma, A. R. Newton, and A. Sangiovanni-Vincentelli. Irredundant Sequential Machines via Optimal Logic Synthesis. *IEEE Transactions on Computer-Aided Design of Integrated Circuits*, 9(1):8–18, January 1990.

[13] S. Devadas, A. R. Wang, A. R. Newton, and A. Sangiovanni-Vincentelli. Boolean Decomposition in Multilevel Logic Optimization. *IEEE Journal of Solid State Circuits*, 24(2):399–408, April 1989.

[14] E. B. Eichelberger and E. Lindbloom. Random-Pattern Coverage Enhancements and Diagnosis for LSSD Logic Self-Test. IBM *Journal of Research and Development*, 27(3):265–272, May 1983.

[15] ExperTest, Inc., 810 East Middlefield Road, Mountain View, CA. *Test Design Expert Users Manual*, 1992.

[16] H. Fujiwara. *Logic Testing and Design for Testability.* MIT Press, Cambridge, MA, 1985.

[17] A. Ghosh, S. Devadas, K. Keutzer, and J. White. Estimation of Average Switching Activity in Combinational and Sequential Circuits. In *Proceedings of the* 29th *Design Automation Conference*, pages 253–259, June 1992.

[18] A. Ghosh, S. Devadas, and A. R. Newton. *Sequential Logic Testing and Verification.* Kluwer Academic Publishers, Norwell, MA, 1991.

[19] C. T. Glover and M. R. Mercer. A Method of Delay Fault Test Generation. In *Proceedings of the* 25th *Design Automation Conference*, pages 90–95, June 1988.

[20] J. P. Hayes and A. D. Friedman. Test Point Placement to Simplify Fault Detection. In *Proceedings of the Fault Tolerant Symposium*, pages 73–78, 1974.

[21] P. Hilfinger. A High-Level Language and Silicon Compiler for Digital Signal Processing. In *Proceedings of the Custom Integrated Circuits Conference*, pages 213–216, May 1985.

[22] V. S. Iyengar and D. Brand. Synthesis of Pseudo-Random Pattern Testable Designs. In *Proceedings of the International Test Conference*, pages 501–508, September 1989.

[23] J. Jain, J. Bitner, D. Fussell, and J. Abraham. Probabilistic Design Verification. In *Proceedings of the International Conference on Computer-Aided Design*, pages 468–471, November 1991.

[24] S-W. Jeong, B. Plessier, G. Hachtel, and F. Somenzi. Extended BDDs: Trading-Off Canonicity for Structure in Verification Algorithms. In *Proceedings of the International Conference on Computer-Aided Design*, pages 464–467, November 1991.

[25] J. Lee and J. H. Patel. A Signal-Driven Discrete Relaxation Technique for Architectural Level Test Generation. In *Proceedings of the International Conference on Computer-Aided Design*, pages 458–461, November 1991.

[26] J. Lee and J. H. Patel. ARTEST: An Architectural Level Test Generator for Data Path Faults and Control Faults. In *Proceedings of the International Test Conference*, pages 729–738, October 1991.

[27] C. E. Leiserson, F. M. Rose, and J. B. Saxe. Optimizing Synchronous Circuitry by Retiming. In *Proceedings of the Third CalTech Conference on VLSI*, pages 23–36, March 1983.

[28] Y. K. Malaiya and R. Narayanswamy. Testing for Timing Failures in Synchronous Sequential Integrated Circuits. In *Proceedings of the International Test Conference*, pages 560–571, October 1983.

[29] F. Najm. Transition Density, A Stochastic Measure of Activity in Digital Circuits. In *Proceedings of the* 28th *Design Automation Conference*, pages 644–649, June 1991.

[30] S. Pateras and J. Rajski. Generation of Correlated Random Patterns for the Complete Testing of Synthesized Multilevel Circuits. In *Proceedings of the International Test Conference*, pages 347–352, October 1991.

[31] Racal-Redac, Inc., 238 Littleton Road, Westford, MA. *INTELLIGEN 2: Users Manual*, 1992.

[32] H. Savoj and R. Brayton. Observability Relations and Observability Don't-Cares. In *Proceedings of the International Conference on Computer-Aided Design*, pages 518–521, November 1991.

[33] H. Savoj, R. Brayton, and H. Touati. Extracting Local Don't-Cares for Network Optimization. In *Proceedings of the International Conference on Computer-Aided Design*, pages 514–517, November 1991.

[34] A. Shen, S. Devadas, A. Ghosh, and K. Keutzer. On Average Power Dissipation and Random Pattern Testability of Combinational Logic Circuits. In *Proceedings of the International Conference on Computer-Aided Design*, pages 402–407, November 1992.

[35] C. B. Shung, R. Jain, K. Rimey, R. W. Brodersen, E. Wang, M. B. Srivastava, B. Richards, E. Lettang, L. Thon, S. K. Azim, P. N. Hilfinger, and J. Rabaey. An Integrated CAD System for Algorithmic-Specific IC Design. *IEEE Transactions on Computer-Aided Design of Integrated Circuits*, 10(4):447–463, April 1991.

Index

ABOUT THE AUTHORS

Srinivas Devadas is an Associate Professor of Electrical Engineering and Computer Science at the Massachusetts Institute of Technology, Cambridge. He is an active researcher in the field of Computer-Aided Design of Integrated Circuits and is a co-author of *Sequential Logic Synthesis* and *Sequential Logic Testing and Verification.*

Abhijit Ghosh is a researcher at Mitsubishi Electric Research Laboratories, Sunnyvale. He is a co-author of *Sequential Logic Testing and Verification.*

Kurt Keutzer is the manager of the Advanced Technology Group at Synopsys, Mountain View. He has authored several articles in the area of logic synthesis and Computer-Aided Design.